a M Rash

AN INTRODUCTION
TO MATRICES
AND LINEAR TRANSFORMATIONS

AN INTRODUCTION TO MATRICES AND LINEAR TRANSFORMATIONS

JOHN H. STAIB, Drexel Institute of Technology

ADDISON-WESLEY PUBLISHING COMPANY, INC.
Reading, Massachusetts · Menlo Park, California · London · Don Mills, Ontario

This book is in the
Addison-Wesley Series in Mathematics

Consulting Editor:
LYNN H. LOOMIS

Copyright © 1969 by Addison-Wesley Publishing Company, Inc. Philippines copyright 1969 by Addison-Wesley Publishing Company, Inc.
All rights reserved. No part of this publication may be reproduced, stored in a retrieval system, or transmitted, in any form or by any means, electronic, mechanical, photocopying, recording, or otherwise, without the prior written permission of the publisher. Printed in the United States of America. Published simultaneously in Canada. Library of Congress Catalog Card No. 69-16405.

PREFACE

Some general remarks

This book is introductory in nature, and is designed primarily for a one-semester, undergraduate course in *linear algebra*. Here at Drexel a preliminary version of the present text has been used by my colleagues and myself on three successive classes of sophomore engineering and science students, about 600 in all. It could also serve easily in a first course for computer scientists, high school mathematics teachers, economists, etc., regardless of their educational level.

The structure of the book is rather typical of today's mathematics texts; I would describe it as a mathematical narrative into which has been embedded a succession of definitions, theorems, examples, and exercise sets. Thus, it has a rather formal appearance. But at the same time it lacks the terseness and elegance of presentation (mathematically speaking) customary with such texts. Rather, one finds that introductions to new topics are often chatty and even clumsy. Ultimately, however, the mathematical content of a new topic is distilled, and the reader is brought to the definitions and theorems. In short, I have tried to bring to the reader a classroom style of presentation, with all its informality and experimentation.

To the teacher

Linear algebra is not an easy course for sophomores—the main reason probably being that the application of linear algebra to "real problems" is not so immediate as in the case of calculus, for example. The student is bound to be more impressed by a mathematical tool that enables him to compute areas and volumes than by one that establishes for him the fact that a certain quadratic form is positive definite. Thus, the teacher of a first course in linear algebra must do more than present clear explanations: he must make a conspicuous effort to sell the course to a skeptical audience. This fact was uppermost in my mind as I wrote this text.

There were also several more specific goals: (1) To show the power of mathematical notation. In particular, to show how a variety of problems are solved by appealing to matrix manipulations. (2) To present the basic concepts of linear algebra, concepts that are increasingly taken to be prerequisites for a diversity of courses. (3) Finally, to instill in the reader some appreciation for the mathematician's seeming obsession with abstract structures; in particular, to give evidence of the economy of thought and the increase in understanding that attend that process whereby a variety of mathematical models are related to a certain neutral (abstract) archetype.

The book begins with a study of matrix arithmetic and the natural application of that arithmetic to the solution of linear systems. Some would postpone the introduction of matrices until after linear transformations have been studied, thus developing a natural environment for their introduction. However, our experience at Drexel has been that students are not at all antagonistic toward matrix computations, even in the absence of a "natural environment". Such a start means that we can move into the study of vector spaces with a ready source of examples (spaces of m by n matrices) and a useful technique (row reduction of matrices) for answering questions relative to the dependence or independence of a set of vectors. Similarly, the notion of a linear transformation can be immediately illustrated with examples such as $Y = AX$, where A is a matrix. Thus, the analogy with $y = ax$ is there at the start. It is against this background that the discovery is made that any linear transformation (between finite spaces) can be realized in a matrix product.

Two other points about the development are worth mentioning here. First, Euclidean ideas are saved for the last chapter. At that time the notions of length and angle are generalized and an *abstract dot product* emerges. This chapter is probably the most sophisticated; here, an attempt is made to justify the mathematician's practice of bringing geometrical terminology and problem-solving techniques into spaces of arbitrary dimension.

The other point is with respect to the treatment of determinants. The determinant function is defined inductively, using a first-column expansion. This allows us to omit the usual discussion of "inversion counting." There is also a very important by-product: such a definition gives us the opportunity to introduce the notion of a *proof by mathematical induction* at a time when it can be put to good use.

A word about the exercise sets: I have tried to arrange the exercises on a priority basis; that is, the earlier exercises in a given set are so placed because they seem essential to an adequate understanding of the section preceding. The later exercises should be interpreted not as being more difficult by design but simply as adding breadth and depth of coverage. Thus, the planning of an assignment may be reduced to making a decision about where to draw the line in a given exercise set. On occasion the earlier exercises will include definitions and theorems that are vital to the development of the theory. I have done this for the following reason: Students should have some opportunity to develop confidence in their ability to read and apply such material without the help of a teacher. Answers to over half of the exercises are included; the remaining answers are available in a separate booklet which may be obtained from the publisher on request.

In all, the text consists of 42 sections and 5 appendixes. Here at Drexel we have only one quarter (4 hours a week for 10 weeks) at our disposal; thus, we are obliged to make some deletions and do some skimming. For example, one of our compromises was to omit entirely Sections 1.8, 4.9, 4.10, 5.2, and 5.3, and then to save additional time by moving more rapidly through Sections 3.5, 3.6, 4.1, 4.4, 4.5, and 4.6. Otherwise, we maintained a pace of one section per class meeting.

To the student

Not so many years ago (about 20) the study of linear algebra was relegated to a graduate-level course in modern algebra. Consequently, only those students working toward an advanced degree in mathematics (or perhaps physics) were exposed to such a study. This state of affairs has changed drastically: it is now not at all unusual to find that a course in linear algebra is required (or at least recommended) for undergraduates majoring in such fields as mathematics, engineering, physics, chemistry, biology, economics, and even psychology.

Granted that mathematical methods are increasingly being applied in all fields professing to use the *scientific method*, it is still necessary to justify the particular need for a course in linear algebra. Why not, for example, a required course in number theory, or modern geometry? Such a justification is not easy to formulate in terms that the reader can fully appreciate. Yet, since motivation of the reader is extremely important, I shall list just a few of the factors that have forced linear algebra into so many undergraduate curricula. (1) In the application of mathematics to problems encountered in other fields, it is frequently the case that the problem is essentially solved once it has been successfully translated into the symbols of mathematics, for the actual solution often results from a routine manipulation of the symbols. A symbol that has been exceedingly useful in this respect is the *matrix*. The study of matrices and the various manipulations involving matrices forms part of a standard course in linear algebra. (2) The study (and invention) of equation-solving techniques has —besides an obvious importance to people who use mathematics—its own special fascination. Particularly interesting are those equations in which the unknown represents a function rather than a number. (A differential equation, for example.) Such *functional equations* are in general difficult to solve. However, there is an easy case: the case where the equation is *linear*. It is the province of courses in linear algebra to define the term "linear equation" and to lay the foundation for other courses in which particular linear equations are met and solved. (3) The study of calculus can be conveniently broken into two courses: the *single-variable calculus* and the *multi-variable calculus*. In the study of single-variable calculus, functions defined by formulas of the form $y = mx$ play a very special role. Indeed, the very notion of *slope* (upon which the notion of *derivative* is based) arises from the graphical study of such functions. The following question arises: With respect to the study of multi-variable calculus, is there also a special class of functions that play a central role, a role analogous to that of the functions defined by $y = mx$? There is. Such functions are called *linear transformations*, and the basic facts about such functions are presented in courses on linear algebra.

The point is this: A knowledge of linear algebra is prerequisite to a variety of other studies. Unfortunately, the full power of that knowledge is not appreciated until one is actually engaged in those "other studies." The situation roughly parallels that of a high school algebra course. There the student learns to add algebraic fractions, factor polynomials, rationalize denominators, and so on. Some of these topics engender their own interest; others do not. But that body of knowledge ultimately proves to be indispensable in subsequent courses in physics, calculus, etc.

Finally, some words of advice with respect to the use of this text: (1) The new terms and symbols of linear algebra are introduced in the *Definitions*. It is crucial that these definitions be mastered. (2) The mathematical content of the course is, in capsule form, found in the *Theorems*. Here lies the real challenge. The goal should be not only to understand these theorems but to appreciate how each one contributes to the developing theory. Such a goal is unapproachable without studying the proofs of the theorems. (3) Most of the theorems and definitions are clarified, hopefully, by accompanying *Examples*. The reader will profit most by reworking each example for himself, filling in any computations or steps in reasoning that have been omitted in the text. (4) At the end of each section there is a set of *Exercises;* at the end of the book there are answers for approximately half of the exercises. Most exercise sets contain more exercises than even the most conscientious student will have time for. However, the reader should take the time to read all of them; frequently, the reading alone will constitute a learning experience.

Acknowledgments

Many people have helped and encouraged me in the writing of this text. In particular, I want to acknowledge the constructive criticism that was offered by Professors Joseph B. Deeds, Miltiades S. Demos, Lynn H. Loomis, Alexander Tartler, and Thomas L. Wade. I am also grateful for the assistance of Miss Amelia Dulisse, Mrs. Sandra Jobson, and Mrs. Doris Wagner, who shared the typing chores; to Mr. Robert Koseluk, who helped with the exercise solutions; and to the staff of Addison-Wesley for their expert technical assistance. Finally, I wish to thank my wife for her patience, encouragement, and confidence.

Philadelphia, Pa.   J. H. S.
October 1968

CONTENTS

Chapter 1 Matrices

1.1 The Arithmetic of Matrices 1
1.2 Properties of Matrix Arithmetic 6
1.3 Elementary Matrices 11
1.4 A Method for Finding A^{-1} 17
1.5 Solving Systems of Linear Equations 22
1.6 The First Diagonalization Problem 31
1.7 Quadratic Forms 36
1.8 Equivalence Relations 42

Chapter 2 Vectors

2.1 Vector Spaces 49
2.2 Subspaces 55
2.3 Isomorphic Vector Spaces 60
2.4 Linear Independence 67
2.5 Another Approach to Dimension 75
2.6 More on Bases 80
2.7 The Row Space of a Matrix 86
2.8 Change-of-Basis Matrix 91
2.9 Coordinate Systems 96
2.10 Changing Coordinate Systems 104
2.11 Lines and Planes 113

Chapter 3 Linear Transformations

3.1 Linear Transformations 121
3.2 The Kernel of a Linear Transformation 127
3.3 Linear Equations 133
3.4 Matrix Representations 140
3.5 Arithmetic Operations for Linear Transformations . . 148
3.6 Linear Algebras 153

Chapter 4 The Determinant Function

4.1	Mathematical Induction	161
4.2	The Determinant Function	168
4.3	Basic Properties	173
4.4	Theory of Determinants	178
4.5	More Theory	184
4.6	A Formula for A^{-1}	191
4.7	The Second Diagonalization Problem	197
4.8	An Application to Linear Differential Equations	208
4.9	Jordan Matrices	215
4.10	The Vandermonde Determinant	224

Chapter 5 Euclidean Spaces

5.1	Length and Angle	233
5.2	Abstract Euclidean Spaces	242
5.3	Isomorphic Euclidean Spaces	250
5.4	Orthonormal Bases	256
5.5	The Characteristic Value Problem for Symmetric Matrices	263
5.6	Applications to Geometry	270

Appendix A	Relations	279
Appendix B	What is a Matrix, Really?	282
Appendix C	The Traditional Approach to the Determinant	285
Appendix D	A Geometric Interpretation for the Determinant Function	288

Answers to Selected Exercises 294

Glossary of Symbols 329

Suggested References 331

Index 335

CHAPTER 1

MATRICES

1.1 THE ARITHMETIC OF MATRICES

In reaching this point in his mathematical studies, the reader should have made contact with a number of arithmetical structures: the arithmetic of natural numbers (positive integers), the arithmetic of real numbers, and the arithmetic of complex numbers, to list a few. Such number systems have many attributes in common. For example, in all three of the listed examples it is the case that $a + b = b + a$ for every pair of numbers a and b. On the other hand, there are also distinguishing features. For example, given any real number a different from 0, there will always be a second real number, say b, such that $ab = 1$. This statement is no longer true if the word "real" is replaced by the word "natural." Or, given any complex number α, there will always be a second complex number, say β, such that $\beta^2 = \alpha$. This statement is no longer true if the word "complex" is replaced by the word "real."

In this chapter we shall begin the study of yet another arithmetical structure: the *arithmetic of matrices*. We begin by defining what a matrix is and explaining how such "matrix-numbers" are to be added and multiplied.

Definition 1.1 An *m by n matrix* is a rectangular array of real numbers having m rows and n columns. We denote it by

$$\begin{bmatrix} a_{11} & a_{12} & \cdots & a_{1n} \\ a_{21} & a_{22} & \cdots & a_{2n} \\ \vdots & \vdots & & \vdots \\ a_{m1} & a_{m2} & \cdots & a_{mn} \end{bmatrix};$$

or, more briefly, we write $[a_{ij}]_{mn}$. (Here a_{ij} denotes a typical element of the matrix: the element to be found in the ith row and the jth column.)

The numbers m and n are called the *dimensions* of the matrix. Two matrices are said to be *equal* when they have the same dimensions and when corresponding entries are identical. If $m = n$, then we speak of an *nth-order matrix*. The elements $a_{11}, a_{22}, \ldots, a_{nn}$ of an nth-order matrix are referred to as the elements along the *main diagonal* of the matrix.

Example 1. Construct the matrix defined by

$$A = \left[\sum_{k=i}^{i+j} k^2\right]_{32}.$$

Solution. Here the element in the ith row and the jth column is given by

$$a_{ij} = \sum_{k=i}^{i+j} k^2.$$

Thus,

$$a_{11} = \sum_{k=1}^{2} k^2 = 1^2 + 2^2 = 5,$$

$$a_{12} = \sum_{k=1}^{3} k^2 = 1^2 + 2^2 + 3^2 = 14,$$

$$a_{21} = \sum_{k=2}^{3} k^2 = 2^2 + 3^2 = 13,$$

and so on. The reader should verify that, finally,

$$A = \begin{bmatrix} 5 & 14 \\ 13 & 29 \\ 25 & 50 \end{bmatrix}.$$

Example 2. Construct the matrix $[[j/i]]_{44}$, where $[x]$ denotes the largest integer not exceeding x.

Solution. Here $a_{ij} = [j/i]$. Thus, $a_{11} = [1/1] = 1$, $a_{12} = [2/1] = 2$, and so on. The reader should verify that

$$A = \begin{bmatrix} 1 & 2 & 3 & 4 \\ 0 & 1 & 1 & 2 \\ 0 & 0 & 1 & 1 \\ 0 & 0 & 0 & 1 \end{bmatrix}.$$

Definition 1.2. The *addition of* $[a_{ij}]_{mn}$ *to* $[b_{ij}]_{mn}$ is defined by

$$[a_{ij}]_{mn} + [b_{ij}]_{mn} = [a_{ij} + b_{ij}]_{mn},$$

i.e., to add two matrices of the same "shape," we simply add corresponding elements; matrices not of the same "shape" are not added.

Definition 1.3. The *multiplication of* $[a_{ij}]_{mn}$ *by the real number* c is defined by

$$c[a_{ij}]_{mn} = [ca_{ij}]_{mn},$$

i.e., to multiply a matrix by a number, we simply multiply all of its elements by that number.

Definition 1.4. The *multiplication of* $[a_{ij}]_{mp}$ *times* $[b_{ij}]_{pn}$ is defined by

$$[a_{ij}]_{mp} \cdot [b_{ij}]_{pn} = \left[\sum_{k=1}^{p} a_{ik} b_{kj} \right]_{mn}.$$

Example 3. Find the product $A \cdot B$, where

$$A = \begin{bmatrix} 1 & 3 & 7 \\ 2 & 4 & 3 \end{bmatrix} \quad \text{and} \quad B = \begin{bmatrix} 3 & 2 & 1 & 2 \\ 1 & 4 & 3 & 1 \\ 6 & 2 & 5 & 2 \end{bmatrix}.$$

Solution. The multiplication of matrices is carried out in practice by employing an informal version of the formula given under Definition 1.4. The informal formula is this: To find the (i,j)-element of the matrix $A \cdot B$ (or AB) we *multiply* the ith row of A by the jth column of B. For example, the second row of A is [2, 4, 3] and the third column of B is

$$\begin{bmatrix} 1 \\ 3 \\ 5 \end{bmatrix}.$$

To *multiply* this row by this column shall mean to perform the calculation

$$(2 \cdot 1) + (4 \cdot 3) + (3 \cdot 5) = 29,$$

i.e., we multiply corresponding entries and then add the results. A diagram here may be helpful:

$$\begin{bmatrix} 1 & 3 & 7 \\ 2 & 4 & 3 \end{bmatrix} \begin{bmatrix} 3 & 2 & 1 & 2 \\ 1 & 4 & 3 & 1 \\ 6 & 2 & 5 & 2 \end{bmatrix} = \begin{bmatrix} \square & \square & \square & \square \\ \square & \square & 29 & \square \end{bmatrix}$$

Note in particular the relative dimensions of the three matrices (compare with Definition 1.4).

Let us return to our formal definition of matrix multiplication. Definition 1.4 tells us that the (i,j)-element of $C = AB$ is given by

$$c_{ij} = \sum_{k=1}^{p} a_{ik} b_{kj}.$$

Here p is the "row length" of A and, at the same time, the "column length" of B. (Only when these two dimensions agree can the matrices

4 MATRICES

be multiplied.) Thus, in this case, $p = 3$. It follows that

$$c_{23} = \sum_{k=1}^{3} a_{2k}b_{k3}$$
$$= a_{21}b_{13} + a_{22}b_{23} + a_{23}b_{33}$$
$$= \text{(second row of } A\text{)(third column of } B\text{).}$$

Similarly,
$$c_{ij} = (i\text{th row of } A)(j\text{th column of } B).$$

Therefore, our row-by-column multiplication scheme is valid.

Finally, the reader should verify that

$$\begin{bmatrix} 1 & 3 & 7 \\ 2 & 4 & 3 \end{bmatrix} \begin{bmatrix} 3 & 2 & 1 & 2 \\ 1 & 4 & 3 & 1 \\ 6 & 2 & 5 & 2 \end{bmatrix} = \begin{bmatrix} 48 & 28 & 45 & 19 \\ 28 & 26 & 29 & 14 \end{bmatrix}.$$

Example 4. Find $(2A + B)C$, where

$$A = \begin{bmatrix} 1 & 3 & 5 \\ 2 & 4 & 6 \end{bmatrix}, \quad B = \begin{bmatrix} 0 & 7 & 2 \\ 3 & 0 & -1 \end{bmatrix}, \quad \text{and} \quad C = \begin{bmatrix} 4 & 0 & 1 \\ 2 & 0 & 1 \\ 1 & 2 & 1 \end{bmatrix}.$$

Solution

$$2A + B = \begin{bmatrix} 2 & 6 & 10 \\ 4 & 8 & 12 \end{bmatrix} + \begin{bmatrix} 0 & 7 & 2 \\ 3 & 0 & -1 \end{bmatrix} = \begin{bmatrix} 2 & 13 & 12 \\ 7 & 8 & 11 \end{bmatrix},$$

$$(2A + B)C = \begin{bmatrix} 2 & 13 & 12 \\ 7 & 8 & 11 \end{bmatrix} \begin{bmatrix} 4 & 0 & 1 \\ 2 & 0 & 1 \\ 1 & 2 & 1 \end{bmatrix} = \begin{bmatrix} 46 & 24 & 27 \\ 55 & 22 & 26 \end{bmatrix}.$$

EXERCISE SET 1.1

1 Find (construct) the matrix defined by the given formula.
  a) $A = [(i + 1)(j - 2)]_{33}$ b) $B = [|i - j|]_{35}$

2 Let A and B be as in Exercise 1. Of the following indicated calculations, perform those that make sense: $A + B$, AB, BA, $A(AB)$, $(AB)B$, $AB + 2B$.

3 Let $A = [i]_{33}$ and $B = [j]_{33}$. Find $A(2A + B)$ and $2A^2 + AB$. [*Hint:* $a_{ij} = i + 0 \cdot j$ and $b_{ij} = 0 \cdot i + j$.]

4 Definition 1.5. We define δ_{ij} by

$$\delta_{ij} = \begin{cases} 1, & i = j, \\ 0, & i \neq j. \end{cases}$$

This zero-or-one symbol is called the *Kronecker delta*.

  Construct the given matrix.
  a) $[\delta_{ij}]_{34}$ b) $[j\delta_{ij}]_{44}$ c) $[5\delta_{2j}]_{33}$

1.1 THE ARITHMETIC OF MATRICES

5 Definition 1.6. The nth-order matrix $I_n = [\delta_{ij}]_{nn}$ is called an *identity matrix*. The m by n matrix $\Theta_{mn} = [0]_{mn}$ is called a *zero matrix*. We also write I and Θ if the shape is known from the context.

Let
$$A = \begin{bmatrix} 1 & 2 & 7 \\ -3 & 1 & 2 \\ 0 & 2 & 5 \end{bmatrix}.$$

a) Show that $IA = AI = A$.   b) Show that $A + \Theta = A$.
c) Show that $A + (-1)A = \Theta$. [Because of this, we write $-A$ in place of $(-1)A$.]

6 Definition 1.7. The matrix $(-1)A$ is called the *negative of* A. It is denoted by $-A$. To *subtract B from A* means to add to A the negative of B. We write $A - B = A + (-B)$.

a) Find $A - B$, where A and B are the matrices of Exercise 3.
b) Prove that $A - A = \Theta$.

7 *On the peculiarities of matrix multiplication.*

a) Show that
$$\begin{bmatrix} 1 & 2 \\ 3 & 4 \end{bmatrix} \begin{bmatrix} 5 & 6 \\ 7 & 8 \end{bmatrix} \neq \begin{bmatrix} 5 & 6 \\ 7 & 8 \end{bmatrix} \begin{bmatrix} 1 & 2 \\ 3 & 4 \end{bmatrix}.$$

Compare with $xy = yx$.

b) Show that
$$\begin{bmatrix} 1 & 2 \\ -4 & -8 \end{bmatrix} \begin{bmatrix} -2 & 6 \\ 1 & -3 \end{bmatrix} = \begin{bmatrix} 0 & 0 \\ 0 & 0 \end{bmatrix}.$$

Compare with $xy = 0$.

c) Show that
$$\begin{bmatrix} -1 & \tfrac{1}{4} \\ -8 & 2 \end{bmatrix}^2 = \begin{bmatrix} -1 & \tfrac{1}{4} \\ -8 & 2 \end{bmatrix}.$$

Compare with $x^2 = x$.

d) Let
$$A = [1, 3, -4] \quad \text{and} \quad B = \begin{bmatrix} 0 \\ 6 \\ 1 \end{bmatrix}.$$

Find AB and BA.

8 Let A, B, C, and D be nth-order matrices.
a) Given that $d_{ij} = a_{ij} + 2b_{ij} + 3c_{ij}$, express D in terms of A, B, and C.
b) Same as (a), where
$$d_{ij} = \sum_{k=1}^{n} (a_{ik} + b_{ik}) c_{kj}.$$

c) Same as (a), where
$$d_{ij} = \sum_{k=1}^{n} \left(\sum_{h=1}^{n} c_{ih} a_{hk} \right) b_{kj}.$$

9 Let $A = [a_{ij}]_{mn}$.
   a) Find (write a formula for) the matrix B such that BA is the pth row of A.
   b) Find the matrix C such that AC is the qth column of A.

10 Prove: If A is any m-rowed matrix, then $I_m A = A$. [*Hint:* Let $C = I_m A$, and begin by writing out the summation formula for c_{ij}.]

11 Prove: If the pth row of A is a "zero row," then so is the pth row of the product AB. [*Hint:* Let $C = AB$, and begin by writing out the summation formula for c_{pj}.]

12 Let $C = AB$, where $A = [i + 2j]_{23}$ and $B = [3i - j]_{34}$. Find a formula for C that is analogous to those given for A and B.

13 Find all second order matrices that satisfy the equation

$$A^2 = \begin{bmatrix} 1 & 2 \\ 0 & 4 \end{bmatrix}.$$

Hint: Let $A = \begin{bmatrix} w & x \\ y & z \end{bmatrix}$.

1.2 PROPERTIES OF MATRIX ARITHMETIC

The arithmetic of matrices is in a number of ways rather strange. In particular, it should be noted that the products AB and BA are usually different. On the other hand, $A(B + C)$ and $AB + AC$ are always the same. In other words, some properties of ordinary arithmetic are preserved and some are not. Now it is essential that we know exactly what can and cannot be done in matrix calculations. Thus, we gather together in the following three theorems the basic calculation rules of matrix arithmetic. (In expressing these rules we use capital letters to denote matrices and lowercase letters to denote real numbers. This practice will be followed throughout the text.)

Theorem 1.1. *Commutative and associative laws:*
1. $A + B = B + A$.
2. $(A + B) + C = A + (B + C)$ and $(AB)C = A(BC)$.

Theorem 1.2. *Distributive laws:*
1. $a(B + C) = aB + aC$ and $(a + b)C = aC + bC$.
2. $A(B + C) = AB + AC$ and $(A + B)C = AC + BC$.

Theorem 1.3. *Mixed multiplication laws:*
1. $(ab)C = a(bC)$.
2. $a(BC) = (aB)C = B(aC)$.

Only one of these properties is at all difficult to prove: the associative law for matrix multiplication.

Proof. Let $A = [a_{ij}]_{mp}$, $B = [b_{ij}]_{pq}$, and $C = [c_{ij}]_{qn}$. It should first be noted that the dimensions of A, B, and C have been chosen so that the product $(AB)C$ is possible. It then turns out, fortunately, that these same dimensions make possible the product $A(BC)$. Moreover, in each case, the dimensions of the final product are m by n.

We now let $D = (AB)C$ and $E = A(BC)$. What we must prove is that $d_{ij} = e_{ij}$ for all i and j. However, we shall show only—to simplify things a bit—that $d_{23} = e_{23}$. If the reader then replaces (in our argument) the 2 by i and the 3 by j, he will have a proof that $d_{ij} = e_{ij}$.

We begin by writing

$$d_{23} = \text{(second row of } AB\text{)(third column of } C\text{)}.$$

Now, a typical entry in the second row of AB, say the kth, is given by $\sum_{h=1}^{p} a_{2h} b_{hk}$. Thus,

$$d_{23} = \sum_{k=1}^{q} \left\{ \left(\sum_{h=1}^{p} a_{2h} b_{hk} \right) (c_{k3}) \right\}.$$

The kth term of this sum—the expression within the braces—is itself a sum multiplied by c_{k3}. This multiplication can be carried out by multiplying each term in the sum by c_{k3}, that is,

$$\left(\sum_{h=1}^{p} a_{2h} b_{hk} \right) (c_{k3}) = \sum_{h=1}^{p} a_{2h} b_{hk} c_{k3}.$$

It follows that

$$d_{23} = \sum_{k=1}^{q} \left\{ \sum_{h=1}^{p} a_{2h} b_{hk} c_{k3} \right\}.$$

Now, if the reader will but mimic the above discussion starting with e_{23}, he will arrive at

$$e_{23} = \sum_{h=1}^{p} \left\{ \sum_{k=1}^{q} a_{2h} b_{hk} c_{k3} \right\}.$$

Despite the difference in the order of summation, these two *double sums* are the same. To see this, suppose that we are asked to find the sum of all of the entries in the matrix $[\alpha_{hk}]$, where $\alpha_{hk} = a_{2h} b_{hk} c_{k3}$. There are two natural ways to proceed: (1) Add the columns of the matrix and then add the subtotals. (The formula for d_{23} does just this.) (2) Add the rows of the matrix and then add the subtotals. (The formula for e_{23} does this.)

Example 1. Solve the matrix equation

$$5A + 4I = \begin{bmatrix} 3 & 1 \\ -7 & 6 \end{bmatrix}.$$

Solution. We note first—because of the right-hand side—that I and the unknown matrix A are both of the second order. We then proceed in the following "natural" manner:

$$5A = \begin{bmatrix} 3 & 1 \\ -7 & 6 \end{bmatrix} - 4I,$$

$$5A = \begin{bmatrix} -1 & 1 \\ -7 & 2 \end{bmatrix},$$

$$A = \begin{bmatrix} -\frac{1}{5} & \frac{1}{5} \\ -\frac{7}{5} & \frac{2}{5} \end{bmatrix}.$$

What properties of matrix arithmetic have been used here?

Example 2. Solve
$$A \begin{bmatrix} 2 & 3 \\ 4 & 5 \end{bmatrix} = I_2.$$

Solution. It would be natural to "divide" both sides by
$$\begin{bmatrix} 2 & 3 \\ 4 & 5 \end{bmatrix},$$
but, so far, at least, we have no division. However, we may proceed as follows: If there is such an A, then it is a second-order matrix. Thus, we let
$$A = \begin{bmatrix} w & x \\ y & z \end{bmatrix}.$$

Then our equation may be written as
$$\begin{bmatrix} w & x \\ y & z \end{bmatrix} \begin{bmatrix} 2 & 3 \\ 4 & 5 \end{bmatrix} = \begin{bmatrix} 1 & 0 \\ 0 & 1 \end{bmatrix}$$
or,
$$\begin{bmatrix} (2w + 4x) & (3w + 5x) \\ (2y + 4z) & (3y + 5z) \end{bmatrix} = \begin{bmatrix} 1 & 0 \\ 0 & 1 \end{bmatrix}.$$

By identifying entries in corresponding positions, we are led to two systems of numerical equations:

$$\begin{cases} 2w + 4x = 1, \\ 3w + 5x = 0, \end{cases} \text{ and } \begin{cases} 2y + 4z = 0, \\ 3y + 5z = 1. \end{cases}$$

The reader may discover for himself that $w = -\frac{5}{2}$, $x = \frac{3}{2}$, $y = 2$, and $z = -1$. Thus,
$$A = \begin{bmatrix} -\frac{5}{2} & \frac{3}{2} \\ 2 & -1 \end{bmatrix}.$$

The reader should verify that this A does indeed satisfy the original equation.

EXERCISE SET 1.2

1 Use the "values"
$$a = 3, \quad b = 5,$$
$$A = \begin{bmatrix} 1 & 2 \\ 3 & 4 \end{bmatrix}, \quad B = \begin{bmatrix} 5 & 3 \\ 2 & 1 \end{bmatrix}, \quad \text{and} \quad C = \begin{bmatrix} 1 & 2 \\ 3 & 0 \end{bmatrix}$$
to illustrate each of the properties considered in Theorems 1.1, 1.2, and 1.3.

2 Solve the given equation.

a) $2A - 3I = \begin{bmatrix} 5 & 2 & 3 \\ 6 & 1 & 7 \\ 4 & 1 & 2 \end{bmatrix}$
b) $3(A + 5I_2) = 2(A - I_2)$

3 Solve the following systems of matrix equations.

a) $\begin{cases} 2A + B = \begin{bmatrix} 1 & 2 \\ 3 & 4 \end{bmatrix}, \\ A - 3B = \begin{bmatrix} 5 & 6 \\ 7 & 8 \end{bmatrix} \end{cases}$
b) $\begin{cases} A\left(B + \begin{bmatrix} 1 & 2 \\ 3 & 4 \end{bmatrix}\right) = \begin{bmatrix} 4 & 2 \\ 1 & 3 \end{bmatrix}, \\ A\left(B + \begin{bmatrix} 3 & 7 \\ 1 & 0 \end{bmatrix}\right) = \begin{bmatrix} 5 & 3 \\ 1 & 2 \end{bmatrix} \end{cases}$

4 Solve the given equation.

a) $A \begin{bmatrix} 1 & 2 \\ 2 & 5 \end{bmatrix} = I$
b) $\begin{bmatrix} 1 & 2 \\ 2 & 5 \end{bmatrix} B = I$

c) $\begin{bmatrix} 1 & 2 \\ 2 & 5 \end{bmatrix} C = \begin{bmatrix} 3 & 4 \\ 7 & 1 \end{bmatrix}$ [*Hint:* Multiply both sides by the matrix obtained in part (a).]

5 Verify that
$$\begin{bmatrix} -3 & \frac{3}{2} \\ 4 & -2 \end{bmatrix} \begin{bmatrix} 1 & 2 \\ 2 & 4 \end{bmatrix} = \Theta.$$
Then do the following.

a) Solve $A \begin{bmatrix} 1 & 2 \\ 2 & 4 \end{bmatrix} = \Theta_{22}$. *Answer:* All matrices of the form $\begin{bmatrix} -2x & x \\ -2z & z \end{bmatrix}$.

b) Solve $A \begin{bmatrix} 3 & 7 \\ -2 & 5 \end{bmatrix} = A \begin{bmatrix} 2 & 5 \\ -4 & 1 \end{bmatrix}$, where A is 2 by 2.

c) Solve $A \begin{bmatrix} 1 & 2 \\ 3 & 4 \end{bmatrix} = \Theta_{22}$.

d) Prove or disprove: If $AB = \Theta$ and $B \neq \Theta$, then $A = \Theta$.

6 Let $A = [a_{ij}]_{mn}$, $B = [b_{ij}]_{np}$, and $C = [c_{ij}]_{np}$.

a) Let $D = A(B + C)$. Write the summation formula for d_{ij}, taking k as the index (variable) of summation.
b) Let $E = AB$. Same as (a) for e_{ij}.
c) Let $F = AC$. Same as (a) for f_{ij}.
d) Show that $d_{ij} = e_{ij} + f_{ij}$.
e) What property of matrix arithmetic has been proven in parts (a) through (d)?

7 Definition 1.8. Given numbers $d_1, d_2, \ldots, d_n$, a matrix of the form $[d_j \delta_{ij}]_{nn}$ is called a *diagonal matrix*. If the d_j are all alike, it is called a *scalar matrix*.
   a) Construct $A = [d_j \delta_{ij}]_{33}$, where $d_1 = 2$, $d_2 = 5$, and $d_3 = 7$.
   b) Solve for B: $AB = I$.
   c) Let
$$C = \begin{bmatrix} 1 & 3 & 2 \\ 4 & 0 & 6 \\ 7 & 1 & 1 \end{bmatrix}.$$

   Find AC and CA, and then make a conjecture about the multiplication of matrices by diagonal matrices.
   d) Prove that the product of two diagonal matrices is again a diagonal matrix. [*Hint:* Let $A = [d_j \delta_{ij}]_{nn}$ and $B = [e_j \delta_{ij}]_{nn}$.]

8 Definition 1.9. Let $A = [a_{ij}]_{nn}$. If $a_{ij} = a_{ji}$ for all i and j, then A is said to be a *symmetric matrix*.
   a) Construct $A = [ij]_{33}$ and $B = [(-1)^{ij}(i+j)]_{33}$.
   b) Find $A + B$ and AB.
   c) Do there exist symmetric matrices A and B such that $A + B$ is not symmetric? Prove your answer.

9 Definition 1.10. Let $A = [a_{ij}]_{nn}$ and suppose that a_{ij} has the property $a_{ij} = 0$ for $i < j$. Then A is said to be a *lower triangular matrix*.
   a) Construct $T = [\gamma_{ij}]_{33}$, where $\gamma_{ij} = \sum_{k=j}^{3} \delta_{ik}$.
   b) Construct $A = [(i + 2j)\gamma_{ij}]_{33}$.
   c) Construct $B = [(i^2 j)\gamma_{ij}]_{33}$.
   d) Find $A + B$ and AB.
   e) Part (d) should suggest a theorem about lower triangular matrices. What is it?
   f) Write a definition for an upper triangular matrix.

10 By following the suggested outline, prove the following theorem: If A and B are nth-order, lower triangular matrices, then so is AB.
   a) Let $\gamma_{ij} = \sum_{k=j}^{n} \delta_{ik}$ and then prove the following: If $p < q$, then $\gamma_{pq} = 0$; if $p \geq q$, then $\gamma_{pq} = 1$.
   b) Prove: If $i < j$, then $\gamma_{ik}\gamma_{kj} = 0$ for $k = 1, 2, \ldots, n$. [*Hint:* First consider $k \leq i$.]
   c) Let $A = [\alpha_{ij}\gamma_{ij}]_{nn}$, $B = [\beta_{ij}\gamma_{ij}]_{nn}$, and let $C = AB$. Then prove that $c_{ij} = 0$ for $i < j$.

11 Find second-order matrices A and B such that $(AB)^2 \neq A^2 B^2$. [This shows that the law of exponents $(xy)^n = x^n y^n$ is not valid in matrix arithmetic.]

12 a) Prove: $(A + I)(A - I) = A^2 - I$.
   b) By exhibiting a suitable counterexample, show that
$$(A + B)(A - B) = A^2 - B^2$$

   is *not* a matrix identity.

13 Which of the following 3 by 1 matrices are solutions of the matrix equation
$$\begin{bmatrix} 1 & 0 & 2 \\ 0 & 1 & 1 \\ 1 & 2 & 4 \end{bmatrix} A = \begin{bmatrix} 1 \\ 3 \\ 7 \end{bmatrix}?$$

a) $\begin{bmatrix} -1 \\ 2 \\ 1 \end{bmatrix}$
b) $\begin{bmatrix} 4 \\ 1 \\ 2 \end{bmatrix}$
c) $\begin{bmatrix} 3 \\ 4 \\ -1 \end{bmatrix}$

14 Find all matrices of the form
$$\begin{bmatrix} a & b \\ -b & a \end{bmatrix}$$
that satisfy the equation $A^2 + I_2 = \Theta_{22}$.

1.3 ELEMENTARY MATRICES

We have learned so far how to add, subtract, and multiply matrices; it remains for us to consider the operation of division. First, recall that to divide a number x by a number y means to multiply x by the *reciprocal* of y. Thus, to proceed analogously with matrices we have first to consider the matrix analog of the reciprocal of a number.

Now it is clear that the identity matrix plays a role in matrix arithmetic that is analogous to the role played by the number 1 in ordinary arithmetic. Thus, the reciprocal of a given matrix A should be a matrix B such that $AB = I$. But already we have a problem: because matrix multiplication is not commutative, we might just as easily decree that the reciprocal of A is a matrix C such that $CA = I$. Thus, we may have to consider both *left reciprocals* and *right reciprocals*. Now before making any definitions at all, it is best that we consider some examples.

Example 1. Let
$$A = \begin{bmatrix} 1 & 2 & 3 \\ 4 & 5 & 6 \end{bmatrix}.$$
Find a matrix (or matrices) B such that $AB = I$.

Solution. Since A is 2 by 3, B must be 3 by something, and since the product AB is to be an identity matrix, hence square, it follows that the something must be 2. Thus, we let
$$B = \begin{bmatrix} u & v \\ w & x \\ y & z \end{bmatrix}$$
and try to solve the matrix equation
$$\begin{bmatrix} 1 & 2 & 3 \\ 4 & 5 & 6 \end{bmatrix} \begin{bmatrix} u & v \\ w & x \\ y & z \end{bmatrix} = \begin{bmatrix} 1 & 0 \\ 0 & 1 \end{bmatrix}.$$

Multiplying out the left-hand side and equating corresponding entries leads to the system
$$\begin{cases} u + 2w + 3y = 1, \\ 4u + 5w + 6y = 0, \\ v + 2x + 3z = 0, \\ 4v + 5x + 6z = 1, \end{cases}$$
which contains four equations in six unknowns. We shall not solve this system. However, it is easy to see that there are infinitely many solutions; arbitrarily choosing values for any two of the unknowns gives rise to a system of four equations in four unknowns that has a solution. For example, taking $u = v = 0$ leads to the system having the solution $w = 2$, $x = 1$, $y = \frac{5}{3}$, and $z = \frac{2}{3}$. Then, constructing the corresponding matrix B allows us to check that AB does indeed equal I:
$$\begin{bmatrix} 1 & 2 & 3 \\ 4 & 5 & 6 \end{bmatrix} \begin{bmatrix} 0 & 0 \\ -2 & 1 \\ \frac{5}{3} & -\frac{2}{3} \end{bmatrix} = \begin{bmatrix} 1 & 0 \\ 0 & 1 \end{bmatrix}.$$
However, reversing the order of multiplication gives
$$\begin{bmatrix} 0 & 0 \\ -2 & 1 \\ \frac{5}{3} & -\frac{2}{3} \end{bmatrix} \begin{bmatrix} 1 & 2 & 3 \\ 4 & 5 & 6 \end{bmatrix} = \begin{bmatrix} 0 & 0 & 0 \\ 2 & 1 & 0 \\ -1 & 0 & 1 \end{bmatrix}.$$

Example 2. Let A be as in Example 1. Find a matrix (or matrices) C such that $CA = I$.

Solution. Reasoning as in the solution of Example 1, we see that I must be a third-order matrix and that C must be 3 by 2. Thus, we are led to the matrix equation
$$\begin{bmatrix} u & v \\ w & x \\ y & z \end{bmatrix} \begin{bmatrix} 1 & 2 & 3 \\ 4 & 5 & 6 \end{bmatrix} = \begin{bmatrix} 1 & 0 & 0 \\ 0 & 1 & 0 \\ 0 & 0 & 1 \end{bmatrix}.$$
Multiplying out the left-hand side and equating corresponding entries leads to a system having nine equations but only six unknowns. Such a system rarely has a solution. For example, in the present case we have three equations involving just u and v:
$$\begin{cases} u + 4v = 1, \\ 2u + 5v = 0, \\ 3u + 6v = 0. \end{cases}$$
The only values satisfying both the second and third equations are $u = 0$ and $v = 0$, but these values do not satisfy the first equation. Thus, our system evidently has no solution, from which we conclude that there is no matrix C such that $CA = I$. ∎

The point of Examples 1 and 2 is this: In the case of a nonsquare matrix, there seems to be no natural way of introducing the notion of a matrix reciprocal. Thus, we turn our attention to square matrices; here we meet with greater success.

Example 3. Let
$$A = \begin{bmatrix} 1 & 2 \\ 3 & 4 \end{bmatrix}.$$
Find B such that $AB = I$.

Solution. The reader may easily solve the matrix equation
$$\begin{bmatrix} 1 & 2 \\ 3 & 4 \end{bmatrix} \begin{bmatrix} w & x \\ y & z \end{bmatrix} = \begin{bmatrix} 1 & 0 \\ 0 & 1 \end{bmatrix}$$
to obtain the unique solution $w = -2$, $x = 1$, $y = \frac{3}{2}$, and $z = -\frac{1}{2}$. Thus, our problem has a single solution:
$$B = \begin{bmatrix} -2 & 1 \\ \frac{3}{2} & -\frac{1}{2} \end{bmatrix}.$$
The reader should verify that $AB = I$. But also (now we seem to be on the right track) he may verify that $BA = I$. In other words, for the given matrix A there exists a unique matrix B such that $AB = BA = I$. The matrix B seems truly to merit being called the *reciprocal of A*, or, using the standard language of matrix algebra, B is the *inverse of A*.

Definition 1.11. Given an nth-order matrix A, let B be a second nth-order matrix such that $AB = BA = I$. Then B is called an *inverse* of A. We usually denote an inverse of A by A^{-1}. If a matrix has no inverse, then we call it a *singular matrix*. Otherwise, it is *nonsingular*.

Our definition does not presume that a matrix can have only one inverse. However, this is indeed the case (see Exercise 8a). Thus, it becomes proper to speak of "the inverse" rather than "an inverse." Also, our definition allows for the existence of square matrices that do not have inverses. To see that there are such things, the reader should consider Exercise 5.

It would now seem appropriate to define the *division of C by a nonsingular matrix A*. We might write: To *divide C by A* means to multiply C by the inverse of A. But, multiply how? On the left, or on the right? In other words, we must anticipate that $A^{-1}C$ and CA^{-1} will, in general, be different. It seems, then, that two definitions are required, one for *left division* and one for *right division*. Because of this complication, and also because not all matrices have inverses, we simply refrain from introducing a division operation into our matrix arithmetic. However, we do multiply—on either side, as the need arises—by the inverses of matrices.

The method we used in Example 3 to find the inverse of a matrix—that of solving a system of equations—is most unsatisfactory for matrices of high order. Thus, it is important that the reader have a more efficient method. We shall attend to this in the next section; the method that we consider there is based on the notion of an *elementary matrix*.

Definition 1.12. An *elementary matrix of order n* is a matrix fitting any of the following descriptions:

1. The matrix that results on interchanging any two rows of I_n. We denote this type by $[i \leftrightarrow k]_n$, where the rows interchanged are the ith and kth.
2. The matrix that results on multiplying a single row of I_n by the number $c \neq 0$. We denote this type by $[i \to (c)i]_n$, where it is the ith row that is multiplied by c.
3. The matrix that results on replacing a given row of I_n by the sum of that row and c times some other row. We denote this type by
$$[i \to i + (c)k]_n,$$
where it is the ith row that is increased by c times the kth.

Example 4. Let $A = [i + 2j]_{33}$. Find $E_1 A$, $E_2 A$, and $E_3 A$, where $E_1 = [2 \leftrightarrow 3]_3$, $E_2 = [1 \to (5)1]_3$, and $E_3 = [3 \to 3 + (2)2]_3$.

Solution. We first construct the four matrices

$$A = \begin{bmatrix} 3 & 5 & 7 \\ 4 & 6 & 8 \\ 5 & 7 & 9 \end{bmatrix}, \quad E_1 = \begin{bmatrix} 1 & 0 & 0 \\ 0 & 0 & 1 \\ 0 & 1 & 0 \end{bmatrix},$$

$$E_2 = \begin{bmatrix} 5 & 0 & 0 \\ 0 & 1 & 0 \\ 0 & 0 & 1 \end{bmatrix}, \quad \text{and} \quad E_3 = \begin{bmatrix} 1 & 0 & 0 \\ 0 & 1 & 0 \\ 0 & 2 & 1 \end{bmatrix}.$$

The reader should now carry out each of the multiplications to see that

$$E_1 A = \begin{bmatrix} 3 & 5 & 7 \\ 5 & 7 & 9 \\ 4 & 6 & 8 \end{bmatrix}, \quad E_2 A = \begin{bmatrix} 15 & 25 & 35 \\ 4 & 6 & 8 \\ 5 & 7 & 9 \end{bmatrix},$$

and

$$E_3 A = \begin{bmatrix} 3 & 5 & 7 \\ 4 & 6 & 8 \\ 13 & 19 & 25 \end{bmatrix}. \quad \square$$

A careful study of these results motivates the following theorem.

Theorem 1.4. *On premultiplication by elementary matrices.* Let E be an elementary matrix. Then the product matrix EA is the same as that resulting on manipulating the rows of A in the same manner as those of I were manipulated to produce E in the first place.

Proof. We will consider here only the case where E is an elementary matrix of type 3 (see Exercises 11 and 12). Let E be an elementary matrix of type 3; that is, $E = [i \to i + (c)k]_n$. Then we may write

$$E = I + cU_{ik},$$

where U_{ik} is the matrix having a 1 in the (i, k)-position and 0's elsewhere. It follows that

$$EA = (I + cU_{ik})A = A + cU_{ik}A.$$

Now, let $B = cU_{ik}A$ and consider the rows of B. The pth row, where $p \neq i$, has entries of the form

$$b_{pj} = c \sum_{h=1}^{n} u_{ph}a_{hj} = c \sum_{h=1}^{n} 0 \cdot a_{hj} = 0.$$

The ith row has entries of the form

$$b_{ij} = c \sum_{h=1}^{n} u_{ih}a_{hj} = cu_{ik}a_{kj} = ca_{kj}.$$

Thus, adding B to A affects only the ith row of A; and in exactly the way that it should, i.e., it adds on to the ith row c times the kth row. □

Now suppose that there are good reasons—we shall show that there are—for wanting to modify a given matrix in the ways that multiplication by an elementary matrix does. Then we must be able to produce on command the elementary matrix which will do a specific task. An informal paraphrasing of Theorem 1.4 produces the following "golden rule":

Do unto the rows of the identity matrix that which you would have the resulting elementary matrix do (as a premultiplier) unto the rows of A.

EXERCISE SET 1.3

1 Find (construct) the elementary matrices defined by the following formulas.
  a) $E_1 = [2 \leftrightarrow 4]_4$
  b) $E_2 = [3 \to (2)3]_4$
  c) $E_3 = [2 \to 2 + (-3)4]_4$

2 Let

$$A = \begin{bmatrix} 1 & 2 \\ 4 & -3 \\ 2 & 0 \\ 1 & 3 \end{bmatrix}$$

and let E_1, E_2, and E_3 be as in Exercise 1. Find E_1A, E_2A, and E_3A.

3 Let A be any 3 by 3 matrix. Find the elementary matrix that (as a premultiplier) (a) multiplies the third row of A by -13, (b) interchanges the first and third rows of A, and (c) adds 5 times the second row of A to its third row.

4 Let
$$A = \begin{bmatrix} 2 & 1 & -2 \\ 3 & 0 & 5 \end{bmatrix}$$
and let E_1, E_2, and E_3 be the three elementary matrices described in Exercise 3.
a) Find AE_1, AE_2, and AE_3.
b) Study the above result and then guess the "golden rule" for postmultiplying a matrix A by an elementary matrix.
c) Find a matrix E that (as a postmultiplier) adds 4 times the third column of A to its first column.

5 Let
$$A = \begin{bmatrix} 1 & 2 \\ 4 & 8 \end{bmatrix} \quad \text{and} \quad B = \begin{bmatrix} -2 & -6 \\ 1 & 3 \end{bmatrix}.$$
a) Show that $AB = \Theta$.
b) Try to find an inverse for A using the method of Example 3.
c) *Another point of view:* Assume that A has an inverse and then produce a contradiction by premultiplying each side of $AB = \Theta$ by A^{-1}.

6 Find by inspection the inverse of the given matrix

a) $\begin{bmatrix} 1 & 0 & 0 \\ 0 & 2 & 0 \\ 0 & 0 & 3 \end{bmatrix}$
b) $\begin{bmatrix} 1 & 0 & 0 \\ 0 & 1 & 1 \\ 0 & 0 & 1 \end{bmatrix}$
c) $\begin{bmatrix} 1 & 0 & 0 \\ 0 & 0 & 1 \\ 0 & 1 & 0 \end{bmatrix}$

7 Theorem 1.5. Every elementary matrix is nonsingular.

Using the special elementary matrix notation, write out the three kinds of elementary matrices along with their inverses.

8 Theorem 1.6. A nonsingular matrix has exactly one inverse.

a) Prove Theorem 1.6 by completing the following argument: Suppose that A has two distinct inverses, say B and C. Then we may write
$$B = IB = (CA)B = \cdots.$$
b) Prove: If $BA = I$ and $AC = I$, then $B = C$ and B is the inverse of A.

9 Prove: If A is a square matrix and B is a nonzero matrix such that $AB = \Theta$, then A is singular. [*Hint:* Consider a proof by contradiction.]

10 Let
$$A = \begin{bmatrix} 1 & 0 & 2 & 1 \\ 0 & 1 & 0 & 1 \\ 1 & 1 & 2 & 2 \end{bmatrix}.$$
Find matrices E and F such that the matrix EAF has the following properties: (1) For some number p the entries $a_{11}, a_{22}, \ldots, a_{pp}$ are all 1's. (2) All other entries are 0's.

11 Prove Theorem 1.4 for the case where $E = [i \to (c)i]_n$.

12 Let U_{pq} be the nth-order matrix having 1 in the (p, q)-position and 0's elsewhere, and let A be an arbitrary nth-order matrix.

a) Let $B = U_{pq}A$. Prove that the pth row of B is the qth row of A and that all other rows of B are zero rows.

b) Prove Theorem 1.4 for the case where $E = [i \leftrightarrow k]_n$. [*Hint:* You may use the fact that $E = I - U_{ii} + U_{ik} - U_{kk} + U_{ki}$.]

13 Prove: If an nth-order matrix A commutes with every other nth-order matrix, then A is necessarily a scalar matrix. [*Hint:* Let U_{pq} be the nth-order matrix having 1 in the (p, q)-position and 0's elsewhere. Then, to show that $a_{rs} = 0$ for $r \neq s$, equate the $(1, s)$-entries of AU_{1r} and $U_{1r}A$. To show that $a_{rr} = a_{11}$ for all r, equate their $(1, r)$-entries.]

14 a) Find the product

$$[2 \to 2 + (1)3]_4 \cdot [2 \to (-1)2]_4 \cdot [3 \to 3 + (1)2]_4 \cdot [2 \to 2 + (-1)3]_4.$$

b) The calculation carried out under part (a) suggests that one of the three types of elementary matrices is actually expendable. Explain.

1.4 A METHOD FOR FINDING A^{-1}

In this section we shall consider an *algorithm* for finding the inverse of a matrix. (An algorithm is, to a mathematician, a recipe for obtaining a desired result that requires a succession of steps that are repetitive in nature.) Our approach will be informal; i.e., rather than present theorems and proofs to justify our algorithm, we will simply present it by way of an illustrative example. First, there are some preliminaries.

Given a square matrix A, suppose that we have found elementary matrices $E_1, E_2, \ldots, E_n$ such that

$$(E_n \cdots (E_2(E_1 A)) \cdots) = I.$$

Then, because of associativity, we may write

$$(E_n \cdots E_2 E_1)A = I.$$

This suggests that the matrix $B = E_n \cdots E_2 E_1$ is the inverse of A (suggests, not proves, because it may be that $AB \neq I$). But this is not the case, for if $E_n \cdots E_2 E_1 A = I$, then we may multiply successively (on the left) by the inverses of the given elementary matrices as follows:

$$E_1^{-1}(E_2^{-1} \cdots (E_n^{-1}(E_n \cdots E_2 E_1 A)) \cdots) = E_1^{-1}(E_2^{-1} \cdots (E_n^{-1} I) \cdots).$$

Successive cancellations lead to

$$A = E_1^{-1} E_2^{-1} \cdots E_n^{-1}.$$

It is supposed here that the reader has discovered (see Exercise 7 of Exercise Set 1.3) that all elementary matrices have inverses. It then follows that

$$AB = (E_1^{-1}E_2^{-1} \cdots E_n^{-1})(E_n \cdots E_2 E_1) = \cdots = E_1^{-1}E_1 = I.$$

Thus, the question of finding A^{-1} may now be replaced by: How do we find elementary matrices having the property described in the foregoing discussion? A numerical example will best serve to illustrate the procedure. Let

$$A = \begin{bmatrix} 0 & 2 & 3 \\ 1 & 4 & 7 \\ 2 & 3 & 6 \end{bmatrix}.$$

We must think now in terms of *row operations* that move A toward I. First of all, we need a 1 in the (1, 1)-position. One way to accomplish this is to interchange rows 1 and 2 by premultiplying A by the appropriate elementary matrix:

$$\begin{bmatrix} 0 & 1 & 0 \\ 1 & 0 & 0 \\ 0 & 0 & 1 \end{bmatrix} \begin{bmatrix} 0 & 2 & 3 \\ 1 & 4 & 7 \\ 2 & 3 & 6 \end{bmatrix} = \begin{bmatrix} 1 & 4 & 7 \\ 0 & 2 & 3 \\ 2 & 3 & 6 \end{bmatrix} \quad (E_1 = [1 \leftrightarrow 2]).$$

Next, we want 0's below the 1:

$$\begin{bmatrix} 1 & 0 & 0 \\ 0 & 1 & 0 \\ -2 & 0 & 1 \end{bmatrix} \begin{bmatrix} 1 & 4 & 7 \\ 0 & 2 & 3 \\ 2 & 3 & 6 \end{bmatrix} = \begin{bmatrix} 1 & 4 & 7 \\ 0 & 2 & 3 \\ 0 & -5 & -8 \end{bmatrix}$$

$$(E_2 = [3 \to 3 + (-2)1]).$$

We now move to the second column, transforming the 2 into a 1 and putting 0's above and below the 1:

$$\begin{bmatrix} 1 & 0 & 0 \\ 0 & \frac{1}{2} & 0 \\ 0 & 0 & 1 \end{bmatrix} \begin{bmatrix} 1 & 4 & 7 \\ 0 & 2 & 3 \\ 0 & -5 & -8 \end{bmatrix} = \begin{bmatrix} 1 & 4 & 7 \\ 0 & 1 & \frac{3}{2} \\ 0 & -5 & -8 \end{bmatrix}$$

$$(E_3 = [2 \to (\tfrac{1}{2})2]),$$

$$\begin{bmatrix} 1 & -4 & 0 \\ 0 & 1 & 0 \\ 0 & 0 & 1 \end{bmatrix} \begin{bmatrix} 1 & 4 & 7 \\ 0 & 1 & \frac{3}{2} \\ 0 & -5 & -8 \end{bmatrix} = \begin{bmatrix} 1 & 0 & 1 \\ 0 & 1 & \frac{3}{2} \\ 0 & -5 & -8 \end{bmatrix}$$

$$(E_4 = [1 \to 1 + (-4)2]),$$

$$\begin{bmatrix} 1 & 0 & 0 \\ 0 & 1 & 0 \\ 0 & 5 & 1 \end{bmatrix} \begin{bmatrix} 1 & 0 & 1 \\ 0 & 1 & \frac{3}{2} \\ 0 & -5 & -8 \end{bmatrix} = \begin{bmatrix} 1 & 0 & 1 \\ 0 & 1 & \frac{3}{2} \\ 0 & 0 & -\frac{1}{2} \end{bmatrix}$$

$$(E_5 = [3 \to 3 + (5)2]).$$

We then proceed to the third column, putting a 1 at the bottom and 0's above:

$$\begin{bmatrix} 1 & 0 & 0 \\ 0 & 1 & 0 \\ 0 & 0 & -2 \end{bmatrix} \begin{bmatrix} 1 & 0 & 1 \\ 0 & 1 & \frac{3}{2} \\ 0 & 0 & -\frac{1}{2} \end{bmatrix} = \begin{bmatrix} 1 & 0 & 1 \\ 0 & 1 & \frac{3}{2} \\ 0 & 0 & 1 \end{bmatrix}$$

$$(E_6 = [3 \to (-2)3]),$$

$$\begin{bmatrix} 1 & 0 & 0 \\ 0 & 1 & -\frac{3}{2} \\ 0 & 0 & 1 \end{bmatrix} \begin{bmatrix} 1 & 0 & 1 \\ 0 & 1 & \frac{3}{2} \\ 0 & 0 & 1 \end{bmatrix} = \begin{bmatrix} 1 & 0 & 1 \\ 0 & 1 & 0 \\ 0 & 0 & 1 \end{bmatrix}$$

$$(E_7 = [2 \to 2 + (-\tfrac{3}{2})3]).$$

Finally,

$$\begin{bmatrix} 1 & 0 & -1 \\ 0 & 1 & 0 \\ 0 & 0 & 1 \end{bmatrix} \begin{bmatrix} 1 & 0 & 1 \\ 0 & 1 & 0 \\ 0 & 0 & 1 \end{bmatrix} = \begin{bmatrix} 1 & 0 & 0 \\ 0 & 1 & 0 \\ 0 & 0 & 1 \end{bmatrix}$$

$$(E_8 = [1 \to 1 + (-1)3]).$$

Thus,

$$E_8 \cdots E_2 E_1 A = I$$

and

$$A^{-1} = E_8 \cdots E_2 E_1 = \begin{bmatrix} 3 & -3 & 2 \\ 8 & -6 & 3 \\ -5 & 4 & -2 \end{bmatrix}.$$

(The reader should verify that $AA^{-1} = A^{-1}A = I$.)

The above procedure involves a lot of wasted effort; it must be refined. First of all, the multiplication required to produce A^{-1} need not be postponed until the end; rather, we could be constructing A^{-1} as we go along. Consider the following pattern:

$$\begin{array}{cc} A & I \\ E_1 A & E_1 I \\ E_2 E_1 A & E_2 E_1 I \\ \vdots & \vdots \\ E_n \cdots E_2 E_1 A & E_n \cdots E_2 E_1 I \\ I & A^{-1} \end{array}$$

In other words, the same multipliers (and in the same order) that transform A into I will also turn I into A^{-1}. But each such multiplication corresponds to a row operation. This means that we can pass from A to I and I to A^{-1} just by performing an identical sequence of row operations on A and I, not once writing down—or even thinking of—an elementary matrix. The accompanying diagram shows how little need actually be written down.

	A			I	
0	2	3	1	0	0
1	4	7	0	1	0
2	3	6	0	0	1
1	4	7	0	1	0
0	2	3	1	0	0
2	3	6	0	0	1
1	4	7	0	1	0
0	2	3	1	0	0
0	-5	-8	0	-2	1
1	4	7	0	1	0
0	1	$\frac{3}{2}$	$\frac{1}{2}$	0	0
0	-5	-8	0	-2	1
1	0	1	-2	1	0
0	1	$\frac{3}{2}$	$\frac{1}{2}$	0	0
0	0	$-\frac{1}{2}$	$\frac{5}{2}$	-2	1
1	0	1	-2	1	0
0	1	$\frac{3}{2}$	$\frac{1}{2}$	0	0
0	0	1	-5	4	-2
1	0	0	3	-3	2
0	1	0	8	-6	3
0	0	1	-5	4	-2
	I			A^{-1}	

Example 1. *An application.* Solve the system

$$\begin{cases} 2x_2 + 3x_3 = 5, \\ x_1 + 4x_2 + 7x_3 = -2, \\ 2x_1 + 3x_2 + 6x_3 = 3. \end{cases}$$

Solution. We note that our peculiar definition of matrix multiplication—this is why that definition exists in the first place—permits us to rewrite the given system as a single matrix equation:

$$\begin{bmatrix} 0 & 2 & 3 \\ 1 & 4 & 7 \\ 2 & 3 & 6 \end{bmatrix} \begin{bmatrix} x_1 \\ x_2 \\ x_3 \end{bmatrix} = \begin{bmatrix} 5 \\ -2 \\ 3 \end{bmatrix}.$$

Multiplying each side by the inverse of the *coefficient matrix* gives

$$\begin{bmatrix} x_1 \\ x_2 \\ x_3 \end{bmatrix} = \begin{bmatrix} 3 & -3 & 2 \\ 8 & -6 & 3 \\ -5 & 4 & -2 \end{bmatrix} \begin{bmatrix} 5 \\ -2 \\ 3 \end{bmatrix} = \begin{bmatrix} 27 \\ 61 \\ -39 \end{bmatrix}.$$

(Since a solution of a system having n unknowns is an n-tuple of numbers, it is convenient to display such solutions as n by 1 column matrices.) ☐

In closing, one observation should be made: The algorithm described above for finding A^{-1} can fail. We would like it to be the case that when this algorithm does fail, then the given matrix simply has no inverse. We have no right to conclude this at the present time; perhaps there are some special matrices whose inverses can be found only by using some more elaborate technique not yet considered. That this is not the case is demonstrated in Exercise 7(c). Thus, the given algorithm always finds the inverse of a matrix that has one.

EXERCISE SET 1.4

1 Theorem 1.7. If A and B are both nonsingular, then so is AB. Moreover, $(AB)^{-1} = B^{-1}A^{-1}$.

a) Verify Theorem 1.7 with respect to the matrices

$$A = \begin{bmatrix} 1 & 2 \\ 3 & 5 \end{bmatrix} \quad \text{and} \quad B = \begin{bmatrix} -2 & 1 \\ 1 & -1 \end{bmatrix}.$$

b) Prove Theorem 1.7. [*Hint:* What two conditions must the matrix $B^{-1}A^{-1}$ fulfill in order to be the inverse of AB?]

2 Find the inverse of the given matrix.

a) $A = [(i+j)!/(i!j!)]_{33}$ b) $B = [[j/i]]_{44}$

c) Make some conjectures about the inverses of symmetric and triangular matrices.

3 Write the given system of linear equations as a single matrix equation in the form $AX = B$. Then solve for X by multiplying through by A^{-1}.

a) $\begin{cases} x_1 + x_2 + 2x_3 = -3, \\ 2x_1 + x_2 = 0, \\ 3x_1 + 2x_2 - x_3 = 4 \end{cases}$ b) $\begin{cases} x_1 = 1 - x_2, \\ x_2 = 2 - x_3, \\ x_3 = 4 + x_4, \\ x_4 = 4 + x_1 \end{cases}$

4 *Another point of view.* Solve the system under Exercise 3(a) by first writing it in the form $[x_1, x_2, x_3]A = [-3, 0, 4]$.

5 Definition 1.13. If there exist elementary matrices $E_1, E_2, \ldots, E_n$ such that $B = (E_n \cdots E_2 E_1)A$, then B is said to be *row equivalent* to A. We write $B \stackrel{\text{(row)}}{=} A$.

Show that the second given matrix is row equivalent to the first one. [*Hint:* It is enough to show that the first matrix may, by row operations, be transformed into the second one.]

a) $\begin{bmatrix} 1 & -2 & 3 \\ 2 & 7 & -1 \\ 4 & 3 & 5 \end{bmatrix}, \begin{bmatrix} 1 & -2 & 3 \\ 0 & 11 & -7 \\ 0 & 0 & 0 \end{bmatrix}$

b) $\begin{bmatrix} 1 & 1 & 3 \\ 2 & 1 & 5 \end{bmatrix}, \begin{bmatrix} 0 & 1 & 1 \\ 1 & 0 & 2 \end{bmatrix}$

c) $\begin{bmatrix} 1 & 2 & 1 & 5 & 3 \\ -1 & 0 & 2 & -7 & -10 \\ 1 & 2 & 4 & -1 & -6 \end{bmatrix}, \begin{bmatrix} 0 & 0 & 1 & -2 & -3 \\ 0 & 1 & 0 & 2 & 1 \\ 1 & 0 & 0 & 3 & 4 \end{bmatrix}$

6 Write a definition for the phrase "*B is column equivalent to A*", and then show that

$$\begin{bmatrix} 1 & 2 & 1 \\ 5 & 7 & 5 \\ 9 & 3 & -1 \end{bmatrix} \stackrel{\text{(col)}}{=} \begin{bmatrix} 1 & 0 & 0 \\ 2 & 3 & 0 \\ 4 & 5 & 10 \end{bmatrix}.$$

7 a) Prove: If the kth row of A consists entirely of zeros, then the same is true about any product matrix AB.

b) Prove: If A is an nth-order matrix having a zero row, then A is singular.

c) Prove: If A is an nth-order matrix that is row equivalent to a matrix having a zero row, then A is singular.

8 a) Find two second-order singular matrices whose sum is nonsingular.

b) Find two second-order nonsingular matrices whose sum is singular.

c) Find second-order nonsingular matrices A and B such that $A + B$ is also nonsingular, but such that $(A + B)^{-1} \neq A^{-1} + B^{-1}$.

9 a) Suppose that the inverses of many second-order matrices are required. Then it makes sense to apply the method of this section to the "general" second-order matrix

$$\begin{bmatrix} a & b \\ c & d \end{bmatrix},$$

thus deriving a formula for the inverse of such a matrix. Do so.

b) What conditions must be met by a, b, c, and d to guarantee the applicability of the derived formula?

10 If $A^{-1} = A$, then A is said to be *self-inverse*. Find all second-order matrices that are self-inverse.

1.5 SOLVING SYSTEMS OF LINEAR EQUATIONS

We have seen in the last section how matrix techniques may be put to good use in solving a system of n equations in n unknowns. The idea of treating a system of equations as a single matrix equation is an appealing one and it continues to be applicable even when the number of unknowns

1.5 SOLVING SYSTEMS OF LINEAR EQUATIONS

is different from the number of equations. For example, the system
$$\begin{cases} x_1 + 2x_2 + 8x_4 = 7, \\ x_1 + 3x_2 + x_3 + 7x_4 = 13, \\ 2x_1 + 4x_2 + x_3 + 12x_4 = 8 \end{cases}$$
may be written
$$\begin{bmatrix} 1 & 2 & 0 & 8 \\ 1 & 3 & 1 & 7 \\ 2 & 4 & 1 & 12 \end{bmatrix} \begin{bmatrix} x_1 \\ x_2 \\ x_3 \\ x_4 \end{bmatrix} = \begin{bmatrix} 7 \\ 13 \\ 8 \end{bmatrix}.$$

However, the idea of multiplying both sides by the inverse of the *coefficient matrix*—since it is rectangular—no longer makes sense. What equation-solving technique would make sense?

We begin by multiplying both sides by elementary matrices that transform the first column of the coefficient matrix into
$$\begin{bmatrix} 1 \\ 0 \\ 0 \end{bmatrix}.$$

Hence we write
$$\begin{bmatrix} 1 & 0 & 0 \\ -1 & 1 & 0 \\ 0 & 0 & 1 \end{bmatrix} \begin{bmatrix} 1 & 2 & 0 & 8 \\ 1 & 3 & 1 & 7 \\ 2 & 4 & 1 & 12 \end{bmatrix} \begin{bmatrix} x_1 \\ x_2 \\ x_3 \\ x_4 \end{bmatrix} = \begin{bmatrix} 1 & 0 & 0 \\ -1 & 1 & 0 \\ 0 & 0 & 1 \end{bmatrix} \begin{bmatrix} 7 \\ 13 \\ 8 \end{bmatrix},$$

$$\begin{bmatrix} 1 & 0 & 0 \\ 0 & 1 & 0 \\ -2 & 0 & 1 \end{bmatrix} \begin{bmatrix} 1 & 2 & 0 & 8 \\ 0 & 1 & 1 & -1 \\ 2 & 4 & 1 & 12 \end{bmatrix} \begin{bmatrix} x_1 \\ x_2 \\ x_3 \\ x_4 \end{bmatrix} = \begin{bmatrix} 1 & 0 & 0 \\ 0 & 1 & 0 \\ -2 & 0 & 1 \end{bmatrix} \begin{bmatrix} 7 \\ 6 \\ 8 \end{bmatrix},$$

$$\begin{bmatrix} 1 & 2 & 0 & 8 \\ 0 & 1 & 1 & -1 \\ 0 & 0 & 1 & -4 \end{bmatrix} \begin{bmatrix} x_1 \\ x_2 \\ x_3 \\ x_4 \end{bmatrix} = \begin{bmatrix} 7 \\ 6 \\ -6 \end{bmatrix}.$$

We now transform the second column into
$$\begin{bmatrix} 0 \\ 1 \\ 0 \end{bmatrix}.$$

But why bother to write down the elementary matrices? We know what must be done: -2 times the second row must be added to the first. Do it! And to both sides:
$$\begin{bmatrix} 1 & 0 & -2 & 10 \\ 0 & 1 & 1 & -1 \\ 0 & 0 & 1 & -4 \end{bmatrix} \begin{bmatrix} x_1 \\ x_2 \\ x_3 \\ x_4 \end{bmatrix} = \begin{bmatrix} -5 \\ 6 \\ -6 \end{bmatrix}.$$

Finally, we move to the third column. We transform it into

$$\begin{bmatrix} 0 \\ 0 \\ 1 \end{bmatrix}.$$

Hence

$$\begin{bmatrix} 1 & 0 & 0 & 2 \\ 0 & 1 & 0 & 3 \\ 0 & 0 & 1 & -4 \end{bmatrix} \begin{bmatrix} x_1 \\ x_2 \\ x_3 \\ x_4 \end{bmatrix} = \begin{bmatrix} -17 \\ 12 \\ -6 \end{bmatrix}.$$

We have gone as far as we can, for any attempt to bring zeros into the fourth column will destroy zeros elsewhere. (Try it!) Now what? Well, let us return to system form:

$$\begin{cases} x_1 & & & + 2x_4 = -17, \\ & x_2 & & + 3x_4 = 12, \\ & & x_3 & - 4x_4 = -6. \end{cases}$$

Transposing the terms in the unknown x, we have

$$\begin{cases} x_1 & & & = -2x_4 - 17, \\ & x_2 & & = -3x_4 + 12, \\ & & x_3 & = 4x_4 - 6. \end{cases}$$

Now this last system evidently has infinitely many solutions. We may choose any value we please for x_4 and then compute corresponding values for x_1, x_2, x_3. More formally, let t be any real number; then all solutions of this last system are generated by

$$X = \begin{bmatrix} -2t - 17 \\ -3t + 12 \\ 4t - 6 \\ t \end{bmatrix}, \quad \text{where} \quad X = \begin{bmatrix} x_1 \\ x_2 \\ x_3 \\ x_4 \end{bmatrix}.$$

But are these columns the solutions of the original system? In other words, could our matrix multiplication have led us to a system whose solutions are different from those of the original system? No, this formula gives all solutions of the original system as well.

This we shall prove shortly. But first, before leaving this example, it should be observed that a further simplification of our method is achieved by eliminating even the matrix equation and working instead with a single matrix that stores all of the information that distinguishes the original system.

Consider the accompanying diagram:

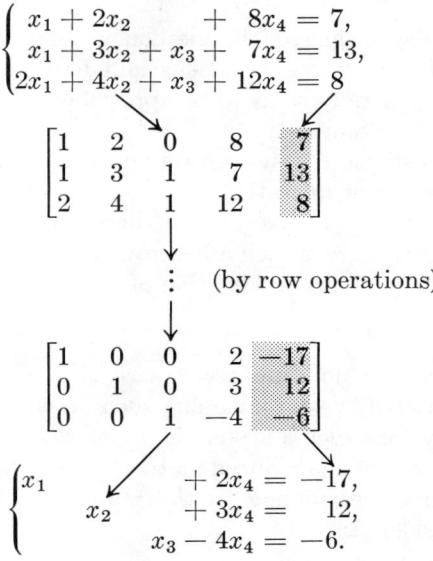

In other words, rather than multiply both sides of a matrix equation by certain elementary matrices, we bring the two sides together into a single matrix. The desired multiplications are then carried out by performing row operations on this matrix.

We close this section by presenting the theory and technical language that supports our discussion.

Definition 1.14. A system of equations that can be brought into the form $AX = B$ is called a *linear system*. The matrix A is called the *coefficient matrix*, and the matrix formed by adjoining B to A in the order A, B is called the *augmented matrix*.

Definition 1.15. If the set of column matrices satisfying $AX = B$ is identical with that satisfying $CX = D$, then the two matrix equations (and the corresponding linear systems) are said to be *equivalent*.

Definition 1.16. Let A be a matrix having the following properties:

1. If the ith row contains only zeros, then this also is true for all subsequent rows (i.e., the zero rows are at the "bottom" of the matrix).
2. Every row that contains other than all zeros has the number 1 as its first nonzero entry.
3. The entries above and below the initial 1 of a nonzero row are all zeros.
4. If the ith and the $(i+1)$st rows are both nonzero rows, then the initial 1 of the lower row occurs in a later column than that in the upper row.

Then we say that A is a *row echelon matrix* or that A is in *row echelon form*. (Some texts call this form the *reduced row echelon form*.)

With this new language, the method we employed earlier to solve a linear system having three equations and four unknowns may now be described quite generally so as to be applicable to a linear system having m equations in n unknowns.

We first construct the system's augmented matrix and then reduce that matrix (using row operations) to row echelon form. Next, we check that reduced matrix for *bad rows*, a bad row being one of the form $[0, 0, \ldots, 0, 1]$. If there is such a bad row, then the system corresponding to the reduced matrix has an equation of the form

$$0 \cdot x_1 + 0 \cdot x_2 + \cdots + 0 \cdot x_n = 1.$$

Evidently there are no numbers $x_1, x_2, \ldots, x_n$ that satisfy such an equation. Thus, the system including such an equation has no solutions. (We usually say that such a system is *inconsistent*.)

Now suppose that the reduced matrix has no bad rows. Then we write out the system corresponding to the reduced matrix. There are two cases to be considered:

1. Each of the n unknowns appears as a *beginner variable* in one of the equations, a beginner variable being the first variable in the equation not having the coefficient zero. Then the system has exactly one solution; it is given by the column of numbers derived from the right-hand sides of the first n equations in the system. (It should be noted that there may be—at the "bottom" of the system—equations of the form $0 = 0$. Such equations impose no conditions on the unknowns; hence, they may be ignored.)

2. If each of the n unknowns does not appear as a beginner variable, then some of the variables may be classified as *nonbeginner variables*. In this case the solution—a formula for an infinite set of column matrices—is obtained as follows: Set each nonbeginner variable equal to one of the letters $r, s, t, \ldots$. Then solve for the beginner variables in terms of these new variables. In this way we obtain a formula for each of the n unknowns in terms of the variables $r, s, t, \ldots$. The column composed of these formulas, taken in the proper order, is the desired solution. (The parenthetical remark at the end of case 1 is equally applicable here.)

Example 1. The augmented matrix of a certain system has been reduced by row operations to the given matrix. Solve the system.

1. $\begin{bmatrix} 1 & 0 & 0 \\ 0 & 1 & 0 \\ 0 & 0 & 1 \end{bmatrix}$, 2. $\begin{bmatrix} 1 & 0 & 5 \\ 0 & 1 & 6 \\ 0 & 0 & 0 \end{bmatrix}$, 3. $\begin{bmatrix} 1 & 3 & 0 & 2 & 2 \\ 0 & 0 & 1 & 5 & 1 \\ 0 & 0 & 0 & 0 & 0 \end{bmatrix}$.

Solution

1. The last row of the reduced matrix is a bad row. Thus, the system has no solutions, i.e., it is inconsistent.
2. The corresponding system is

$$\begin{cases} x_1 = 5, \\ x_2 = 6, \\ 0 = 0. \end{cases}$$

Its only solution is

$$X = \begin{bmatrix} 5 \\ 6 \end{bmatrix}.$$

3. The corresponding system is

$$\begin{cases} x_1 + 3x_2 \phantom{{}+x_3} + 2x_4 = 2, \\ x_3 + 5x_4 = 1, \\ 0 = 0. \end{cases}$$

The nonbeginner variables are x_2 and x_4. We set $x_2 = s$ and $x_4 = t$, and then solve for the beginner variables. We obtain

$$\begin{cases} x_1 = -3s - 2t + 2, \\ x_3 = -5t + 1. \end{cases}$$

It follows that all solutions (infinitely many) are "captured" by the formula

$$X = \begin{bmatrix} (-3s - 2t + 2) \\ (s) \\ (-5t + 1) \\ (t) \end{bmatrix}.$$

An algorithm that promises to solve *all* systems of linear equations needs justification. There are two basic concerns: (1) Starting with the matrix equation $AX = B$, our method, in its original form, requires that we premultiply both sides of this equation with a succession of elementary matrices, each multiplication bringing the equation closer to its final, simplest form. Thus, if the string of multipliers is given by $C = E_n \cdots E_2 E_1$—a nonsingular matrix—then we must be assured that the final equation, $(CA)X = (CB)$, is equivalent to the original one. This assurance is given by Theorem 1.8 (below). (2) The second concern could almost be overlooked; it is this: Does there always exist a succession of row operations that will carry a given matrix into row echelon form, and will that final form be unique? The affirmative answer to this question is given by Theorem 1.9 (below).

Theorem 1.8. If C is a nonsingular matrix, then the equations $AX = B$ and $(CA)X = CB$ are equivalent.

Proof

1. Let X^* be a solution of the equation $AX = B$. Then $AX^* = B$. It follows that $C(AX^*) = CB$. By associativity, this is the same as $(CA)X^* = CB$. Thus, X^* is also a solution of the equation $(CA)X = CB$.
2. The argument is reversible. Starting with $(CA)X^* = CB$, we conclude, using now the hypothesis that C^{-1} exists, that $AX^* = B$.

Corollary. If $A \stackrel{\text{(row)}}{=} B$, then the linear systems having A and B as their augmented matrices are equivalent.

Theorem 1.9. Corresponding to every matrix A there is a unique matrix B such that:

1. B is in row echelon form.
2. $B \stackrel{\text{(row)}}{=} A$ (that is, B may be obtained from A by a sequence of row operations).

Theorem 1.9 can be proved by literally writing down a set of rules for transforming an arbitrary m by n matrix into row echelon form. It is an uninteresting proof and difficult to make rigorous. We shall omit it.

EXERCISE SET 1.5

1 Consider the system $\begin{cases} x_1 + 2x_2 + 3x_4 = -1, \\ 4x_1 + 8x_2 + x_3 + 14x_4 = -3, \\ 3x_1 + 6x_2 + x_3 + 11x_4 = -2. \end{cases}$

  a) Construct the augmented matrix associated with this system and reduce it (by row operations) to row echelon form.
  b) Write out the system corresponding to the matrix obtained under part (a).
  c) Solve the original system by appealing to the equivalent system obtained under part (b).
  d) Check your solution by carrying out a certain matrix product.

2 Repeat Exercise 1 for each of the following systems.

  a) $\begin{cases} x_1 + 2x_2 + 3x_3 = 4, \\ 2x_1 + 5x_2 + 7x_3 = 8, \\ x_1 + 2x_2 + 4x_3 = 16 \end{cases}$

b) $\begin{cases} x_1 + 2x_2 + 3x_3 = 4, \\ 2x_1 + 5x_2 + 7x_3 = 8, \\ 4x_1 + 9x_2 + 13x_3 = 16 \end{cases}$
c) $\begin{cases} x_1 + 2x_2 + 3x_3 = 4, \\ 2x_1 + 4x_2 + 6x_3 = 8, \\ 3x_1 + 6x_2 + 9x_3 = 12 \end{cases}$

In Exercises 3 through 5, solve the system whose augmented matrix has been reduced to the given row echelon matrix.

3 *Systems where* $m = n$.

a) $\begin{bmatrix} 1 & 0 & 0 & 2 \\ 0 & 1 & 0 & 3 \\ 0 & 0 & 1 & 4 \end{bmatrix}$
b) $\begin{bmatrix} 1 & 0 & 0 & 0 \\ 0 & 1 & 0 & 0 \\ 0 & 0 & 0 & 1 \end{bmatrix}$
c) $\begin{bmatrix} 1 & 0 & 0 & 2 \\ 0 & 1 & 0 & 3 \\ 0 & 0 & 0 & 0 \end{bmatrix}$

4 *Systems where* $m > n$.

a) $\begin{bmatrix} 1 & 0 & 0 & 0 \\ 0 & 1 & 0 & 0 \\ 0 & 0 & 1 & 0 \\ 0 & 0 & 0 & 1 \end{bmatrix}$
b) $\begin{bmatrix} 1 & 0 & 0 & 2 \\ 0 & 1 & 0 & 3 \\ 0 & 0 & 1 & 4 \\ 0 & 0 & 0 & 0 \end{bmatrix}$
c) $\begin{bmatrix} 1 & 0 & 2 & 4 \\ 0 & 1 & 3 & 5 \\ 0 & 0 & 0 & 0 \\ 0 & 0 & 0 & 0 \end{bmatrix}$

5 *Systems where* $m < n$.

a) $\begin{bmatrix} 1 & 0 & 0 & 2 & 5 \\ 0 & 1 & 0 & 3 & 6 \\ 0 & 0 & 1 & 4 & 7 \end{bmatrix}$
b) $\begin{bmatrix} 1 & 0 & 2 & 4 & 0 \\ 0 & 1 & 3 & 5 & 0 \\ 0 & 0 & 0 & 0 & 1 \end{bmatrix}$
c) $\begin{bmatrix} 1 & 2 & 3 & 0 & 4 \\ 0 & 0 & 0 & 1 & 5 \\ 0 & 0 & 0 & 0 & 0 \end{bmatrix}$

6 Theorem 1.10. Suppose that a linear system has the following two characteristics:

1. There are more unknowns than equations.
2. The right-hand side of each equation is 0.

Then such a system always has infinitely many solutions.

Prove this theorem. [*Hint:* Two questions must be answered: Why is the system evidently consistent? Why must there be nonbeginner variables in the simplified system?]

7 Let

$$A = \begin{bmatrix} 1 & 1 & -1 & 0 \\ 0 & 1 & 2 & 1 \\ 1 & 2 & 3 & 3 \end{bmatrix}, \quad B = \begin{bmatrix} 1 \\ 0 \\ 3 \end{bmatrix}, \quad \text{and} \quad C = \begin{bmatrix} 1 & 0 & 2 \\ 0 & -1 & -1 \\ 1 & -1 & 1 \end{bmatrix}.$$

a) Solve $AX = B$.
b) Solve $(CA)X = CB$.
c) Find a solution of the second equation that is *not* a solution of the first one.
d) Reread (or read) Theorem 1.8 and its proof.

8 Solve the given matrix equation for A.

a) $A \begin{bmatrix} 1 & 2 & 1 \\ 2 & 0 & 1 \end{bmatrix} = I_3.$
b) $\begin{bmatrix} 1 & 2 & 1 \\ 2 & 0 & 1 \end{bmatrix} A = I_2.$

9 The following information is given about a certain function f:
1. $f(0) = -1, f(1) = -5, f(-1) = \frac{1}{3}$.
2. There exist numbers a, b, c, and d that permit f to be described by the formula
$$f(x) = \frac{ax+b}{cx+d}.$$

Show that this information completely characterizes f, that is, that there is exactly one such function f. [*Hint:* Begin by constructing a system of three equations in the unknowns a, b, c, and d.]

10 A certain problem involves four unknowns: x_1, x_2, x_3, and x_4. A certain problem solver discovers six ways of relating these unknowns:

$$\begin{cases} x_1 - x_2 + x_4 = 3, \\ x_2 + x_3 = 5, \\ x_1 + x_2 + 2x_3 + 2x_4 = 17, \\ x_1 + x_2 + x_3 = 6, \\ 3x_1 + x_2 + 3x_3 + 3x_4 = 26, \\ 2x_1 + 2x_2 + 3x_3 + 2x_4 = 23. \end{cases}$$

a) Show that two of these equations are expendable, that is, that the values of x_1, x_2, x_3, and x_4 are completely determined by four of the equations.
b) Are the first two equations expendable?

11 Four numbers a_1, a_2, a_3, and a_4 satisfy the following conditions:
1. The sum of the four numbers is four more than the fourth.
2. The sum of the first three is three more than the third.
3. The sum of the first two is two more than the second.

a) Verify that one such quadruple of numbers is [2, 1, 1, 2].
b) Find all such quadruples.

12 Prove: If the equation $AX = B$ has more than one solution, then it has infinitely many solutions. [*Hint:* Let X^1 and X^2 both be solutions of $AX = B$. Then show that $X^1 + c(X^2 - X^1)$ is also a solution.]

13 The matrix technique that we have learned for solving linear systems is useful for solving certain nonlinear systems as well.

a) Solve the system
$$\begin{cases} x_1^2 + 2x_2^2 + 3x_3^2 = 7, \\ 2x_1^2 + 3x_2^2 - 2x_3^2 = -2, \\ x_1^2 - x_2^2 + x_3^2 = 3. \end{cases}$$

[*Hint:* Set $u_1 = x_1^2$, $u_2 = x_2^2$, and $u_3 = x_3^2$.]

b) Solve the system
$$\begin{cases} x_1^2 + 2\sin x_2 + 3e^{x_3} = 7, \\ 2x_1^2 + 3\sin x_2 - 2e^{x_3} = -2, \\ x_1^2 - \sin x_2 + e^{x_3} = 3. \end{cases}$$

14 Suppose that we are required to solve the r linear systems whose matrix equations are $AX = B^1$, $AX = B^2, \ldots, AX = B^r$. Describe a technique for solving these systems that is more efficient than solving each system individually.

1.6 THE FIRST DIAGONALIZATION PROBLEM

The application of matrices to solving linear systems is surely impressive as a labor-saving device. On the other hand, it must be admitted that our method—that of reducing the augmented matrix of a system to row echelon form—has little to distinguish it from the traditional high school approach of eliminating unknowns by adding multiples of one equation to another. Thus, so that the reader may immediately be reassured that there is more to matrix theory than this, we move directly to a second application of matrices involving notions that are probably less familiar to the reader. We present this application in the next section; in the present section we make certain preparations.

Definition 1.17. Let A be an m by n matrix and let B be an n by m matrix such that $b_{ij} = a_{ji}$ for all i and j. Then we say that B is the *transpose* of A. We will denote the transpose of A by A'.

Example 1. Find A', where
$$A = \begin{bmatrix} 2 & 3 & 7 \\ 4 & 6 & 1 \end{bmatrix}.$$

Solution. The reader should first verify that the definition of A' reduces to the following informal statement: To find the transpose of A, we simply construct the matrix whose first column is the first row of A, whose second column is the second row of A, and so on. Thus,
$$A' = \begin{bmatrix} 2 & 4 \\ 3 & 6 \\ 7 & 1 \end{bmatrix}.$$

Theorem 1.11.
1. $(A + B)' = A' + B'$.
2. $(cA)' = cA'$.
3. $(AB)' = B'A'$.

Proof. We consider only part 3. Let $A = [a_{ij}]_{mp}$ and $B = [b_{ij}]_{pn}$. Also, let $A' = [\alpha_{ij}]_{pm}$ and $B' = [\beta_{ij}]_{np}$. Then $\alpha_{ij} = a_{ji}$ and $\beta_{ij} = b_{ji}$. We now compute the (i,j)-element of each side of our identity:

1. (i,j)-element of $(AB)' = (j,i)$-element of AB

$$= \sum_{k=1}^{p} a_{jk}b_{ki}$$

2. (i,j)-element of $B'A' = \sum_{k=1}^{p} \beta_{ik}\alpha_{kj}$

$$= \sum_{k=1}^{p} b_{ki}a_{jk}$$

$$= \sum_{k=1}^{p} a_{jk}b_{ki}$$

Theorem 1.12. If A is symmetric (Definition 1.9), then $A' = A$. And conversely.

Proof

1. Let $B = A'$. Then $b_{ij} = a_{ji}$. But $a_{ji} = a_{ij}$. (Why?) Therefore, $b_{ij} = a_{ij}$. That is, $A' = A$.
2. The proof of the converse is left to the reader.

Theorem 1.13. If A is symmetric, then so is BAB' (for any B for which the product makes sense).

Proof. Let $C = BAB'$. Then, using Theorem 1.11 (part 3) as applied to three factors, $C' = (BAB')' = (B')'A'B'$. Now surely $(B')' = B$. And by hypothesis (plus Theorem 1.12), $A' = A$. Therefore, $C' = BAB' = C$. The result follows by a second application of Theorem 1.12, the converse part. □

We now consider a *diagonalization problem*. Given a symmetric matrix A, find a nonsingular matrix B such that BAB' is diagonal. This problem always has a solution; indeed, it has infinitely many solutions (see Exercise 5). (A reason for wanting to solve such a problem will be given in the next section.) Finding a solution again involves a technique that is based on elementary matrices.

As an illustration, we take as our symmetric matrix

$$A = \begin{bmatrix} 1 & 2 & 3 \\ 2 & 6 & 5 \\ 3 & 5 & 10 \end{bmatrix}.$$

Our plan is this: We introduce zeros under the main diagonal by premultiplying A with the appropriate elementary matrices; we introduce

1.6 THE FIRST DIAGONALIZATION PROBLEM

zeros above the main diagonal by postmultiplying A with the transposes of the premultiplication matrices. The pre- and postmultipliers are applied alternately; in this way symmetric positions about the main diagonal are in pairs filled by zeros, and the symmetry of A is restored after each pair of multiplications.

Let us see how it works:

$$\begin{bmatrix} 1 & 0 & 0 \\ -2 & 1 & 0 \\ 0 & 0 & 1 \end{bmatrix} \begin{bmatrix} 1 & 2 & 3 \\ 2 & 6 & 5 \\ 3 & 5 & 10 \end{bmatrix}$$

$$\downarrow$$

$$\begin{bmatrix} 1 & 2 & 3 \\ 0 & 2 & -1 \\ 3 & 5 & 10 \end{bmatrix} \begin{bmatrix} 1 & -2 & 0 \\ 0 & 1 & 0 \\ 0 & 0 & 1 \end{bmatrix}$$

$$\downarrow$$

$$\begin{bmatrix} 1 & 0 & 0 \\ 0 & 1 & 0 \\ -3 & 0 & 1 \end{bmatrix} \begin{bmatrix} 1 & 0 & 3 \\ 0 & 2 & -1 \\ 3 & -1 & 10 \end{bmatrix}$$

$$\downarrow$$

$$\begin{bmatrix} 1 & 0 & 3 \\ 0 & 2 & -1 \\ 0 & -1 & 1 \end{bmatrix} \begin{bmatrix} 1 & 0 & -3 \\ 0 & 1 & 0 \\ 0 & 0 & 1 \end{bmatrix}$$

$$\downarrow$$

$$\begin{bmatrix} 1 & 0 & 0 \\ 0 & 1 & 0 \\ 0 & \frac{1}{2} & 1 \end{bmatrix} \begin{bmatrix} 1 & 0 & 0 \\ 0 & 2 & -1 \\ 0 & -1 & 1 \end{bmatrix}$$

$$\downarrow$$

$$\begin{bmatrix} 1 & 0 & 0 \\ 0 & 2 & -1 \\ 0 & 0 & \frac{1}{2} \end{bmatrix} \begin{bmatrix} 1 & 0 & 0 \\ 0 & 1 & \frac{1}{2} \\ 0 & 0 & 1 \end{bmatrix}$$

$$\downarrow$$

$$\begin{bmatrix} 1 & 0 & 0 \\ 0 & 2 & 0 \\ 0 & 0 & \frac{1}{2} \end{bmatrix}.$$

Now, letting E_1, E_2, and E_3 be the three successive premultipliers, we have shown that

$$(E_3((E_2((E_1 A)E_1'))E_2'))E_3' = D = \begin{bmatrix} 1 & 0 & 0 \\ 0 & 2 & 0 \\ 0 & 0 & \frac{1}{2} \end{bmatrix}.$$

Equivalently, using associativity,

$$(E_3 E_2 E_1) A (E_1' E_2' E_3') = D,$$

or, using a three-factor extension of Theorem 1.11 (part 3), we may write

$$(E_3 E_2 E_1) A (E_3 E_2 E_1)' = D.$$

Thus, we have in $B = E_3 E_2 E_1$ a matrix that diagonalizes A in the sense that $BAB' = D$. The reader should verify that

$$B = \begin{bmatrix} 1 & 0 & 0 \\ -2 & 1 & 0 \\ -4 & \frac{1}{2} & 2 \end{bmatrix}.$$

The foregoing procedure can be refined. As with our algorithm for finding A^{-1}, we can replace our matrix multiplications with row operations and column operations, assembling B as we go by performing the sequence of row operations on I (see the accompanying illustration).

A			I			
1	2	3	1	0	0	
2	6	5	0	1	0	
3	5	10	0	0	1	
1	2	3	1	0	0	(Two row operations)
0	2	−1	−2	1	0	
0	−1	1	−3	0	1	
1	0	0	1	0	0	(Corresponding column operations—but only on the A-side.)
0	2	−1	−2	1	0	
0	−1	1	−3	0	1	
1	0	0	1	0	0	
0	2	−1	−2	1	0	(One row operation.)
0	0	$\frac{1}{2}$	−4	$\frac{1}{2}$	1	
1	0	0	1	0	0	(Corresponding column operation—but only on the A-side.)
0	2	0	−2	1	0	
0	0	$\frac{1}{2}$	−4	$\frac{1}{2}$	1	
D			B			

The fact that we apply only the row operations to I is based on the realization that the desired matrix B is obtained by collecting just the premultiplying elementary matrices.

The symmetry-restoring column operations, applied only to the A side, seem, at first, unnecessary. For example, in the case of the given A

we could have proceeded by row operations alone from

$$\left[\begin{array}{ccc|ccc} 1 & 2 & 3 & 1 & 0 & 0 \\ 2 & 6 & 5 & 0 & 1 & 0 \\ 3 & 5 & 10 & 0 & 0 & 1 \end{array}\right] \quad \text{to} \quad \left[\begin{array}{ccc|ccc} 1 & 2 & 3 & 1 & 0 & 0 \\ 0 & 2 & -1 & -2 & 1 & 0 \\ 0 & 0 & \frac{1}{2} & -4 & \frac{1}{2} & 1 \end{array}\right],$$

and then concluded that

$$D = \begin{bmatrix} 1 & 0 & 0 \\ 0 & 2 & 0 \\ 0 & 0 & \frac{1}{2} \end{bmatrix} \quad \text{and} \quad B = \begin{bmatrix} 1 & 0 & 0 \\ -2 & 1 & 0 \\ -4 & \frac{1}{2} & 1 \end{bmatrix}.$$

This shortcut will work if A can be reduced to upper triangular form by adding multiples of upper rows to lower ones. But if, say, a row interchange is needed, then the corresponding column interchange must be employed, on the A side, before continuing. This will become more clear to the reader after he has considered some examples.

EXERCISE SET 1.6

1 Let

$$A = \begin{bmatrix} 1 & 7 & 3 \\ 2 & 0 & -4 \end{bmatrix} \quad \text{and} \quad B = \begin{bmatrix} 4 & 0 & 1 & 2 \\ 3 & 1 & 3 & 1 \\ 2 & 2 & 5 & -1 \end{bmatrix}.$$

Verify that $(AB)' = B'A'$.

2 Prove: For any matrix A, it is always true that both AA' and $A'A$ are symmetric. [*Hint:* Use Theorem 1.12.]

3 a) Prove: If A is a symmetric, nonsingular matrix, then its inverse is also symmetric. [*Hint:* $I = I' = (AA^{-1})' = \cdots = (A^{-1})'A$.]
b) Prove: If A is nonsingular, then so is A'. Moreover, $(A')^{-1} = (A^{-1})'$.

4 Find a nonsingular matrix B such that BAB' is diagonal, where A is the given matrix.

a) $\begin{bmatrix} 1 & 7 \\ 7 & 3 \end{bmatrix}$

b) $\begin{bmatrix} 1 & 2 & 1 \\ 2 & 22 & 14 \\ 1 & 14 & 34 \end{bmatrix}$

c) $\begin{bmatrix} 1 & 1 & 1 \\ 1 & 1 & 3 \\ 1 & 3 & 6 \end{bmatrix}$

d) $\begin{bmatrix} 4 & 5 & 1 \\ 5 & 10 & 4 \\ 1 & 4 & 3 \end{bmatrix}$

e) $\begin{bmatrix} 0 & 2 & 1 \\ 2 & 3 & 4 \\ 1 & 4 & 5 \end{bmatrix}$

f) $\begin{bmatrix} 0 & 1 & 2 \\ 1 & 0 & 3 \\ 2 & 3 & 0 \end{bmatrix}$

g) $\begin{bmatrix} 1 & 1 & 1 & 0 & 0 \\ 1 & 3 & 1 & -2 & 0 \\ 1 & 1 & 1 & 1 & 0 \\ 0 & -2 & 1 & 4 & 0 \\ 0 & 0 & 0 & 0 & 3 \end{bmatrix}$

5 Let A and B be as in Exercise 4(b). Let $E = [2 \to (5)2]_3$.
  a) Verify that $EBAB'E'$ is diagonal.
  b) Prove or disprove: Given a symmetric matrix A, let S be the set of all matrices B such that BAB' is diagonal. Then S is an infinite set.

6 Let A be as in Exercise 4(b). Find a nonsingular matrix C such that $C'AC$ is diagonal.

7 Prove the parts of Theorem 1.11 that are not proven in the text.

8 A square matrix A is said to be *skew symmetric* if $A' = -A$.
  a) Prove: A is skew symmetric if and only if $a_{ij} = -a_{ji}$ for all i and j.
  b) Prove: If A is skew symmetric, then A^2 is symmetric.
  c) Let A be an arbitrary nth-order matrix. Let $S = \frac{1}{2}(A + A')$ and $K = \frac{1}{2}(A - A')$. Prove that S is symmetric and that K is skew symmetric.
  d) Let $A = S + K$, where it is given that S is symmetric and that K is skew symmetric. Prove that $S = \frac{1}{2}(A + A')$ and that $K = \frac{1}{2}(A - A')$.
  e) Parts (c) and (d) taken together establish the following theorem: Every square matrix can be decomposed in exactly one way into the sum of a symmetric matrix and a skew symmetric matrix. Find this decomposition for $A = [i + 2j]_{44}$.
  f) Let
  $$K = \begin{bmatrix} 0 & 1 & 0 \\ -1 & 0 & -2 \\ 0 & 2 & 0 \end{bmatrix} \quad \text{and} \quad B = (I - K)(I + K)^{-1}.$$
  Find B^{-1} and then verify that $B^{-1} = B'$.
  g) State and prove the theorem suggested by part (f). [*Hint:* It will be useful to note that the matrices $I - K$ and $I + K$ commute.]

1.7 QUADRATIC FORMS

In this section we shall be interested in certain polynomial expressions, expressions in which all terms are of the second degree. An example of a three-variable expression of this type (with all possible terms) is

$$x_1^2 + 4x_1x_2 + 6x_2^2 + 10x_2x_3 + 10x_3^2 + 6x_1x_3.$$

It would be unexpected, perhaps, that such expressions should in any way be related to matrices. However, we shall now see that this is the case. Indeed, every such n-variable expression can be identified in exactly one way with an n by n symmetric matrix.

Let us return to the polynomial given above. We seek a 3 by 3 symmetric matrix that identifies this polynomial, that is, we want a matrix that distinguishes this polynomial from all others. Or, being more precise, we seek a matrix whose entries will lead us, in some manner or other, back to the original polynomial.

To discover this identification process, we must suppose ourselves clever enough to relate the subscripts of the variables with the positions of a matrix. For example, if we write

$$x_1^2 = (1)x_1x_1,$$

then we might reasonably elect to place 1 in the (1, 1)-position. Similarly,

$$6x_2^2 = (6)x_2x_2 \quad \text{and} \quad 10x_3^2 = (10)x_3x_3.$$

Thus, we place 6 in the (2, 2)-position and 10 in the (3, 3)-position. We have in this way completed the main diagonal of a third-order matrix. There are six other positions to fill in, and only three terms of our polynomial to draw from. With considerable inspiration we write

$$4x_1x_2 = (2)x_1x_2 + (2)x_2x_1,$$
$$10x_2x_3 = (5)x_2x_3 + (5)x_3x_2,$$

and

$$6x_1x_3 = (3)x_1x_3 + (3)x_3x_1.$$

Finally, by matching each coefficient of the "expanded polynomial" with the position corresponding to the subscript listing of that term, we are led to the symmetric matrix

$$A = \begin{bmatrix} 1 & 2 & 3 \\ 2 & 6 & 5 \\ 3 & 5 & 10 \end{bmatrix}.$$

This construction is significant for, as the reader may verify, it turns out that

$$[x_1, x_2, x_3] \begin{bmatrix} 1 & 2 & 3 \\ 2 & 6 & 5 \\ 3 & 5 & 10 \end{bmatrix} \begin{bmatrix} x_1 \\ x_2 \\ x_3 \end{bmatrix}$$
$$= [x_1^2 + 4x_1x_2 + 6x_2^2 + 10x_2x_3 + 10x_3^2 + 6x_1x_3].$$

Observe that the last matrix is a 1 by 1 matrix whose single entry is the expression we started with. With this motivation we are led to the following definition.

Definition 1.18. Let A be a symmetric matrix. The function f defined by

$$f(X) = X'AX, \quad \text{where} \quad X = \begin{bmatrix} x_1 \\ x_2 \\ \vdots \\ x_n \end{bmatrix},$$

is called a *quadratic form*. (Actually, this term is applied either to the function itself or to the formula defining the function. Also, to be more precise, we should write $[f(X)]$ instead of $f(X)$. Why?)

Quadratic forms play an important role in the study of maximum-minimum problems. In that study it becomes important to be able to answer certain questions about the range of a quadratic form. And the answers to these questions are most easily obtained by considering the matrix representation of the form in question. Now, it would take us too far afield to discuss this application of quadratic forms in detail; however, for motivational purposes, we shall consider briefly the main idea.

Suppose that a function f is defined by a formula of the following type:

$$f(x_1, x_2) = 5 + 2x_1^2 + 3x_2^2 + \text{(terms of higher degree)}.$$

Then $f(0, 0) = 5$. Furthermore, it is not hard to see that 5 is a *relative minimum* for the function f, for consider the following reasoning: To establish that $f(0, 0) = 5$ is a relative minimum for f, we have only to show that 5 is the smallest value taken on by f in the immediate vicinity of the point $(0, 0)$. Now, we observe first that if x_1 and x_2 are both in absolute value less than 1, then higher degree products in x_1 and x_2 are smaller (in absolute value) than those of lower degree. It follows—we do not give here a rigorous argument—that if we restrict (x_1, x_2) to a small enough circular neighborhood about the origin, say N, then the contribution of the terms of higher degree in our formula will be smaller (in absolute value) than that of the quadratic part. But this means that the sign of

$$2x_1^2 + 3x_2^2 + \text{(terms of higher degree)}$$

is the same as the sign of

$$2x_1^2 + 3x_2^2.$$

Now it is evident that $2x_1^2 + 3x_2^2 \geq 0$ for all (x_1, x_2). Thus, we may write

$$f(x_1, x_2) = 5 + \text{(nonnegative quantity)},$$

provided that (x_1, x_2) is in N. In other words, in the neighborhood N the value of f never drops below 5.

The point is this: Whether or not 5 is a relative minimum (or maximum) for f is seen to depend on the behavior of the quadratic part of the formula for f. Consider now a second function g, where

$$g(x_1, x_2) = 5 + x_1^2 + 4x_1x_2 + 3x_2^2 + \text{(terms of higher degree)}.$$

Then we may argue as before that 5 is a relative minimum for g if and only if the quadratic part of our formula is nonnegative for all choices of x_1 and x_2. But now it is not clear that such is the case. It is conceivable that x_1 and x_2 might be picked with opposite signs in such a manner that the negativeness of the term $4x_1x_2$ would overpower the positiveness of

$x_1^2 + 3x_2^2$. In other words, given a quadratic formula in two variables, we cannot always deduce simply by inspection that it is nonnegative for all x_1 and x_2. The situation is even more complex in the case of the three-variable analog to the problem that we have been considering. Happily there is an easy way to deduce the nonnegativeness (or nonpositiveness) of quadratic forms in two, three, or any number of variables: it depends on the matrix diagonalizing scheme considered in the last section.

Example 1. Let
$$f(x_1, x_2, x_3) = x_1^2 + 4x_1x_2 + 6x_2^2 + 10x_2x_3 + 10x_3^2 + 6x_1x_3.$$

Prove that $f(x_1, x_2, x_3) > 0$ provided that $(x_1, x_2, x_3) \neq (0, 0, 0)$.

Solution. We rewrite our function in matrix form:
$$f(X) = X' \begin{bmatrix} 1 & 2 & 3 \\ 2 & 6 & 5 \\ 3 & 5 & 10 \end{bmatrix} X.$$

(This is the matrix A that was diagonalized in the last section; the matrix called B' there will be called C below.) Then we make a "change of variable." We consider the function g that arises on replacing the X of $f(X)$ by CU, where
$$C = \begin{bmatrix} 1 & -2 & -4 \\ 0 & 1 & \frac{1}{2} \\ 0 & 0 & 1 \end{bmatrix}.$$

In other words,
$$g(U) = f(CU) = (CU)' \begin{bmatrix} 1 & 2 & 3 \\ 2 & 6 & 5 \\ 3 & 5 & 10 \end{bmatrix} (CU).$$

Now this formula for g has a remarkable simplification:
$$g(U) = U'C' \begin{bmatrix} 1 & 2 & 3 \\ 2 & 6 & 5 \\ 3 & 5 & 10 \end{bmatrix} CU$$
$$= U' \begin{bmatrix} 1 & 0 & 0 \\ 0 & 2 & 0 \\ 0 & 0 & \frac{1}{2} \end{bmatrix} U$$
$$= u_1^2 + 2u_2^2 + \tfrac{1}{2}u_3^2.$$

Look at this formula: a sum of squares. Evidently, $g(U)$ never takes on negative values and is zero only once, at $(0, 0, 0)$. Can we conclude that the same is true for $f(X)$? That we can is the message of the following theorem.

Theorem 1.14. Given a quadratic form $f(X)$ and any nonsingular matrix C, let $g(U)$ be the quadratic form defined by

$$g(U) = f(CU).$$

Then the range of f is identical with that of g. Moreover, f takes on the value zero more than once if and only if the same is true for g.

Proof. Let a be in the range of f. This means that there is a column X^* such that $f(X^*) = a$. Now, let $U^* = C^{-1}X^*$. Then we may write

$$g(U^*) = f(CU^*) = f(C(C^{-1}X^*)) = f(X^*) = a.$$

Thus, there is a U^* such that $g(U^*) = a$. This means that a is in the range of g.

So far we have proved only that the range of f is a subset of the range of g. The foregoing argument may be reversed to verify the converse statement. Then the two ranges will be proved identical. (We leave this to the reader.)

Finally, suppose that $f(X^1) = 0$ and $f(X^2) = 0$. Then also $g(U^1) = 0$ and $g(U^2) = 0$, where $U^1 = C^{-1}X^1$ and $U^2 = C^{-1}X^2$. This argument may also be reversed.

Corollary. Given a symmetric matrix A, suppose that there is a nonsingular matrix B such that BAB' is a diagonal matrix having only positive entries along its main diagonal. Then the quadratic form $f(X) = X'AX$ takes on only positive values, except for $f(\Theta) = 0$.

Definition 1.19. If a quadratic form takes on positive values for all X other than $X = \Theta$, then it is said to be *positive definite*. A symmetric matrix is said to be *positive definite* if there is a nonsingular matrix B such that BAB' is a diagonal matrix having only positive entries along its main diagonal.

Our corollary may now be rewritten: *If a symmetric matrix is positive definite, then the corresponding quadratic form is also positive definite.*

EXERCISE SET 1.7

1 Prove that the given function is a quadratic form. [One has only to find, recall Definition 1.18, a symmetric matrix A such that $f(X) = X'AX$.]

a) $f(X) = x_1^2 + 3x_1x_2 + 5x_2^2 - 2x_2x_3 - x_3^2$

b) $f(X) = (x_1 + 2x_2)(3x_1 - x_2)$

c) $f(X) = X' \begin{bmatrix} 1 & 4 & 6 \\ 2 & 5 & 3 \\ 3 & 0 & 0 \end{bmatrix} X$

2 Let

$$f(X) = X' \begin{bmatrix} 1 & 2 & 3 \\ 2 & 6 & 7 \\ 3 & 7 & 10 \end{bmatrix} X.$$

a) Find a quadratic form $g(U) = d_1 u_1^2 + d_2 u_2^2 + d_3 u_3^2$ such that $g(U) = f(CU)$, where C is nonsingular.
b) Show that $g(U)$ is positive definite.
c) May we conclude that $f(X)$ is positive definite? On what authority?

3 Show that

$$A = \begin{bmatrix} 6 & 0 & -1 \\ 0 & 5 & 2 \\ -1 & 2 & 1 \end{bmatrix}$$

is positive definite.

4 Let $f(x, y) = ax^2 + bxy + cy^2$. Prove that f is positive definite if and only if $a > 0$ and $4ac - b^2 > 0$.

5 Definition 1.20. If a quadratic form takes on nonnegative values for all X, then it is said to be *positive semidefinite*.

Let

$$f(X) = X' \begin{bmatrix} 2 & 4 & 2 \\ 4 & 9 & 3 \\ 2 & 3 & 3 \end{bmatrix} X.$$

Proceed as in Exercise 2, replacing "positive definite" by "positive semidefinite."

6 Write definitions for *negative definite* and *negative semidefinite* that parallel Definitions 1.19 and 1.20.

7 Definition 1.21. If a quadratic form takes on both negative and positive values, then it is said to be *indefinite*.

Let

$$f(X) = X' \begin{bmatrix} 3 & 4 & 11 \\ 4 & 4 & 12 \\ 11 & 12 & 35 \end{bmatrix} X.$$

Proceed as in Exercise 2, replacing "positive definite" by "indefinite."

8 Let

$$f(X) = X' \begin{bmatrix} 1 & 2 \\ 2 & 3 \end{bmatrix} X \quad \text{and} \quad g(U) = f(CU),$$

where

$$C = \begin{bmatrix} 0 & 1 \\ 2 & -3 \end{bmatrix}.$$

a) Let $X^* = [1, 1]'$. Verify that $f(X^*) = 8$ and then find a column U^* such that $g(U^*) = 8$.
b) Let $U^* = [8, 7]'$. Verify that $g(U^*) = -16$ and then find a column X^* such that $f(X^*) = -16$.
c) Reread (or read) the proof of Theorem 1.14.

9 Let $X = [x_1, x_2, \ldots, x_n]'$ and let

$$f(X) = \sum_{i=1}^{n} \sum_{j=1}^{n} a_{ij} x_i x_j,$$

where $a_{ij} = a_{ji}$. Prove that

$$f(X) = X'[a_{ij}]_{nn} X.$$

1.8 EQUIVALENCE RELATIONS

The words "equal" and "equivalent" are not, even in nonmathematical usage, synonyms. A popular dictionary makes the following distinction:

equal: identical in size or extent.
equivalent: having the same force, meaning, or effect.

In mathematics the difference in meaning attributed to these two words is along similar lines, but more precise. Two mathematical objects are said to be *equal* only if they are identical. If x and y are equal, we write $x = y$. For example, we write

$$\sin^2 t + \cos^2 t = 1$$

because the function that assigns to each number t the number $\sin^2 t + \cos^2 t$ is *identical* with the function that assigns to each number t the number 1. (Here the formulas given for the functions are not identical, but the functions defined by the formulas are. Thus, the use of the equality sign implies that we are talking about functions and not formulas.) Consider the statement

$$\begin{bmatrix} 2 & 1 & 3 \\ 4 & 2 & 6 \end{bmatrix} + \begin{bmatrix} 0 & 2 & 0 \\ 3 & 1 & 3 \end{bmatrix} = \begin{bmatrix} 2 & 3 & 3 \\ 7 & 3 & 9 \end{bmatrix}.$$

This means that the matrix sum of the matrices on the left is *identical*—entry by entry the same—with the matrix on the right.

The notion of *equivalence* is more subtle. Before we consider the formal definition adopted by mathematicians, let us first consider some examples.

Example 1. Let I be the set of all integers. We denote its members by $i, j, k, \ldots$. We introduce a *relation* (a scheme for relating objects) into I as follows: If the difference $i - j$ is divisible by 2, then we say that i is parity equivalent to j. We write $i \stackrel{(p)}{=} j$. For example, $7 \stackrel{(p)}{=} 13$, for $7 - 13 = -6$ and -6 is divisible by 2.

Consider now the following application of parity equivalence. Suppose that there is a sequence $a_1, a_2, a_3, \ldots$ for which we have derived the

formula
$$a_n = (-1)^{n+7} n^2.$$

Observing that $n + 7 \stackrel{(p)}{=} n + 1$ for all n, we may replace our formula for a_n by a simpler version:
$$a_n = (-1)^{n+1} n^2.$$

This simplification is based on the following easy-to-prove theorem: If $i \stackrel{(p)}{=} j$, then $(-1)^i = (-1)^j$. The point is this: although the formulas $n + 7$ and $n + 1$ are not equal, they are nevertheless *equivalent* in their roles as powers of -1.

Example 2. Let M be the set of all matrices. We have already (Definition 1.13) introduced into M the relation of *row equivalence*. Recall that we write $A \stackrel{\text{(row)}}{=} B$ when A is row equivalent to B.

Consider now the following (already familiar) application of row equivalence. Suppose that we are required to solve the system
$$\begin{cases} x_1 + 2x_2 = 7, \\ 2x_1 + 6x_2 = 20, \\ x_1 - x_2 = -2. \end{cases}$$

We first let
$$A = \begin{bmatrix} 1 & 2 & 7 \\ 2 & 6 & 20 \\ 1 & -1 & -2 \end{bmatrix},$$

and then reduce A to row echelon form. We obtain
$$B = \begin{bmatrix} 1 & 0 & 1 \\ 0 & 1 & 3 \\ 0 & 0 & 0 \end{bmatrix}.$$

Now, there is an important theorem concerning row equivalence: If $A \stackrel{\text{(row)}}{=} B$, then the linear systems having A and B as their augmented matrices have identical solution sets. (Corollary to Theorem 1.8.) Because of this theorem we may replace the original system by the simpler system
$$\begin{cases} x_1 = 1, \\ x_2 = 3, \end{cases}$$

and then deduce that the original system has $X = [1, 3]'$ as its only solution. In other words, although the matrices A and B are *not equal*, they are nevertheless *equivalent* in the sense that they represent systems having the same solution sets.

Example 3. Let α be a real number and let S_α be the set of all real-valued functions that are defined (at least) in some deleted neighborhood of α.

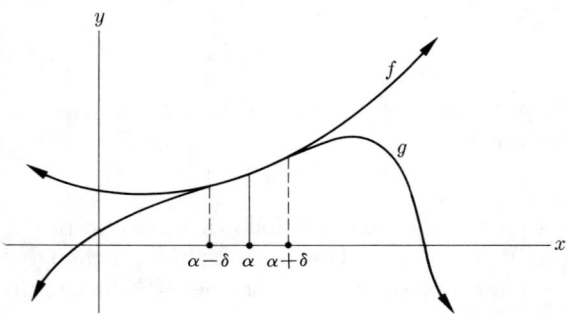

Figure 1.1

[A *deleted neighborhood* of α is an interval of the form $(\alpha - \delta, \alpha + \delta)$, $\delta > 0$, from which α itself has been removed.] For example, the functions having the following formulas are all members of S_2:

$$t^2 + 3t - 1, \quad \frac{1}{t}, \quad \frac{1}{t-2}, \quad \sqrt{5-t^2}, \quad \frac{t-2}{t^2-4},$$

but the function defined by $\sqrt{t-2}$ is not in S_2. Why?

We introduce a relation into S_α as follows: We say that f is α-*equivalent* to g, writing $f \stackrel{(\alpha)}{=} g$, if there exists a deleted neighborhood of α throughout which $f(t) = g(t)$ (see Fig. 1.1).

Now, there is an important theorem concerning α-equivalent functions which is sometimes proved in calculus courses: If $f \stackrel{(\alpha)}{=} g$ and

$$\lim_{t \to \alpha} g(t) = \beta,$$

then

$$\lim_{t \to \alpha} f(t) = \beta.$$

Suppose, for example, that we are trying to find

$$\lim_{t \to \alpha} f(t), \quad \text{where} \quad f(t) = \frac{t-2}{t^2-4}.$$

Then we might proceed as follows: Let g be defined by

$$g(t) = \frac{1}{t+2}.$$

Then $f \neq g$. (Why?) However, since $f(t) = g(t)$ for all $t \neq 2$, we can certainly say that $f \stackrel{(2)}{=} g$. Now g is continuous at 2. Thus,

$$\lim_{t \to 2} g(t) = g(2) = \tfrac{1}{4}.$$

Using our theorem, we may then conclude that
$$\lim_{t \to 2} f(t) = \tfrac{1}{4}.$$
In other words, although the functions f and g are not *equal*, they are nevertheless *equivalent* in the sense that they both converge to the same limit as t approaches 2. □

The point of Examples 1 through 3 is this: There are many arguments in mathematics in which an object may be replaced, without changing the conclusion, by a different, but related, object. When objects are interchangeable in this way, it is natural to refer to them as equivalent. However, this observation does not bring us closer to a proper definition for *equivalent;* we cannot simply define *equivalent objects* as *objects that are interchangeable under certain circumstances*. This is too vague; for a suitable definition we must try a different approach.

A further study of the foregoing equivalences reveals that they share with equality a "common denominator": each of these relations satisfies three basic laws. In terms of parity equivalence, these laws assume the following forms:

1. For every integer i we have $i \stackrel{(p)}{=} i$. (This is true since $i - i = 0$ is surely divisible by 2. Thus, we may say that every integer is parity equivalent to itself.)

2. If $i \stackrel{(p)}{=} j$, then $j \stackrel{(p)}{=} i$. (If $i - j$ is divisible by 2, then surely $j - i$ is also.)

3. If $i \stackrel{(p)}{=} j$ and $j \stackrel{(p)}{=} k$, then $i \stackrel{(p)}{=} k$. [If $i - j$ and $j - k$ are both divisible by 2, then so is their sum. But $(i - j) + (j - k) = i - k$.]

It is this fact, the fact that various equivalences share with equality these three properties, that led mathematicians to take these properties as characterizing the notion of *equivalence*.

Definition 1.22. Suppose that we have introduced a relation into a set S, writing $a \sim b$ when a is related to b. This relation is said to be an *equivalence relation* if it has the following properties:

1. $a \sim a$ (reflexive property).
2. If $a \sim b$, then $b \sim a$ (symmetric property).
3. If $a \sim b$ and $b \sim c$, then $a \sim c$ (transitive property).

Objects related with respect to an equivalence relation are said to be *equivalent*.

Introducing an equivalence relation into a set has the effect of decomposing that set into a unique collection of nonintersecting subsets.

For example, let us return to parity equivalence. Take any integer, say 17, and consider the set of all integers that are parity equivalent to 17. It is easy to see that this set contains all of the odd integers and none of the even ones. Moreover, the role played by 17 could just as well have been played by any odd number. In other words, given any odd number, the set of all integers parity equivalent to that odd number is simply the set of all odd numbers. Similarly, the set of all integers that are parity equivalent to some arbitrarily chosen even number is simply the set of all even numbers. Thus, the relation of parity equivalence serves to decompose the set of all integers into two classes: the odd ones and the even ones.

Definition 1.23. Given an equivalence relation in a set S and an element x in S, we shall call the set of all elements that are equivalent to x the *equivalence class generated by x*. We denote this set by S_x.

Theorem 1.15. Given an equivalence relation in a set S, the resulting equivalence classes meet the following conditions:

1. The union of all the equivalence classes generated by the members of S is S itself.
2. The equivalence classes generated by x and y, where $x \neq y$, are either identical or else they have no element in common.

Proof

1. By the reflexive property, we have $x \sim x$ for any x in S. It follows that x is in S_x. That is, each x in S belongs (at least) to the equivalence class that it generates. Thus, the union of all equivalence classes must contain every member of S. But then that union must be exactly S. (Why?)
2. Take $x \neq y$ and consider the equivalence classes S_x and S_y. There are two cases: (a) Suppose that $x \sim y$. Then, take any element z from S_x. This means that $z \sim x$. Thus, we have

$$z \sim x \quad \text{and} \quad x \sim y.$$

By the transitive property, we are assured that $z \sim y$. It then follows that z is in S_y. This argument proves that every element in S_x is also in S_y. But the same argument can be used to show that every element in S_y is also in S_x. We have only to observe first that the hypothesis $x \sim y$ implies immediately (the symmetry property) that $y \sim x$. Thus, we conclude finally that $S_x = S_y$. (b) Suppose that x is *not* equivalent to y and that there is an element of S, say z, that belongs to both S_x and S_y. Then

$$z \sim x \quad \text{and} \quad z \sim y.$$

1.8 EQUIVALENCE RELATIONS

Using again the symmetry property, we may write

$$x \sim z \quad \text{and} \quad z \sim y,$$

from which it follows (transitivity) that $x \sim y$. This last result contradicts our opening statement. Thus, we must conclude that there is no such element z.

Example 4. Let $S = \{2, 3, 4, 5, 6, \ldots\}$. We introduce an equivalence relation into S as follows: We say that i is *factor equivalent* to j, writing $i \stackrel{(f)}{=} j$, if i has the same prime factors as j. For example, since $30 = 2 \cdot 3 \cdot 5$ and $300 = 2 \cdot 2 \cdot 3 \cdot 5 \cdot 5$, we may write $30 \stackrel{(f)}{=} 300$. Now, what do the equivalence classes of S look like?

The equivalence classes generated by prime numbers have an especially simple form. For example,

$$S_2 = \{2, 2^2, 2^3, 2^4, \ldots\},$$
$$S_3 = \{3, 3^2, 3^3, 3^4, \ldots\},$$

and so on. Evidently, there are infinitely many equivalence classes.

The equivalence classes generated by composite numbers are not so easily exhibited. For example, S_6 consists of all those numbers having the form $2^\alpha 3^\beta$, where $\alpha, \beta > 0$. The reader may verify that the first 10 members of S_6 are as follows:

$$S_6 = \{6, 12, 18, 24, 36, 48, 54, 72, 96, 108, \ldots\}.$$

EXERCISE SET 1.8

1 Determine which of the following relations are equivalence relations. With respect to those that are not, state which of the three conditions under Definition 1.22 are violated.
   a) R is the set of real numbers; the relation is $a < b$.
   b) R as above; $a \sim b$ means that $b - a$ is a rational number.
   c) R as above; $a \sim b$ means that $ab \geq 0$.
   d) I is the set of all integers; $i \sim j$ means that i divides j evenly.
   e) I as above; $i \sim j$ means that $|i - j| < 10$.
   f) T is the set of all triangles in a plane; $\alpha \sim \beta$ means that α is similar to β.
   g) L is the set of all lines in space; $\alpha \sim \beta$ means either that α is parallel to β or that α coincides with β.

2 Let S and factor equivalence be as in Example 4. Find the first six numbers in S_{30}.

3 R is the real number system; an equivalence relation is introduced as follows: $a \sim b$ means that $(a - b)/2\pi$ is an integer.
   a) Find five numbers in the equivalence class R_2.

b) Describe a situation in which equivalent objects (with respect to this equivalence relation) are interchangeable. [*Hint:* Consider the function $\sin t$.]

4 Definition 1.24. Let M be the set of all symmetric matrices; a relation is introduced as follows: $A \sim B$ if there exists a nonsingular matrix C such that $CAC' = B$. We say that A is *congruent* to B.
  a) Prove that congruence in M is an equivalence relation.
  b) Describe a sense in which congruent matrices are interchangeable.
  c) Prove that
  $$\begin{bmatrix} 1 & 1 & 1 \\ 1 & 0 & -1 \\ 1 & -1 & -1 \end{bmatrix}$$
  is congruent to
  $$\begin{bmatrix} 1 & 2 & 0 \\ 2 & -4 & 3 \\ 0 & 3 & -1 \end{bmatrix}.$$

5 Let S_2 and 2-equivalence be as in Example 3. Let
$$f(t) = |t - 3| \cdot \frac{t^2 - 4}{t^2 - 5t + 6}.$$
  a) Prove first that f is in S_2.
  b) Find a function g that is 2-equivalent to f and such that it is continuous at $t = 2$.
  c) Use g to find $\lim_{t \to 2} f(t)$.

6 I is the set of integers; an equivalence relation is introduced as follows: $i \sim j$ means that $i - j$ is evenly divisible by 12. Let k be any integer in the equivalence class I_5. If it is k hours past noon, to what numeral is the hour hand pointing?

7 F is the set of real functions that are continuous over the interval $[0, 1]$; an equivalence relation is introduced as follows: $f \sim g$ means that
$$\int_0^1 f(t)\, dt = \int_0^1 g(t)\, dt.$$
Find three functions other than t^2 in the equivalence class generated by t^2.

8 Prove that row equivalence of matrices is indeed an equivalence relation.

9 Read Appendix A.

CHAPTER 2

VECTORS

2.1 VECTOR SPACES

We begin this chapter with a short digression. Certain nouns are said to be *abstract*, others *concrete*. Consider the noun "happiness" and contrast it with, say, the noun "chair." The word "chair" is a symbol that stands for a certain physical object. On the other hand, although the word "happiness" stands for something, it is not a physical something. Rather, in this word we are symbolizing a certain collection of states of being: happiness is walking barefoot in the rain; happiness is smelling bread baking; happiness is making someone else happy; and so on. The very existence of the word "happiness" presupposes that many states of being have some common denominator. To identify this common denominator is to *abstract* the notion of happiness. Thus, happiness is an *abstraction*, and the word "happiness" is, therefore, an *abstract* noun.

In an analogous fashion, the mathematician surveys a variety of mathematical structures and observes, perhaps, that they have certain attributes in common. In such an event, it is natural for him to try to abstract that commonness. A detailed discussion explaining how mathematical abstractions are abstracted would entail too long a digression. (Recall the preceding section in which we struggled with the notion of *equivalent objects*.) Let it suffice to note that the ultimate goal of the mathematician (in this connection) is to describe an idealized structure that embodies just those features of the structures studied that make an essential contribution to the observed commonness. This ultimate structure, an *abstract mathematical structure*, then serves as an archetype for a variety of *concrete mathematical structures* or *models*.

Now, our interest here is in *arithmetical structures:* sets of objects for which operations analogous to ordinary addition and multiplication are defined. The particular abstract arithmetical structure that we are about to consider is one that derives its inspiration from various sources familiar to the reader. Some examples will help make this clear.

Example 1. Our first example is not truly mathematical in nature, for it deals with a set of objects that are portrayed as being physical things that we can see and move about. Actually, these "physical objects" do have a precise, mathematical characterization; however, for our purpose it will suffice to take a more naive point of view.

Consider the set of all *arrows* emanating from a fixed point in space, along with the point itself, which we shall call the *zero arrow* (or *origin*). Call this set V. We shall now describe schemes for adding the arrows in V and for multiplying them by real numbers; we shall use the symbols $\oplus$ and $\odot$ to denote these operations.

1. Given the arrows X and Y, complete the parallelogram "embraced" by these two arrows. Then $X \oplus Y$ is defined as that arrow which reaches from the origin to the opposite vertex of this parallelogram. In the event that X is the zero arrow, we take $X \oplus Y$ to be Y; similarly, if Y is the zero arrow.
2. If $c > 0$, then $c \odot X$ is defined as the arrow having the same direction as X but c times as long. If $c < 0$, then $c \odot X$ is oppositely directed from X and is $|c|$ times as long. If $c = 0$, then $c \odot X$ is taken as the zero arrow.

The set of arrows V along with these two arithmetic operations constitute an *arithmetic structure*.

It would be proper to pause at this point and study this structure: to see in which ways, if any, it parallels the ordinary arithmetic of real numbers. For example, we might seek answers to questions like:

Is $X \oplus Y = Y \oplus X$?
Is $c \odot (X \oplus Y) = (c \odot X) + (c \odot Y)$?

Instead, we shall move right on to other arithmetical structures, leaving such questions until later.

Example 2. Let V be the set of all 2 by 3 matrices. Again we consider schemes for adding our objects and multiplying them by real numbers:
1. Given the 2 by 3 matrices X and Y, we define $X \oplus Y$ to be the ordinary matrix sum.
2. Similarly, $c \odot X$ is defined as the ordinary numerical multiple of the matrix X.

The point is this: Since V consists only of 2 by 3 matrices, we cannot regard ordinary matrix multiplication as an operation in V. However, the other two matrix operations do make sense. Thus, the set of matrices

V along with the operations of matrix addition and numerical multiplication do constitute an arithmetical structure.

Example 3. Let V be the set of all *real functions* (functions whose domain and range are sets of real numbers) that satisfy the differential equation

$$Y'' - Y' - 2Y = 0.$$

Let f, g, h, etc. denote functions in V. We wish to consider an arithmetic for these functions.

There are a number of natural ways of defining arithmetic operations for real functions. We list just three:

1. By $f \oplus g$ we mean the function h defined by $h(t) = f(t) + g(t)$.
2. By $f \otimes g$ we mean the function h defined by $h(t) = f(t) \times g(t)$.
3. By $c \odot f$ we mean the function h defined by $h(t) = c \cdot f(t)$.

However, although each of these operations makes sense if we are considering, say, the set of all functions defined on the interval $(-\infty, \infty)$, they do not all necessarily make sense as operations for our particular set V. For suppose that f and g belong to V. What assurance do we have that the function $h = f \oplus g$ is actually a function in V? The point is this: If we announce that there is a certain scheme for adding objects in a certain set S, then it is implicit in that announcement that the sum of any two objects in S is actually to be found in S.

Now, we leave it to the reader (see Exercise 1) to establish the following: If f and g are both solutions of the given differential equation, then so also are the functions $f \oplus g$ and $c \odot f$, for any real number c. On the other hand, the function $f \otimes g$ is generally not a solution. Thus, with respect to the particular set of real functions V, we can consider that arithmetical structure which consists of V along with the operations $\oplus$ and $\odot$. □

The arithmetic structures described in Examples 1, 2, and 3 have a certain "commonness" about them. First, in each case we are dealing with a set of objects along with two arithmetic operations: an addition operation for the objects themselves and a second operation for multiplying the objects by real numbers. But there is more to it than that. If we make a detailed study of the three structures—recall the remark at the close of Example 1—we find that they share many properties in common. Furthermore, mathematicians have determined that these common properties are derivable from a select few. Thus, if we take such a selected set of properties as the defining axioms of an *abstract* mathematical structure, then that structure will serve as the archetype for all of the *concrete* structures considered under Examples 1, 2, and 3, as well as for many others that we shall shortly come upon. The list of axioms turns out to be long, but each axiom is easily understood.

Definition 2.1. Let V be any nonempty set, and let R be the real number system. (We denote the members of V with capitals, those of R with lowercase letters.) Suppose that there is a rule for adding objects in V that fulfills the following axioms:

1. $X \oplus Y = Y \oplus X$.
2. $X \oplus (Y \oplus Z) = (X \oplus Y) \oplus Z$.
3. There exists a unique object in V, denoted by Θ, such that $X \oplus \Theta = X$.
4. For each X in V there exists a unique mate, denoted by $-X$, such that $X \oplus (-X) = \Theta$.

Furthermore, suppose that there is a rule for multiplying the objects in V by the members of R, a multiplication rule that fulfills the following axioms:

5. $c \odot (X \oplus Y) = (c \odot X) \oplus (c \odot Y)$.
6. $(c + d) \odot X = (c \odot X) \oplus (d \odot X)$.
7. $c \odot (d \odot X) = (cd) \odot X$.
8. $1 \odot X = X$.

Then this entire structure, the quadruple $(V, R, \oplus, \odot)$, is called a *vector space*. The elements of V are called *vectors*. The elements of R (being real numbers in a special context) are called *scalars*.

Two remarks may serve to clarify this definition. The first has to do with notation: The curious symbols "$\oplus$" and "$\odot$" are used so as to emphasize the fact that the arithmetic operations in V are not, except by coincidence, the usual addition and multiplication of real number arithmetic. In contrast, the appearance of "$+$" and the juxtaposition notation for multiplication—see Axioms 6 and 7—do refer to the usual arithmetic operation denoted by that notation. The second remark has to do with the addition and multiplication rules. The phrase "there is a rule for adding objects in V" means that for every pair of vectors X and Y in V there is a unique Z in V such that $X \oplus Y = Z$. Similarly, the phrase "there is a rule for multiplying the objects in V by the members of R" means that for every scalar c in R and every vector X in V there is a unique Z in V such that $c \odot X = Z$.

The reader should consider each of the foregoing axioms with respect to the structures of Examples 1, 2, and 3, and thus verify that these structures are indeed vector spaces. The *arrow space* of Example 1 can, of course, be regarded as the primary motivation for even considering the notion of a vector space, and our axioms seem, perhaps, most "self-evident" in that space (see Fig. 2.1). The reader may profit by verbalizing each

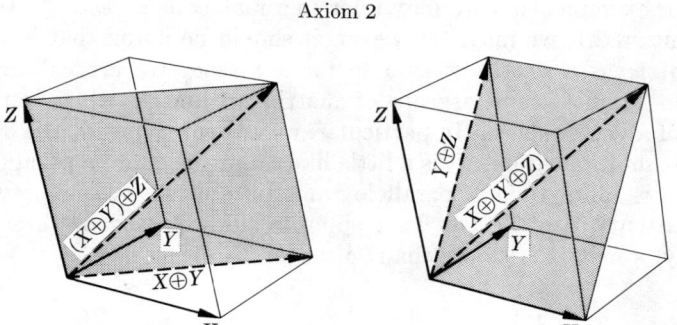

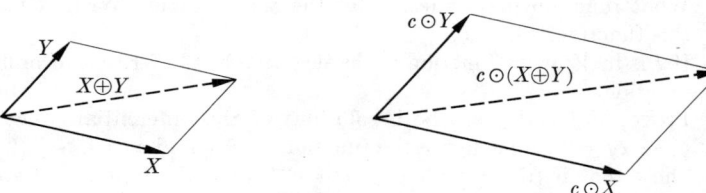

Figure 2.1

of the axioms in the "language of arrows." For example, Axiom 7 says: If an arrow is lengthened by a factor of d, and that arrow is lengthened by a factor of c, then surely the resulting arrow is the same as that achieved by lengthening the original arrow by a factor of cd. On the other hand—and this is the main point—these same axioms are just as valid for the other two structures considered. Thus, in this sense, the set of all 2 by 3 matrices or the set of all solutions of the differential equation $y'' - y' - 2y = 0$ may be likened to the set of all arrows emanating from a point. Another example, a somewhat peculiar one, shows that real numbers may be viewed as vectors.

Example 4. Take V as the set of real numbers. Take $\oplus$ as the usual addition of the real number system and $\odot$ as the usual multiplication. Then all the axioms listed under Definition 2.1 are fulfilled in a trivial way: since both the capital letters and the lowercase letters are to be interpreted as real numbers, the axioms express well known truths about the arithmetic of real numbers. Thus, the structure $(V, R, \oplus, \odot)$ is a vector space, the real numbers being both the vectors and scalars in this case. □

Thus, we have vector spaces of arrows, vector spaces of matrices, vector spaces of functions, and vector spaces of numbers. Does this mean, for example, that we may refer to a matrix as a vector? According to Definition 2.1, we may. However, it should be noted that in referring to a matrix as a vector, we do so for a reason: we are calling special attention to just those aspects of matrix arithmetic which satisfy the axioms of a vector space. In particular, we are putting aside the operation of matrix multiplication. It is a little like calling a square a parallelogram: a square is more than a parallelogram, but by so calling it, we draw special attention to the fact that opposite sides of a square are parallel. Similarly, a matrix is "more than" a vector.

EXERCISE SET 2.1

1. Recall the vector space V of Example 3.
   a) What function plays the role of the zero vector? Write a formula for this function.
   b) If f is in V, what function is the negative of f? Write a formula for this function.
   c) Prove: If f and g are both solutions of the differential equation $y'' - y' - 2y = 0$, then so are the functions $f \oplus g$ and $c \odot f$.
   d) Show that if $f(t) = e^{2t}$ and $g(t) = e^{-t}$, then the functions f and g both satisfy the differential equation of part (c). Also, show that $f \otimes g$ does not satisfy this equation.

2. Let S be the set of all ordered pairs of real numbers. We define sums and numerical multiples of pairs as follows:
   1. $(x_1, x_2) \oplus (y_1, y_2) = (x_1 + y_1, x_2 + y_2)$.
   2. $c \odot (x_1, x_2) = (2cx_1, cx_2)$.

   Show that $(S, R, \oplus, \odot)$ is *not* a vector space. [*Hint:* Find an axiom under Definition 2.1 that is not fulfilled in this arithmetical structure.]

3. Let S be as in Exercise 2, but let $\oplus$ and $\odot$ be defined as follows:
   1. $(x_1, x_2) \oplus (y_1, y_2) = (x_1 y_1, x_2 y_2)$.
   2. $c \odot (x_1, x_2) = (cx_1, cx_2)$.

   Show that $(S, R, \oplus, \odot)$ is *not* a vector space.

4. Let S and $\oplus$ be as in Exercise 2, but let $\odot$ be defined as follows:
   $$c \odot (x_1, x_2) = (x_1, x_2).$$
   Show that $(S, R, \oplus, \odot)$ is *not* a vector space.

5. Prove that the following theorems are valid in any vector space.
   a) $c \odot \Theta = \Theta$ for all c. [*Hint:* Show first that $c \odot \Theta = (c \odot \Theta) \oplus (0 \odot \Theta)$.]
   b) $0 \odot X = \Theta$ for all X. [*Hint:* Show first that $X = X \oplus (0 \odot X)$.]
   c) $(-1) \odot X = -X$ for all X. [*Hint:* $X \oplus [(-1) \odot X] = [(1) \odot X] \oplus [(-1) \odot X] = \cdots$.]

6 a) Let V be a set consisting of a single element, denoted by "loner." We define vector operations as follows:

1. loner $\oplus$ loner = loner.
2. $c \odot$ loner = loner.

Show that $(V, R, \oplus, \odot)$ *is* a vector space. (Thus, a vector space can consist of a single element.)

b) Can a vector space have just n vectors, where $n \neq 1$?

7 Let V be the set of all real functions over the interval $(-\infty, \infty)$, and let $\oplus$ and $\odot$ be the *function operations* defined in Example 3. Show that $(V, R, \oplus, \odot)$ *is* a vector space. (Your only real concern is with respect to Axioms 3 and 4. Why?)

8 Let W be the set of positive numbers, and let $\oplus$ and $\odot$ be defined as follows:

1. $X \oplus Y = XY$ (vector addition = ordinary multiplication).
2. $c \odot X = X^c$ (numerical multiplication = exponentiation).

Show that $(W, R, \oplus, \odot)$ *is* a vector space.

9 *On the perverseness of mathematicians.* Do the objects of a vector space, the things called *vectors*, have length and direction?

10 a) Let S be as in Exercise 2, but let $\oplus$ and $\odot$ be defined as follows:

1. $(x_1, x_2) \oplus (y_1, y_2) = ([x_1^3 + y_1^3]^{1/3}, [x_2^3 + y_2^3]^{1/3})$
2. $c \odot (x_1, x_2) = (cx_1, cx_2)$.

Show that $(S, R, \oplus, \odot)$ is *not* a vector space.

b) Change the multiplication rule to

$$c \odot (x_1, x_2) = (c^{1/3}x_1, c^{1/3}x_2)$$

and then show that $(S, R, \oplus, \odot)$ *is* a vector space.

2.2 SUBSPACES

Suppose that $(V, R, \oplus, \odot)$ is a vector space and that W is a *proper subset* of V, that is, W consists of some but not all of the vectors in V. Suppose next that we discard all of the vectors in V except those in our select subset W, but that we retain our knowledge of the way vectors in V are added and multiplied by scalars. Then we might reasonably ask if W, along with these operations, is a vector space in its own right. Unfortunately, if W has been selected quite arbitrarily, there is no assurance that all sums and multiples of vectors in W are still available. An example will make this clear.

Example 1. Let $(V, R, \oplus, \odot)$ be the vector space of 3 by 1 matrices. And let W be that subset of V consisting of just those columns having the

form
$$\begin{bmatrix} a \\ b \\ a+1 \end{bmatrix}.$$

(That is, to be in W your third entry must exceed your first by exactly 1.) Now let X^1 and X^2 be any two members of W; we write

$$X^1 = \begin{bmatrix} a \\ b \\ a+1 \end{bmatrix} \quad \text{and} \quad X^2 = \begin{bmatrix} c \\ d \\ c+1 \end{bmatrix}.$$

Then

$$X = X^1 \oplus X^2 = \begin{bmatrix} a+c \\ b+d \\ a+c+2 \end{bmatrix}.$$

Thus, X is in V, but not in W. Consequently, if we were to discard all columns except those in W, then the above addition simply could not be performed. Hence W, at least with respect to the usual matrix operations, cannot be regarded as a vector space.

Example 2. Let $(V, R, \oplus, \odot)$ be as in Example 1, and let W be the subset of V consisting of vectors of the form

$$\begin{bmatrix} a \\ b \\ a+b \end{bmatrix}.$$

Then sums and multiples of vectors in W can be found in W:

1. $X^1 \oplus X^2 = \begin{bmatrix} a \\ b \\ a+b \end{bmatrix} \oplus \begin{bmatrix} c \\ d \\ c+d \end{bmatrix} = \begin{bmatrix} a+c \\ b+d \\ a+b+c+d \end{bmatrix}$

$$= \begin{bmatrix} (a+c) \\ (b+d) \\ (a+c)+(b+d) \end{bmatrix}.$$

2. $c \odot X^1 = c \odot \begin{bmatrix} a \\ b \\ a+b \end{bmatrix} = \begin{bmatrix} ca \\ cb \\ c(a+b) \end{bmatrix} = \begin{bmatrix} (ca) \\ (cb) \\ (ca)+(cb) \end{bmatrix}.$

Thus, at least it makes sense to ask whether or not the structure $(W, R, \oplus, \odot)$ is a vector space. Now Axioms 1, 2, 5, 6, 7, and 8 are automatically fulfilled because these properties of $\oplus$ and $\odot$ are valid for all vectors in V; hence, in particular, they are valid for vectors in W. Thus, we need only check out Axioms 3 and 4.

Axiom 3: Since
$$\Theta = \begin{bmatrix} 0 \\ 0 \\ 0 \end{bmatrix} = \begin{bmatrix} 0 \\ 0 \\ 0+0 \end{bmatrix},$$
it follows that Θ is in W.

Axiom 4: If
$$X = \begin{bmatrix} a \\ b \\ a+b \end{bmatrix},$$
then
$$-X = \begin{bmatrix} -a \\ -b \\ -(a+b) \end{bmatrix} = \begin{bmatrix} (-a) \\ (-b) \\ (-a)+(-b) \end{bmatrix}.$$

Evidently, $-X$ is in W.

Thus, $(W, R, \oplus, \odot)$ is a vector space in its own right. It is natural to refer to it as a *subspace* of the original space.

Definition 2.2 Let $(V, R, \oplus, \odot)$ be a vector space and let $(W, R, \oplus', \odot')$ be a second vector space having the following properties:
1. W is a subset of V.
2. The operations $\oplus'$ and $\odot'$ are the same as $\oplus$ and $\odot$, except that they are restricted to the set W.

Then $(W, R, \oplus', \odot')$ is said to be a *subspace* of $(V, R, \oplus, \odot)$.

Theorem 2.1. Let $(V, R, \oplus, \odot)$ be a vector space. Let W be a nonempty subset of V having the following properties:
1. If X and Y are both in W, then so is $X \oplus Y$.
2. If X is in W and c is any number, then $c \odot X$ is in W.

Then $(W, R, \oplus', \odot')$ is a subspace of $(V, R, \oplus, \odot)$. [An abbreviated way of stating properties (1) and (2): *The set W is closed with respect to the operations of V.*]

Proof. For the reasons already cited under Example 2, we have only to consider Axioms 3 and 4.

Axiom 3: Choosing $c = 0$ and using part 2 of our hypothesis, we may conclude that $0 \odot X$ is in W, where X is any vector whatsoever chosen from W. But $0 \odot X = \Theta$. (See Exercise 5(b) of Section 2.1.) Thus, Θ is in W.

Axiom 4: Take X as any vector in W, and consider the vector $(-1) \odot X$. This latter vector—again using part 2 of our hypothesis—is in W. But $(-1) \odot X = -X$. (See Exercise 5(c) of Section 2.1.) Thus, the negative of each X in W is also in W.

Example 3. Let W be the collection of all polynomial functions. Since the sum of two polynomials is a polynomial and a numerical multiple of a polynomial is a polynomial, it follows (by Theorem 2.1) that W is a subspace of the vector space of all functions defined on $(-\infty, \infty)$. Being a subspace, W is itself a vector space. (We have used "W" instead of "$(W, R, \oplus, \odot)$." This ambiguous use of W is a common practice when the vector operations are understood.)

Example 4. Let W be the set of all 3 by 1 matrices that satisfy the equation

$$\begin{bmatrix} 1 & 2 & -3 \\ 2 & 1 & 3 \\ 3 & 3 & 0 \end{bmatrix} \begin{bmatrix} x_1 \\ x_2 \\ x_3 \end{bmatrix} = \begin{bmatrix} 0 \\ 0 \\ 0 \end{bmatrix}.$$

The reader may show that W is the set generated by

$$X = \begin{bmatrix} -3t \\ 3t \\ t \end{bmatrix}.$$

Now, it is immediate—recall the method of Example 2—that sums and multiples of columns of this type are of the same type. Thus, the hypotheses of Theorem 2.1 are fulfilled and, consequently, W is a subspace of the space of all 3 by 1 matrices. It is natural to call W the *solution space* of the given matrix equation.

It is important to observe that this result is obtainable without knowing the precise form of the members of W. Indeed, the important thing here is that the given equation is of the form $AX = \Theta$. For suppose that X^1 and X^2 both satisfy this equation. Then, using only the basic properties of matrix arithmetic, we may write:

1. $A(cX^1) = c(AX^1) = c(\Theta) = \Theta$. Thus, cX^1 is in W.
2. $A(X^1 + X^2) = AX^1 + AX^2 = \Theta + \Theta = \Theta$. Thus, $(X^1 + X^2)$ is in W.

EXERCISE SET 2.2

1 Let V be the vector space of all 3 by 1 matrices, with the usual addition and numerical multiplication. Show that the given subset of V along with V's operations either does or does not constitute a subspace of V.

a) W consists of all columns of the form $\begin{bmatrix} a \\ b \\ 0 \end{bmatrix}$.

b) $W, \begin{bmatrix} a \\ a \\ a^2 \end{bmatrix}$ c) $W, \begin{bmatrix} a \\ b \\ 3a+4b \end{bmatrix}$

2 Let V be the space of all polynomial functions, with the usual addition and numerical multiplication. Proceed as in Exercise 1.
  a) W consists of all polynomials of the form $at^2 + bt + c$. (Here and elsewhere we write "polynomials" when we mean "polynomial functions.")
  b) W, polynomials that take on only positive values.
  c) W, polynomials that take on the value 0 at $t = 5$.
  d) W, polynomials with real zeros only. (A *zero* of a polynomial is a number for which the value of the polynomial is 0.)

3 Let V be the space of all 2 by 2 matrices. Proceed as in Exercise 1.
  a) W consists of all matrices of the form $\begin{bmatrix} a & b \\ -b & a \end{bmatrix}$.
  b) W, symmetric matrices.   c) W, nonsingular matrices.
  d) W, matrices whose entries are either all positive, all negative, or all zero.

4 Let A be an nth order matrix and let B be n by 1. Then consider the equation $AX = B$.
  a) Prove: If $B = \Theta$, then the set of solutions of $AX = B$ is a subspace of the vector space of all n by 1 columns.
  b) If A is nonsingular, then the solution set of $AX = \Theta$ consists of a single vector: the vector Θ. Does this in any way contradict the theorem of part (a)?
  c) Is the proposition of part (a) true when $B \neq \Theta$? (Prove your answer.)

5 Let V be the vector space of all real-valued functions that are defined on the interval $[0, 1]$, with the usual addition and numerical multiplication of functions. Proceed as in Exercise 1.
  a) W consists of all functions f that have equal values at the endpoints of $[0, 1]$, that is, $f(0) = f(1)$.
  b) W, all f such that $f(0) + f(1) = \frac{1}{2}$.
  c) W, all f such that $f(0) + f(1) = f(\frac{1}{2})$.
  d) W, all increasing functions.

6 Let V be the vector space of Exercise 5.
  a) Let W be the set of all continuous functions in V. Quote the theorems from calculus that permit us to conclude that W is a subspace of V.
  b) Let W^* be the set of all differentiable functions in V. Quote the theorems from calculus that permit us to conclude that W^* is a subspace of W.
  c) Let W^{**} be the set of all functions in V that satisfy the differential equation
  $$ay'' + by' + cy = 0,$$
  where a, b, and c are constants. Prove that W^{**} is a subspace of W^*.

7 Let V be the vector space of Example 1, Section 2.1. Let L be a line in the given space, and let W be the subset of V consisting of those vectors whose tips lie on L. For which lines will the resulting W be a subspace of V? (No proof required.)

2.3 ISOMORPHIC VECTOR SPACES

Certain vector spaces will be used sufficiently often for illustrative purposes that it will be convenient to adopt a special notation and terminology for them.

Definition 2.3.
1. The vector space of all m by n matrices will be denoted by R^{mn}. Such spaces will be called *matrix spaces*. The spaces R^{1n} and R^{n1} will be called *row spaces* and *column spaces*, respectively.
2. The space of polynomial functions corresponding to expressions of the form $a_n t^n + \cdots + a_1 t + a_0$ will be called the *polynomial space of degree n*. We denote it by P^n. The space of all polynomial functions will be denoted by P^∞.

Consider now the space P^2. Any member of P^2, say $p(t) = at^2 + bt + c$, is known by knowing its coefficients, and their order. Thus, the polynomial function p is "captured" in the matrix $[a, b, c]$. More interesting, however, is that the vector arithmetic of such polynomial functions is "mirrored" in the vector arithmetic of R^{13}. That is, corresponding to each polynomial computation there is a corresponding matrix computation. Consider, for example, the following parallel computations:

Let $\quad p(t) = 3t^2 + 2t - 6$ $\qquad$ Let $\quad X^1 = [3, 2, -6]$
and $\quad q(t) = 2t^2 - 5t + 7$. $\qquad$ and $\quad X^2 = [2, -5, 7]$.

Then $\qquad\qquad\qquad\qquad\qquad\qquad\qquad$ Then

$\qquad 2\{p(t) + 3q(t)\}$ $\qquad\qquad\qquad\qquad 2\{X^1 + 3X^2\}$
$\qquad = 2\{(3t^2 + 2t - 6)$ $\qquad\qquad\qquad = 2\{[3, 2, -6]$
$\qquad\quad + (6t^2 - 15t + 21)\}$ $\qquad\qquad\quad + [6, -15, 21]\}$
$\qquad = 2(9t^2 - 13t + 15)$ $\qquad\qquad\qquad = 2[9, -13, 15]$
$\qquad = 18t^2 - 26t + 30.$ $\qquad\qquad\qquad\quad = [18, -26, 30].$

(Note that we have used the same operation symbols in both computations. This will be our custom from now on, when there is no confusion as to the kinds of addition and numerical multiplication being used. In theoretical discussions, however, it will still often be necessary to distinguish the operations in distinct spaces by using different symbols.)

2.3 ISOMORPHIC VECTOR SPACES

The point of the foregoing computations is this: Although the spaces P^2 and R^{13} are quite distinct—polynomial functions are certainly not 1 by 3 matrices—the two spaces are alike in that their arithmetic structures duplicate each other.

To make this "arithmetic sameness" more precise, we introduce a one-to-one function having domain P^2 and range R^{13}. Let Id be the function defined by the formula

$$\mathrm{Id}(at^2 + bt + c) = [a, b, c].$$

We say that Id *identifies* P^2 with R^{13}. (It is as if each polynomial function in P^2 were joined with a string to a matrix in R^{13}, no element in either space having more than one string attached to it.) Now the function Id has the following two important properties:

1. $\mathrm{Id}(p + q) = \mathrm{Id}(p) + \mathrm{Id}(q)$.
2. $\mathrm{Id}(cp) = c\,\mathrm{Id}(p)$.

In words:

1. The matrix that is identified with the polynomial function $(p + q)$ is the same as the sum of the matrices that are individually identified with p and q.
2. The matrix identified with the polynomial function cp is the same as that obtained by multiplying the matrix identified with p by c.

Illustrations of these properties:

1. $\mathrm{Id}\{(2t^2 + 3) + (t^2 + t - 2)\} = \mathrm{Id}(3t^2 + t + 1) = [3, 1, 1]$,
   $\mathrm{Id}(2t^2 + 3) + \mathrm{Id}(t^2 + t - 2) = [2, 0, 3] + [1, 1, -2] = [3, 1, 1]$.
2. $\mathrm{Id}\{15(t^2 - 2t + 1)\} = \mathrm{Id}(15t^2 - 30t + 15) = [15, -30, 15]$,
   $15\,\mathrm{Id}\{t^2 - 2t + 1\} = 15[1, -2, 1] = [15, -30, 15]$.

The significance of these two properties is, perhaps, best appreciated by redescribing them in an alternative form obtained by applying the inverse of Id to both sides:

1. $p + q = \mathrm{Id}^{-1}\{\mathrm{Id}(p) + \mathrm{Id}(q)\}$.
2. $cp = \mathrm{Id}^{-1}\{c\,\mathrm{Id}(p)\}$.

It is this last formulation which brings out the role of Id in shifting calculations from P^2 to R^{13}:

1. $(2t^2 + 3) + (t^2 + t - 2) \xrightarrow{\mathrm{Id}} [2, 0, 3] + [1, 1, -2]$
   $3t^2 + t + 1 \xleftarrow{\mathrm{Id}^{-1}} [3, 1, 1]$
2. $15(t^2 - 2t + 1) \xrightarrow{\mathrm{Id}} 15[1, -2, 1]$
   $15t^2 - 30t + 15 \xleftarrow{\mathrm{Id}^{-1}} [15, -30, 15]$

(Except for the small economy involved in replacing the symbol "$at^2 + bt + c$" by "$[a, b, c]$," there is little motivation for shifting calculations from P^2 to R^{13}, but see Exercise 8 in this regard.)

Definition 2.4. Given the vector spaces $(V, R, +, \cdot)$ and $(W, R, \oplus, \odot)$, suppose that there is a one-to-one function Id with domain V and range W having the following properties:

1. $\text{Id}(X + Y) = \text{Id}(X) \oplus \text{Id}(Y)$.
2. $\text{Id}(c \cdot X) = c \odot \text{Id}(X)$.

Then we say that the V-space is *isomorphic* to the W-space and that Id is an *isomorphism* from the V-space onto the W-space. (We shall also use, because it is most suggestive, the following language: Id is an *identification* function which *identifies* the V-space with the W-space.)

Example 1. Prove that R^{22} is isomorphic to R^{41}.

Solution. To meet the requirements of Definition 2.4, we must find an isomorphism Id that identifies R^{22} with R^{41}. This is easy to do. Let Id be defined by

$$\text{Id}\left(\begin{bmatrix} a & b \\ c & d \end{bmatrix}\right) = \begin{bmatrix} a \\ b \\ c \\ d \end{bmatrix}.$$

That Id is a one-to-one function is intuitively obvious, but somewhat tedious to prove formally. Here is an acceptable proof. Let

$$X = \begin{bmatrix} a & b \\ c & d \end{bmatrix} \quad \text{and} \quad Y = \begin{bmatrix} \alpha & \beta \\ \gamma & \delta \end{bmatrix},$$

where $X \neq Y$. Then, by our definition of Id, we have

$$\text{Id}(X) = \begin{bmatrix} a \\ b \\ c \\ d \end{bmatrix} \quad \text{and} \quad \text{Id}(Y) = \begin{bmatrix} \alpha \\ \beta \\ \gamma \\ \delta \end{bmatrix}.$$

Now, the assumption that $X = Y$ implies only that corresponding entries of these two matrices differ in at least one instance. That same pair of differing entries serves to make

$$\begin{bmatrix} a \\ b \\ c \\ d \end{bmatrix} \neq \begin{bmatrix} \alpha \\ \beta \\ \gamma \\ \delta \end{bmatrix}.$$

Thus, Id assigns to different domain elements different mates (values). This is precisely what it means to say that Id is one-to-one. We must now verify that Id has the properties required of an isomorphism:

1. $\text{Id}\left(\begin{bmatrix} a & b \\ c & d \end{bmatrix} + \begin{bmatrix} \alpha & \beta \\ \gamma & \delta \end{bmatrix}\right) = \text{Id}\left(\begin{bmatrix} (a+\alpha) & (b+\beta) \\ (c+\gamma) & (d+\delta) \end{bmatrix}\right)$

$= \begin{bmatrix} (a+\alpha) \\ (b+\beta) \\ (c+\gamma) \\ (d+\delta) \end{bmatrix} = \begin{bmatrix} a \\ b \\ c \\ d \end{bmatrix} + \begin{bmatrix} \alpha \\ \beta \\ \gamma \\ \delta \end{bmatrix}$

$= \text{Id}\left(\begin{bmatrix} a & b \\ c & d \end{bmatrix}\right) + \text{Id}\left(\begin{bmatrix} \alpha & \beta \\ \gamma & \delta \end{bmatrix}\right).$

2. This part is left to the reader. □

Recall the space of geometric vectors considered in Example 1 of Section 2.1; call it V. Suppose that a *coordinate system* is superimposed on the plane of vectors V, placing the origin of the coordinate system over the point from which the vectors emanate. Then this coordinate system provides the means for identifying V with R^{21}. Again, because of our mathematically inadequate description of V, it is impossible to prove that V is isomorphic to R^{21}. However, that this is indeed the case is made evident by the sketches in Fig. 2.2.

In an analogous manner, the space of all vectors emanating from the origin of a "three-dimensional space" is isomorphic to R^{31}. Thus, we have suitable motivation to declare—in an attempt to introduce the notion of "dimension" for arbitrary vector spaces—that R^{n1} has dimension n and that every space isomorphic to R^{n1} has dimension n. However, such a definition would be nonsense if it turned out that it was possible to identify a given vector space V with both R^{n1} and with R^{m1}, where $n \neq m$. But this is not the case.

Theorem 2.2. If V is isomorphic to R^{n1} and $n \neq m$, then V is *not* isomorphic to R^{m1}.

Proof. The notion of isomorphism is such that for V to be simultaneously isomorphic to both R^{n1} and R^{m1} it would be necessary for R^{n1} to be isomorphic to R^{m1}. (We leave this part of the argument to the reader, see Exercise 5.) Thus, the essential thing to be proved is that R^{n1} is not isomorphic to R^{m1}.

To avoid notational problems that might obscure the "spirit" of the proof, we consider only the special case where $n = 3$ and $m = 2$. We begin with the assumption—our proof is by contradiction—that there is a function Id that identifies R^{31} with R^{21}. There are a number of steps

64 VECTORS

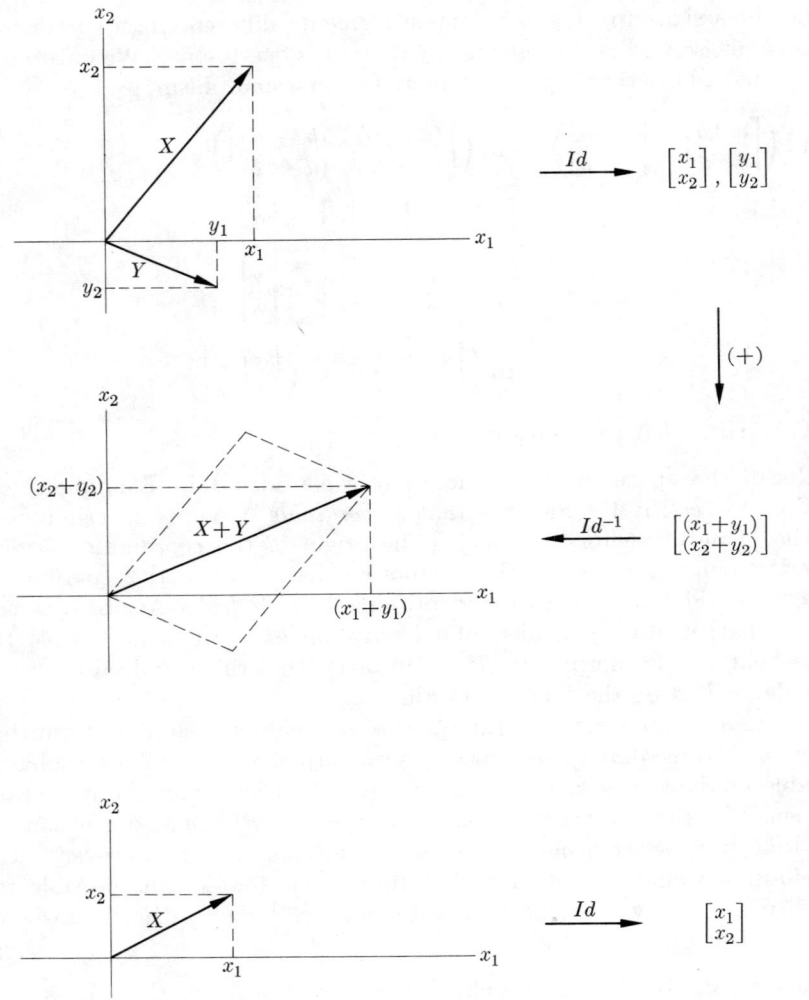

Figure 2.2

2.3 ISOMORPHIC VECTOR SPACES

to our proof:

1. Let

$$N^1 = \begin{bmatrix} 1 \\ 0 \\ 0 \end{bmatrix}, \quad N^2 = \begin{bmatrix} 0 \\ 1 \\ 0 \end{bmatrix}, \quad \text{and} \quad N^3 = \begin{bmatrix} 0 \\ 0 \\ 1 \end{bmatrix}.$$

Since we are given the function Id (by our hypothesis), it makes sense to "compute" $\text{Id}(N^j)$ for $j = 1, 2, 3$. Thus, we let

$$\text{Id}(N^1) = \begin{bmatrix} a_{11} \\ a_{21} \end{bmatrix},$$

$$\text{Id}(N^2) = \begin{bmatrix} a_{12} \\ a_{22} \end{bmatrix},$$

and

$$\text{Id}(N^3) = \begin{bmatrix} a_{13} \\ a_{23} \end{bmatrix}.$$

2. Next, we compute $\text{Id}(X)$, where

$$X = [x_1, x_2, x_3]' = x_1 N^1 + x_2 N^2 + x_3 N^3.$$

$$\begin{aligned}
\text{Id}(X) &= \text{Id}(x_1 N^1 + x_2 N^2 + x_3 N^3) \\
&= \text{Id}(x_1 N^1) + \text{Id}(x_2 N^2) + \text{Id}(x_3 N^3) \quad \text{(Why?)} \\
&= x_1 \text{Id}(N^1) + x_2 \text{Id}(N^2) + x_3 \text{Id}(N^3) \quad \text{(Why?)} \\
&= x_1 \begin{bmatrix} a_{11} \\ a_{21} \end{bmatrix} + x_2 \begin{bmatrix} a_{12} \\ a_{22} \end{bmatrix} + x_3 \begin{bmatrix} a_{13} \\ a_{23} \end{bmatrix} \\
&= \begin{bmatrix} (a_{11}x_1 + a_{12}x_2 + a_{13}x_3) \\ (a_{21}x_1 + a_{22}x_2 + a_{23}x_3) \end{bmatrix} \\
&= \begin{bmatrix} a_{11} & a_{12} & a_{13} \\ a_{21} & a_{22} & a_{23} \end{bmatrix} \begin{bmatrix} x_1 \\ x_2 \\ x_3 \end{bmatrix}.
\end{aligned}$$

3. Finally, consider the equation $\text{Id}(X) = \Theta$, or

$$\begin{bmatrix} a_{11} & a_{12} & a_{13} \\ a_{21} & a_{22} & a_{23} \end{bmatrix} \begin{bmatrix} x_1 \\ x_2 \\ x_3 \end{bmatrix} = \begin{bmatrix} 0 \\ 0 \end{bmatrix}.$$

This equation has infinitely many solutions. (Recall Theorem 1.10.) But this means, recalling part 2, that $\text{Id}(X) = \Theta$ for infinitely many X—a most unseemly behavior for a one-to-one function!

Definition 2.5. If V is a vector space that is isomorphic to R^{n1}, then V is said to have *dimension n*.

EXERCISE SET 2.3

1 Let V be the subspace of R^{22} consisting of all symmetric matrices. An isomorphism from V on to R^{31} is defined by

$$\text{Id}\left(\begin{bmatrix} a & b \\ b & c \end{bmatrix}\right) = \begin{bmatrix} a \\ b \\ c \end{bmatrix}.$$

Also, let

$$X = \begin{bmatrix} 1 & 2 \\ 2 & 3 \end{bmatrix} \quad \text{and} \quad Y = \begin{bmatrix} 7 & 4 \\ 4 & -2 \end{bmatrix}.$$

a) Verify that $\text{Id}(X + Y) = \text{Id}(X) + \text{Id}(Y)$.
b) Verify that $\text{Id}(17X) = 17\,\text{Id}(X)$.
c) Find $\text{Id}^{-1}([0, 1, 2]')$.
d) What is the dimension of V?

2 Let V be the subspace of P^3 consisting of all polynomial functions that take on the value 0 at $t = 5$. An isomorphism from V onto R^{31} is defined by the following two-step rule. To find $\text{Id}(p)$, first express $p(t)$ in the form $(t - 5)(at^2 + bt + c)$. Then take $\text{Id}(p) = [a, b, c]'$. Now, let $p(t) = t^3 - 125$ and $q(t) = t^3 - 5t^2 - t + 5$, and proceed as follows:

a) Verify that $\text{Id}(p + q) = \text{Id}(p) + \text{Id}(q)$.
b) Find $\text{Id}^{-1}([1, 2, 3]')$, writing your answer in standard polynomial form.
c) What is the dimension of V?

3 Let V be the subspace of R^{33} consisting of all diagonal matrices. Then the function Id defined by

$$\text{Id}\left(\begin{bmatrix} a & 0 & 0 \\ 0 & b & 0 \\ 0 & 0 & c \end{bmatrix}\right) = \begin{bmatrix} a \\ b \\ c \end{bmatrix}$$

is an isomorphism from V onto R^{31}. Prove this assertion. (Use Example 1 as a model.)

4 By exhibiting an appropriate identification function, determine the dimension of the given vector space. (No proof required.)

a) All 1 by n matrices.
b) All 3 by 2 matrices.
c) All 1 by 5 matrices of the form $[a, a, a, b, b]$.
d) All polynomial functions of the form $at^5 + bt$.
e) All functions of the form $a \cos t + b \sin t$.
f) The real number space (Example 4, Section 2.1).
g) All members of P^3 that take on the value 0 at $t = 2$.
h) All members of P^3 that take on the value 0 at both $t = 2$ and $t = 3$.

5 Theorem 2.3. The isomorphism relation of vector spaces is an equivalence relation.

a) Theorem 2.3 is proved by proving three separate "subtheorems." State these subtheorems.
b) Prove Theorem 2.3.

6 Prove: If Id is an isomorphism from V_1 onto V_2, then $\text{Id}(\Theta_1) = \Theta_2$. (That is, an isomorphism between vector spaces necessarily identifies the zeros of the two spaces.)

7 a) Let f be defined on R^{22} by

$$f\left(\begin{bmatrix} a & b \\ c & d \end{bmatrix}\right) = \begin{bmatrix} a \\ b \\ c \end{bmatrix}.$$

Show that f is *not* an isomorphism from R^{22} onto R^{31}.

b) Let f be defined on R^{22} by

$$f\left(\begin{bmatrix} a & b \\ c & d \end{bmatrix}\right) = [a^3, b^3, c^3, d^3]'.$$

Show that f is *not* an isomorphism from R^{22} onto R^{41}.

c) Let f be defined on the space V of symmetric second order matrices by

$$f\left(\begin{bmatrix} a & b \\ b & c \end{bmatrix}\right) = [a, b, b, c]'.$$

Show that f is *not* an isomorphism from V onto R^{41}.

8 Let V be the space of Example 4, Section 2.1. Let W be the space of Exercise 8, Section 2.1.

a) Show that W is isomorphic to V. [*Hint:* Let $\text{Id}(X) = \log_e X$.]
b) What is gained by shifting computations from W to V?

9 Let V be the space of polynomials of the form $at^2 + bt$. Let the function D be defined on V by the formula $D(p) = p'$, where p' is the derivative of p. Prove that D is an isomorphism from V onto P^1.

2.4 LINEAR INDEPENDENCE

To determine the dimension of a vector space V it is necessary, according to Definition 2.5, to demonstrate the existence of an isomorphism identifying V with some column space R^{n1}. In the exercises that we have considered so far, such an isomorphism was easy to construct. This is not always the case. (Indeed, we have no assurance yet that such an isomorphism always exists.)

Suppose, for example, that we let V be the subspace of R^{41} generated by

$$X = a \begin{bmatrix} 1 \\ 2 \\ 3 \\ 4 \end{bmatrix} + b \begin{bmatrix} 2 \\ 0 \\ 1 \\ -2 \end{bmatrix} + c \begin{bmatrix} 3 \\ 2 \\ 4 \\ 2 \end{bmatrix}.$$

It would be reasonable to guess that V had dimension 3, and to establish that guess by constructing an isomorphism that identifies each X in V

68   VECTORS

with the triple of coefficients that produces X. That is, we are prompted to let

$$\text{Id}(X) = \begin{bmatrix} a \\ b \\ c \end{bmatrix}.$$

But this would be nonsense! For suppose that

$$X = \begin{bmatrix} 11 \\ 6 \\ 13 \\ 4 \end{bmatrix}.$$

Then X "fits" the formula defining V in more than one way:

$$X = 2\begin{bmatrix} 1 \\ 2 \\ 3 \\ 4 \end{bmatrix} + 3\begin{bmatrix} 2 \\ 0 \\ 1 \\ -2 \end{bmatrix} + 1\begin{bmatrix} 3 \\ 2 \\ 4 \\ 2 \end{bmatrix} \quad \text{or} \quad X = 1\begin{bmatrix} 1 \\ 2 \\ 3 \\ 4 \end{bmatrix} + 2\begin{bmatrix} 2 \\ 0 \\ 1 \\ -2 \end{bmatrix} + 2\begin{bmatrix} 3 \\ 2 \\ 4 \\ 2 \end{bmatrix}.$$

Thus, there is confusion as to whether

$$\text{Id}(X) = \begin{bmatrix} 2 \\ 3 \\ 1 \end{bmatrix} \quad \text{or} \quad \text{Id}(X) = \begin{bmatrix} 1 \\ 2 \\ 2 \end{bmatrix}.$$

That is to say, our seeming definition of Id is not a definition at all! Thus, we have not (in this attempt) succeeded in identifying V with R^{31}; hence, the question of V's dimension remains unanswered.

A proper understanding of what lies behind the difficulties encountered in the foregoing discussion depends on a number of theoretical notions that we shall now consider. In the next section, having acquired the necessary preparation, we shall return to the problem of finding the dimension of a given vector space. The approach then will be quite different from the one that has just failed us.

Definition 2.6. Let $X^1, X^2, \ldots, X^k$, and Y all belong to a vector space V, and suppose that there are scalars $a_1, a_2, \ldots, a_k$ such that

$$Y = a_1 X^1 + a_2 X^2 + \cdots + a_k X^k.$$

Then Y is said to be a *linear combination* of the vectors X^1 through X^k. If at least one a_j is different from 0, then we say that the combination is *nontrivial*.

Example 1. Show that

$$\begin{bmatrix} 2 \\ 3 \end{bmatrix}$$

is a linear combination of the vectors

$$\begin{bmatrix} 1 \\ 2 \end{bmatrix} \quad \text{and} \quad \begin{bmatrix} 1 \\ 3 \end{bmatrix}.$$

Solution. The problem is simply to find—or, at least, show the existence of—scalars a_1 and a_2 such that

$$a_1 \begin{bmatrix} 1 \\ 2 \end{bmatrix} + a_2 \begin{bmatrix} 1 \\ 3 \end{bmatrix} = \begin{bmatrix} 2 \\ 3 \end{bmatrix}.$$

Equivalently, we have the system

$$\begin{cases} a_1 + a_2 = 2, \\ 2a_1 + 3a_2 = 3. \end{cases}$$

The augmented matrix is easily reduced:

$$\begin{bmatrix} 1 & 1 & 2 \\ 2 & 3 & 3 \end{bmatrix} \stackrel{\text{(row)}}{=} \begin{bmatrix} 1 & 0 & 3 \\ 0 & 1 & -1 \end{bmatrix}.$$

It follows that $a_1 = 3$ and $a_2 = -1$. Thus, as may easily be checked,

$$3 \begin{bmatrix} 1 \\ 2 \end{bmatrix} + (-1) \begin{bmatrix} 1 \\ 3 \end{bmatrix} = \begin{bmatrix} 2 \\ 3 \end{bmatrix}.$$

Example 2. Show that $[0, 0, 0, 0]$ is a nontrivial linear combination of the vectors $[1, 2, 3, 4]$, $[2, 0, 1, -2]$, and $[3, 2, 4, 2]$. (The question of trivial versus nontrivial combinations arises only when Θ is the resultant vector.)

Solution. We seek a_1, a_2, and a_3, not all zero, such that

$$a_1[1, 2, 3, 4] + a_2[2, 0, 1, -2] + a_3[3, 2, 4, 2] = [0, 0, 0, 0].$$

This vector equation is equivalent to the system

$$\begin{cases} a_1 + 2a_2 + 3a_3 = 0, \\ 2a_1 + 2a_3 = 0, \\ 3a_1 + a_2 + 4a_3 = 0, \\ 4a_1 - 2a_2 + 2a_3 = 0. \end{cases}$$

The augmented matrix is easily reduced:

$$\begin{bmatrix} 1 & 2 & 3 & 0 \\ 2 & 0 & 2 & 0 \\ 3 & 1 & 4 & 0 \\ 4 & -2 & 2 & 0 \end{bmatrix} \stackrel{\text{(row)}}{=} \begin{bmatrix} 1 & 0 & 1 & 0 \\ 0 & 1 & 1 & 0 \\ 0 & 0 & 0 & 0 \\ 0 & 0 & 0 & 0 \end{bmatrix}.$$

It follows that
$$\begin{bmatrix} a_1 \\ a_2 \\ a_3 \end{bmatrix} = \begin{bmatrix} -t \\ -t \\ t \end{bmatrix}.$$

Taking $t = 5$, for example, we see that
$$-5[1, 2, 3, 4] - 5[2, 0, 1, -2] + 5[3, 2, 4, 2] = [0, 0, 0, 0].$$

Definition 2.7. Let S be a subset of a vector space V, and suppose that every vector in V can be realized as a linear combination of vectors taken from S. Then S is said to be a *spanning set for* V. We also say that S *spans* V.

Example 3. Show that
$$S = \left\{ \begin{bmatrix} 1 & 0 \\ 0 & 0 \end{bmatrix}, \begin{bmatrix} 1 & 1 \\ 0 & 0 \end{bmatrix}, \begin{bmatrix} 1 & 1 \\ 1 & 0 \end{bmatrix}, \begin{bmatrix} 1 & 1 \\ 1 & 1 \end{bmatrix} \right\}$$
spans R^{22}.

Solution. Let X be an arbitrary vector in R^{22}, say
$$X = \begin{bmatrix} \alpha & \beta \\ \gamma & \delta \end{bmatrix}.$$

We seek a_1, a_2, a_3, and a_4 such that
$$a_1 \begin{bmatrix} 1 & 0 \\ 0 & 0 \end{bmatrix} + a_2 \begin{bmatrix} 1 & 1 \\ 0 & 0 \end{bmatrix} + a_3 \begin{bmatrix} 1 & 1 \\ 1 & 0 \end{bmatrix} + a_4 \begin{bmatrix} 1 & 1 \\ 1 & 1 \end{bmatrix} = \begin{bmatrix} \alpha & \beta \\ \gamma & \delta \end{bmatrix}.$$

Equivalently, we have the system
$$\begin{cases} a_1 + a_2 + a_3 + a_4 = \alpha, \\ a_2 + a_3 + a_4 = \beta, \\ a_3 + a_4 = \gamma, \\ a_4 = \delta. \end{cases}$$

(In reality we are dealing here with an infinite set of systems: one system for each choice of α, β, γ, and δ. Our goal is to solve every such system, i.e., to solve for the a_j's *in terms of* α, β, γ, and δ.) Solving this system, we have
$$a_4 = \delta, \qquad a_3 = \gamma - \delta,$$
$$a_2 = \beta - \gamma, \quad \text{and} \quad a_1 = \alpha - \beta.$$

It follows that
$$(\alpha - \beta) \begin{bmatrix} 1 & 0 \\ 0 & 0 \end{bmatrix} + (\beta - \gamma) \begin{bmatrix} 1 & 1 \\ 0 & 0 \end{bmatrix} + (\gamma - \delta) \begin{bmatrix} 1 & 1 \\ 1 & 0 \end{bmatrix} + \delta \begin{bmatrix} 1 & 1 \\ 1 & 1 \end{bmatrix} = \begin{bmatrix} \alpha & \beta \\ \gamma & \delta \end{bmatrix}.$$

The fact that the foregoing system has only the one solution means, of course, that each matrix

$$\begin{bmatrix} \alpha & \beta \\ \gamma & \delta \end{bmatrix}$$

has only this one expression as a linear combination of the given matrices.

Example 4. Determine whether or not the set

$$S = \{t^3,\, t^2 + 2t,\, t^3 - 2t + 1\}$$

spans P^3.

Solution. Let $p(t) = \alpha t^3 + \beta t^2 + \gamma t + \delta$ be any vector in P^3. We must determine whether or not there exist numbers a_1, a_2, and a_3 such that

$$a_1(t^3) + a_2(t^2 + 2t) + a_3(t^3 - t + 1) = \alpha t^3 + \beta t^2 + \gamma t + \delta.$$

Rewriting the left-hand side in *standard polynomial form*, we have

$$(a_1 + a_3)t^3 + (a_2)t^2 + (2a_2 - a_3)t + (a_3) = \alpha t^3 + \beta t^2 + \gamma t + \delta.$$

But polynomials in standard form define the same function only if their coefficients agree. Thus, we are led to the system

$$\begin{cases} a_1 \quad\;\; + a_3 = \alpha, \\ \quad\;\; a_2 \qquad\;\; = \beta, \\ \quad\;\; 2a_2 - a_3 = \gamma, \\ \qquad\qquad\; a_3 = \delta. \end{cases}$$

The augmented matrix is easily reduced:

$$\begin{bmatrix} 1 & 0 & 1 & \alpha \\ 0 & 1 & 0 & \beta \\ 0 & 2 & -1 & \gamma \\ 0 & 0 & 1 & \delta \end{bmatrix} \underset{=}{\text{(row)}} \begin{bmatrix} 1 & 0 & 0 & (\alpha - 2\beta + \gamma) \\ 0 & 1 & 0 & (\beta) \\ 0 & 0 & 1 & (2\beta - \gamma) \\ 0 & 0 & 0 & (-2\beta + \gamma + \delta) \end{bmatrix}.$$

Now, the last row of the reduced matrix is a "bad row" for some choices of α, β, γ, and δ. For example, suppose that we take $\alpha = 1$, $\beta = 1$, $\gamma = 1$, and $\delta = 2$. Then the reduced matrix is

$$\begin{bmatrix} 1 & 0 & 0 & 0 \\ 0 & 1 & 0 & 1 \\ 0 & 0 & 1 & 1 \\ 0 & 0 & 0 & 1 \end{bmatrix},$$

and, consequently, the corresponding system is inconsistent. In other words, the polynomial $t^3 + t^2 + t + 2$ cannot be realized as a linear combination of the vectors in S. On the other hand, suppose that $\alpha = 1$,

$\beta = 1$, $\gamma = 1$, and $\delta = 1$. Then the reduced matrix is

$$\begin{bmatrix} 1 & 0 & 0 & 0 \\ 0 & 1 & 0 & 1 \\ 0 & 0 & 1 & 1 \\ 0 & 0 & 0 & 0 \end{bmatrix},$$

and the corresponding system has a solution: $a_1 = 0$, $a_2 = 1$, and $a_3 = 1$. It is easy to check that

$$0(t^3) + 1(t^2 + 2t) + 1(t^3 - t + 1) = t^3 + t^2 + t + 1.$$

The point is this: Some vectors in P^3 are linear combinations of the vectors in S while others are not. Thus, it is not the case that S spans P^3.

Definition 2.8. Let S be a subset of a vector space V, and suppose that there is a nontrivial linear combination of vectors taken from S that equals Θ. Then S is said to be *linearly dependent*. Otherwise, S is said to be *linearly independent*.

Example 5. Show that the set $\{[1, 2, 3, 4], [2, 0, 1, -2], [3, 2, 4, 2]\}$ is linearly dependent.

Solution. See Example 2, the last line.

Example 6. Show that the set S of Example 3 is linearly independent.

Solution. In this space,

$$\Theta = \begin{bmatrix} 0 & 0 \\ 0 & 0 \end{bmatrix}.$$

Thus, substituting $\alpha = \beta = \gamma = \delta = 0$ into the formulas obtained under Example 3, we see that the only linear combination of the vectors in S that equals Θ is the trivial one.

Theorem 2.4. Let S_1 and S_2 be finite subsets of a vector space V, and suppose that S_1 is a subset of S_2. Then:

1. If S_1 is linearly dependent, then so is S_2. (An "overset" of a dependent set is still dependent.)

2. If S_2 is linearly independent, then so is S_1. (A subset of an independent set is still independent.)

Proof

1. Since S_1 is linearly dependent, there exist vectors $X^1, X^2, \ldots, X^k$ taken from S_1 and numbers $a_1, a_2, \ldots, a_k$, not all zero, such that

$$a_1 X^1 + a_2 X^2 + \cdots + a_k X^k = \Theta.$$

Since the vectors $X^1, X^2, \ldots, X^k$ are also in S_2, it follows that the above linear combination of vectors may be viewed as a nontrivial combination of vectors in S_2 that equals Θ. Thus, S_2 is linearly dependent.

2. *By contradiction.* Suppose that S_1 is not linearly independent. Then it is linearly dependent. But then, by part 1, any overset of S_1, including S_2, must be dependent. But S_2 is linearly independent! (Proposition 2 is the *contrapositive form* of Proposition 1; a proposition and its contrapositive are always both true or both false.)

EXERCISE SET 2.4

1 a) Show that
$$\begin{bmatrix} 1 & 2 \\ 3 & 4 \end{bmatrix}$$
is a linear combination of the vectors in
$$S = \left\{ \begin{bmatrix} 1 & 0 \\ 0 & 0 \end{bmatrix}, \begin{bmatrix} 0 & 1 \\ 0 & 0 \end{bmatrix}, \begin{bmatrix} 0 & 0 \\ 1 & 0 \end{bmatrix}, \begin{bmatrix} 0 & 0 \\ 0 & 1 \end{bmatrix} \right\}.$$
b) Show that S is a spanning set for R^{22}.
c) Show that S is a linearly independent set.

2 a) Show that $[4, 3, 8]$ is a linear combination of the vectors in $S = \{[4, 1, 0], [2, 1, 2], [2, 0, -2]\}$.
b) Show that S is *not* a spanning set for R^{13}.
c) Show that S is a linearly dependent set.

3 a) Show that $3t^2$ is a linear combination of the vectors in
$$S = \{t, t^2 + 1, 2t - 3, 2t - 4\}.$$
b) Show that S is a spanning set for P^2.
c) Show that S is a linearly dependent set.

4 Theorem 2.5. Let $X^1, X^2, \ldots, X^n$ be chosen from a vector space V. Let W be the set of all linear combinations of these n vectors. Then W constitutes a subspace of V.

a) Prove Theorem 2.5. [*Hint:* Easy, just apply Theorem 2.1.]
b) Let W be the subspace of R^{41} spanned by
$$\left\{ \begin{bmatrix} 1 \\ 2 \\ 0 \\ 0 \end{bmatrix}, \begin{bmatrix} 2 \\ 1 \\ 3 \\ 0 \end{bmatrix}, \begin{bmatrix} 3 \\ 2 \\ 0 \\ 4 \end{bmatrix}, \begin{bmatrix} 7 \\ 7 \\ 3 \\ 4 \end{bmatrix} \right\}.$$

Show that $[5, -1, 3, 8]'$ is in W.

5 Theorem 2.6. Any set of vectors containing Θ is linearly dependent.
  a) Show that the set $\{[1, 2, 3], [4, 5, 6], [0, 0, 0]\}$ is linearly dependent.
  b) Prove Theorem 2.6.

6 Definition 2.9. If $X = a_1 X^1 + \cdots + a_n X^n$, then X is said to be *linearly dependent on the set* $\{X^1, X^2, \ldots, X^n\}$.
  a) **Theorem 2.7.** If X is linearly dependent on $\{X^1, X^2, \ldots, X^n\}$, then the set $\{X^1, X^2, \ldots, X^n, X\}$ is linearly dependent.
  Prove Theorem 2.7.
  b) **Theorem 2.8.** If a set is linearly dependent, then at least one of its members is linearly dependent on the set of the remaining members.
  Prove Theorem 2.8.
  c) Let
  $$X^1 = \begin{bmatrix} 1 \\ 2 \\ 3 \end{bmatrix}, \quad X^2 = \begin{bmatrix} 2 \\ 4 \\ 6 \end{bmatrix}, \quad \text{and} \quad X^3 = \begin{bmatrix} 7 \\ -11 \\ 46 \end{bmatrix}.$$
  Show that $\{X^1, X^2, X^3\}$ is linearly dependent, that X^2 is linearly dependent on $\{X^1, X^3\}$, but that X^3 is not linearly dependent on $\{X^1, X^2\}$.

7 Let V be the subspace of R^{41} described in the beginning of this section.
  a) Let W be the subspace of R^{41} spanned by the set
  $$\{[1, 2, 3, 4]', [2, 0, 1, -2]'\}.$$
  Show that W and V are identical.
  b) Find the dimension of V.

8 a) Consider a pair of arrows originating from the same point in space, the angle between them being other than 0 or π. Make a sketch showing the subspace of arrows "spanned" by these two arrows. (This exercise motivates our use of the term "spanning set.")
  b) Consider three arrows originating from the same point in a given plane. Show, by an informal geometric argument, that this set is linearly dependent.

9 Prove: A set consisting of a single nonzero vector is linearly independent.

10 Prove: If the set $S = \{X, Y, Z\}$ is linearly independent, then so is $S^* = \{X, X + Y, X + Y + Z\}$.

11 Student Q is asked to verify that $S = \{t^2, t + 1, 2t + 3\}$ is a spanning set for P^2. He proceeds as follows: He sets
$$a_1(t^2) + a_2(t + 1) + a_3(2t + 1) = \alpha t^2 + \beta t + \gamma$$
and then replaces t successively by the values 0, 1, and -1, obtaining three equations. He solves the system for a_1, a_2, and a_3 in terms of α, β, and γ. Obtaining a solution, he concludes that S is indeed a spanning set. Comment on this method after having completed the following.
  a) Carry out Q's program as described above.
  b) Repeat part (a) after first replacing $t + 1$ by $t^3 + 1$.

2.5 ANOTHER APPROACH TO DIMENSION

Let $S = \{X^1, X^2, \ldots, X^k\}$ be a spanning set for V. A question that arises quite naturally is this: Can we remove certain vectors from S and still have—in the vectors that remain—a spanning set for V? If S is linearly independent, then the answer is: No. For suppose that $S^* = \{X^1, X^2, \ldots, X^{k-1}\}$ is a spanning set for V. Then any X in V is dependent on S^*; in particular, X^k is dependent on S^*. But then, applying Theorem 2.7, we may conclude that S is a dependent set. A contradiction! On the other hand, if S is linearly dependent, then the answer is: Yes. (This follows directly from Theorem 2.8.) Thus, a spanning set that is also linearly independent is a *minimal* spanning set: none of its vectors is expendable. Such sets play an important role in the study of vector spaces; they deserve a special name.

Definition 2.10. Let S be a subset of a vector space V, and suppose that S has the following properties:
1. S spans V.
2. S is linearly independent.

Then S is said to be a *basis* for V. The members of S are called *basis vectors*.

Example 1. Let
$$S = \left\{ \begin{bmatrix} 1 & 0 \\ 0 & 0 \end{bmatrix}, \begin{bmatrix} 1 & 1 \\ 0 & 0 \end{bmatrix}, \begin{bmatrix} 1 & 1 \\ 1 & 0 \end{bmatrix}, \begin{bmatrix} 1 & 1 \\ 1 & 1 \end{bmatrix} \right\}.$$

In Examples 3 and 6 of section 2.4, it was established that S is both a spanning set for R^{22} and a linearly independent set. Thus, S is a basis for R^{22}.

Example 2. Let V be the solution space of the equation

$$\begin{bmatrix} 1 & 0 & 2 & 1 & 1 \\ 1 & 1 & 1 & 2 & 3 \\ 1 & 2 & 0 & 3 & 5 \end{bmatrix} \begin{bmatrix} x_1 \\ x_2 \\ x_3 \\ x_4 \\ x_5 \end{bmatrix} = \begin{bmatrix} 0 \\ 0 \\ 0 \end{bmatrix}.$$

Find a basis for V.

Solution. We first solve this equation in order to get a formula for a typical member of V:

$$\begin{bmatrix} 1 & 0 & 2 & 1 & 1 & 0 \\ 1 & 1 & 1 & 2 & 3 & 0 \\ 1 & 2 & 0 & 3 & 5 & 0 \end{bmatrix} \xrightarrow{\text{(row)}} \begin{bmatrix} 1 & 0 & 2 & 1 & 1 & 0 \\ 0 & 1 & -1 & 1 & 2 & 0 \\ 0 & 0 & 0 & 0 & 0 & 0 \end{bmatrix}.$$

The system corresponding to the reduced matrix is

$$\begin{cases} x_1 \phantom{{}+x_2} + 2x_3 + x_4 + x_5 = 0, \\ \phantom{x_1+{}} x_2 - x_3 + x_4 + 2x_5 = 0, \\ \phantom{x_1 + 2x_3 + x_4 + 2x_5{}} 0 = 0. \end{cases}$$

Thus, taking $x_3 = r$, $x_4 = s$, and $x_5 = t$, we have

$$X = \begin{bmatrix} (-2r - s - t) \\ (r - s - 2t) \\ (r) \\ (s) \\ (t) \end{bmatrix}.$$

But each such X may be "broken down" as follows:

$$X = \begin{bmatrix} -2r - s - t \\ r - s - 2t \\ r \\ s \\ t \end{bmatrix} = r\begin{bmatrix} -2 \\ 1 \\ 1 \\ 0 \\ 0 \end{bmatrix} + s\begin{bmatrix} -1 \\ -1 \\ 0 \\ 1 \\ 0 \end{bmatrix} + t\begin{bmatrix} -1 \\ -2 \\ 0 \\ 0 \\ 1 \end{bmatrix}.$$

Now let

$$S = \left\{ \begin{bmatrix} -2 \\ 1 \\ 1 \\ 0 \\ 0 \end{bmatrix}, \begin{bmatrix} -1 \\ -1 \\ 0 \\ 1 \\ 0 \end{bmatrix}, \begin{bmatrix} -1 \\ -2 \\ 0 \\ 0 \\ 1 \end{bmatrix} \right\}.$$

The foregoing formula shows that each X—each member of V—may be realized as a linear combination of the members of S. Thus, S is a spanning set for V. It is almost as immediate that S is linearly independent. For consider just the last three entries of each vector in S. Each vector has a 1 where the others have 0's. Since 1 cannot arise as a sum of multiples of 0, it follows that no vector in S is dependent on the other two. But then, by Theorem 2.8, the set S is linearly independent. Finally, then, S is a basis for V.

Example 3. Let $S = \{t^2 + 1, 2t - 3, 2t^2 - 5\}$. Show that S is a basis for P^2.

Solution. Take $p(t) = \alpha t^2 + \beta t + \gamma$ as an arbitrary member of P^2. To show that S spans P^2, we must show that there are numbers a_1, a_2, and a_3 such that

$$a_1(t^2 + 1) + a_2(2t - 3) + a_3(2t^2 - 5) = \alpha t^2 + \beta t + \gamma.$$

Proceeding as in Example 4, Section 2.4, we are led to the system

$$\begin{cases} a_1 \phantom{{}+2a_2} + 2a_3 = \alpha, \\ \phantom{a_1+{}}2a_2 \phantom{{}+2a_3} = \beta, \\ a_1 - 3a_2 - 5a_3 = \gamma. \end{cases}$$

This system has a solution; moreover, that solution is unique:

$$\begin{cases} a_1 = \tfrac{5}{7}\alpha + \tfrac{3}{7}\beta + \tfrac{2}{7}\gamma, \\ a_2 = \tfrac{1}{2}\beta, \\ a_3 = \tfrac{1}{7}\alpha - \tfrac{3}{14}\beta - \tfrac{1}{7}\gamma. \end{cases}$$

This means that each polynomial $\alpha t + \beta t + \gamma$ can be realized as a linear combination of the members of S in exactly one way. Thus, S is a spanning set for P^2, and S is linearly independent; for the substitution $\alpha = \beta = \gamma = 0$ produces the solution $a_1 = a_2 = a_3 = 0$. (In other words, the zero polynomial arises only by taking a trivial combination of the members of S.) □

We now turn to some important theorems about the basis of a vector space, theorems that relate the notion of a basis to that of dimension.

Theorem 2.9. Let $S = \{X^1, X^2, \ldots, X^n\}$ be a basis for V and let X be any vector in V. Then the representation of X as a linear combination of the members of S is unique.

Proof. *By contradiction.* Suppose that we have both

$$X = a_1 X^1 + a_2 X^2 + \cdots + a_n X^n$$

and

$$X = b_1 X^1 + b_2 X^2 + \cdots + b_n X^n,$$

where $a_j \neq b_j$ for at least one j. Then, using the arithmetic properties of vector spaces, we may write

$$\Theta = X - X = (a_1 - b_1)X^1 + (a_2 - b_2)X^2 + \cdots + (a_n - b_n)X^n.$$

Since $(a_j - b_j) \neq 0$ for at least one j, we have here a nontrivial combination of the basis vectors that is equal to the zero vector. This is impossible!

Theorem 2.10. If V has a basis containing n vectors, then V has dimension n.

Proof. To prove the theorem we must establish the existence of an isomorphism that identifies V with R^{n1}. Let $S = \{X^1, X^2, \ldots, X^n\}$ be a basis for V. Then, by Theorem 2.9, each X in V may be expressed

in terms of the vectors of S in exactly one way:
$$X = a_1 X^1 + a_2 X^2 + \cdots + a_n X^n.$$
Thus, it makes sense to define a function that assigns to each X the n-tuple of coefficients that appear in its S-vector representation. We write
$$\text{Id}(X) = \begin{bmatrix} a_1 \\ a_2 \\ \vdots \\ a_n \end{bmatrix}.$$

It is also clear that Id is a one-to-one function, for different n-tuples of coefficients surely lead to different n by 1 column matrices. Thus, to establish that Id is actually an isomorphism from V to R^{n1}, we have only to verify that Id has the two arithmetic-preserving properties:

1. Let
$$X = a_1 X^1 + a_2 X^2 + \cdots + a_n X^n$$
and
$$Y = b_1 X^1 + b_2 X^2 + \cdots + b_n X^n.$$
Then we may write
$$\text{Id}(X+Y) = \text{Id}[(a_1+b_1)X^1 + (a_2+b_2)X^2 + \cdots + (a_n+b_n)X^n]$$
$$= \begin{bmatrix} (a_1+b_1) \\ (a_2+b_2) \\ \vdots \\ (a_n+b_n) \end{bmatrix} = \begin{bmatrix} a_1 \\ a_2 \\ \vdots \\ a_n \end{bmatrix} + \begin{bmatrix} b_1 \\ b_2 \\ \vdots \\ b_n \end{bmatrix} = \text{Id}(X) + \text{Id}(Y).$$

2. It is just as easy—and just as uninteresting—to show that $\text{Id}(cX) = c\,\text{Id}(X)$.

Theorem 2.11. If V has a basis containing n vectors, then every basis for V has n vectors.

Proof. *By contradiction.* Suppose that V has a basis containing k vectors, where $k \neq n$. Then, by Theorem 2.10, V has dimension k; but also, applying Theorem 2.10 to the hypothesis of our theorem, V has dimension n. But the dimension of a space is unique. (Recall the discussion preceding Theorem 2.2.)

EXERCISE SET 2.5

1 Prove that the given space has the given set as a basis.

a) $R^{31}, \left\{ \begin{bmatrix} 1 \\ 0 \\ 0 \end{bmatrix}, \begin{bmatrix} 0 \\ 1 \\ 0 \end{bmatrix}, \begin{bmatrix} 0 \\ 0 \\ 1 \end{bmatrix} \right\}.$

b) R^{22}, $\left\{\begin{bmatrix} 1 & 0 \\ 0 & 0 \end{bmatrix}, \begin{bmatrix} 0 & 1 \\ 0 & 0 \end{bmatrix}, \begin{bmatrix} 0 & 0 \\ 1 & 0 \end{bmatrix}, \begin{bmatrix} 0 & 0 \\ 0 & 1 \end{bmatrix}\right\}$.

c) P^2, $\{t^2, t, 1\}$.

2 Find a basis for the given space. (No justification required.)

a) R^{41} b) R^{14} c) R^{23} d) P^4

3 Prove that $\{t^3 + t, t^3 - t, t^2 + 1, t^2 - 1\}$ is a basis for P^3.

4 a) Let
$$S = \left\{\begin{bmatrix} 1 & 1 \\ 0 & 0 \end{bmatrix}, \begin{bmatrix} 0 & 1 \\ 0 & 1 \end{bmatrix}, \begin{bmatrix} 0 & 0 \\ 1 & 1 \end{bmatrix}, \begin{bmatrix} 1 & 0 \\ 1 & 0 \end{bmatrix}\right\}.$$

Make a guess: Is S a basis for R^{22}?

b) Prove or disprove your guess.

5 Let V be the subspace of R^{51} that satisfies
$$\begin{bmatrix} 2 & 0 & 3 & 1 & 2 \\ 4 & 1 & 2 & 3 & -2 \\ 8 & 1 & 8 & 5 & 2 \end{bmatrix} X = \Theta.$$

Find the dimension of V.

6 Find a basis for the given subspace of R^{22}. (No justification required.)

a) W: matrices of the form $\begin{bmatrix} a & a \\ a & a \end{bmatrix}$.

b) W: matrices of the form $\begin{bmatrix} a & 0 \\ 0 & b \end{bmatrix}$.

c) W: matrices of the form $\begin{bmatrix} a & a+b \\ b+c & c \end{bmatrix}$.

7 This exercise concerns the vector space that gave us trouble at the beginning of Section 2.4. Let W be the subspace of R^{41} given by
$$X = a\begin{bmatrix} 1 \\ 2 \\ 3 \\ 4 \end{bmatrix} + b\begin{bmatrix} 2 \\ 0 \\ 1 \\ -2 \end{bmatrix} + c\begin{bmatrix} 3 \\ 2 \\ 4 \\ 2 \end{bmatrix}.$$

a) Find a basis for W.

b) What is the dimension of W?

8 Let V be the space of continuous functions on $(-\infty, \infty)$. Let W be the subspace spanned by $\{\cos^2 t, \sin^2 t, \cos 2t\}$.

a) Show that $3 - 5 \cos 2t$ belongs to W.

b) Show that $\{\cos^2 t, \sin^2 t\}$ is a basis. [Hint: If $a_1 \cos^2 t + a_2 \sin^2 t = 0$ for all t, then in particular it must be true for $t = 0$ and $t = \pi/2$.]

c) Show that $\{\cos^2 t, \sin^2 t, \cos 2t\}$ is not a basis for W.

9 The set of all 2 by 1 matrices having complex number entries, with addition and multiplication by real numbers defined in the expected way, is a vector space. Call it V.

a) Let
$$S = \left\{ \begin{bmatrix} 1 \\ i \end{bmatrix}, \begin{bmatrix} i \\ 1 \end{bmatrix}, \begin{bmatrix} i \\ i \end{bmatrix} \right\}.$$
Show that S is linearly independent.
b) Find a basis for V.

10 The arithmetical structure defined in Definition 2.1 is more properly called a *real vector space*. If in that definition we take the scalars to be complex numbers, then the resulting structure is called a *complex vector space*. Now, let W be the vector space of Exercise 9 except that we shall let the scalars be complex numbers.

a) Let S be as in Exercise 9(a). Show that S is linearly dependent.
b) Find a basis for V.

2.6 MORE ON BASES

We continue the study begun in Section 2.5: a study of the interplay between the notion of a basis and that of the dimension of a vector space. The theorems that we shall consider in this section are all quite "believable," and should need no special motivation or explanation.

Definition 2.11. Let S be a set of vectors and let T be a linearly independent subset of S. In addition, suppose that there is no independent subset of S which properly contains T. Then T is said to be a *maximal independent subset* of S.

Example 1. Let $S = \{X^1, X^2, X^3, X^4\}$, where

$$X^1 = \begin{bmatrix} 1 \\ 0 \\ 1 \\ 2 \end{bmatrix}, \quad X^2 = \begin{bmatrix} 2 \\ 0 \\ 2 \\ 4 \end{bmatrix}, \quad X^3 = \begin{bmatrix} 3 \\ 2 \\ 1 \\ 4 \end{bmatrix}, \quad \text{and} \quad X^4 = \begin{bmatrix} 4 \\ 2 \\ 3 \\ 5 \end{bmatrix}.$$

We seek a maximal independent subset of S.

Solution. First, we note that S is not itself linearly independent, for $X^2 = 2X^1$. It follows that $2X^1 + (-1)X^2 + 0 \cdot X^3 + 0 \cdot X^4 = \Theta$. Thus, we turn to the three-element subsets of S.

1. Let $S_1 = \{X^1, X^2, X^3\}$. S_1 is linearly dependent for the same reason that S is: $X^2 = 2X^1$. Similarly, $S_2 = \{X^1, X^2, X^4\}$ is linearly dependent.

2. Let $S_3 = \{X^1, X^3, X^4\}$. This set (we leave the verification to the reader) is linearly independent. So also is $S_4 = \{X^2, X^3, X^4\}$.

It follows, by Definition 2.11, that S_3 and S_4 are both maximal independent subsets of S.

Theorem 2.12. If the set $S = \{X^1, X^2, \ldots, X^k\}$ spans a vector space V, then any maximal independent subset of S is a basis for V.

Proof. If S is itself linearly independent, then it is its own maximal independent subset; and, as a linearly independent spanning set, it is a basis. Thus, we turn to the case where S has a maximal independent subset containing r vectors, $r < k$.

Let S_1 be a maximal independent subset of S containing r vectors. By relabeling our vectors, if necessary, we may assume that $S_1 = \{X^1, X^2, \ldots, X^r\}$.

1. We show first that each of the vectors deleted in passing from S to S_1 is linearly dependent on the vectors of S_1. Consider the set $S_2 = \{X^1, \ldots, X^r, X^j\}$, where $j > r$. Then S_2 is linearly dependent. (Why?) Thus, we may write

$$a_1 X^1 + \cdots + a_r X^r + a_j X^j = \Theta,$$

where at least one of the coefficients is different from 0. Now suppose that $a_j = 0$. Then we are left with a nontrivial combination of the members of S_1 set equal to Θ, denying that S_1 is independent. Thus, $a_j \neq 0$, and we may solve for X^j in terms of the members of S_1:

$$X^j = -\frac{1}{a_j}(a_1 X^1 + \cdots + a_r X^r) = \left(-\frac{a_1}{a_j}\right) X^1 + \cdots + \left(-\frac{a_r}{a_j}\right) X^r.$$

Thus, X^j is indeed dependent on the members of S_1.

2. Now, let X be any vector in V. Then, because S is a spanning set, we may write

$$X = a_1 X^1 + \cdots + a_r X^r + a_{r+1} X^{r+1} + \cdots + a_k X^k.$$

By part 1, each of the last $(k - r)$ terms may be replaced by linear combinations of the vectors X^1 through X^r. It follows that X may be realized as a linear combination of the members of S_1. This means that S_1 is a spanning set for V. But from the beginning, S_1 has been linearly independent. Thus, S_1 is a basis for V.

Theorem 2.13. If V has a basis containing exactly n vectors, then no linearly independent subset of V has more than n vectors.

Proof. Let $A = \{X^1, X^2, \ldots, X^n\}$ be a basis for V, and let $B = \{Y^1, Y^2, \ldots, Y^m\}$ be a linearly independent subset of V. (We do not mean to imply by our notation that the X^j are necessarily distinct from

the Y^j.) Consider the union of these two sets:
$$S = \{X^1, \ldots, X^n, Y^1, \ldots, Y^m\}.$$
It is a trivial observation that S is a spanning set for V. (Why?)
We next turn our attention to B; there are two cases:

1. If B is itself a maximal independent subset of S, then (Theorem 2.12) B is a basis for V. As such (Theorem 2.11) it must contain the same number of vectors as A. Thus, $m = n$.
2. Suppose that B is not as in part 1. Then we add to B vectors from A until we get a maximal independent subset of S. This resulting set, as in part 1, must contain n vectors. It follows that $m < n$. Combining parts 1 and 2, it follows that $m \leq n$.

Theorem 2.14. Let V be isomorphic to W, and let Id be an isomorphism identifying V with W. Now, suppose that V^* is a subspace of V spanned by $S = \{X^1, X^2, \ldots, X^n\}$ and that W^* is the subspace of W spanned by $T = \{\text{Id}(X^1), \text{Id}(X^2), \ldots, \text{Id}(X^n)\}$. Then V^* is isomorphic to W^*. Moreover, if S is a basis for V^*, then T is a basis for W^*.

Proof. Let Id* be the *restriction* of Id to V^*, that is, $\text{Id}^*(X) = \text{Id}(X)$ for all X in V^*, but Id* is not defined outside of V^*. We will show that Id* is an isomorphism from V^* onto W^*. Now, since Id* inherits the properties of an isomorphism from Id, the only thing we have to prove is that W^* is actually the range of Id*.

1. Let R be the range of Id* and let Y belong to R. This means that there is an X in V^* such that $\text{Id}^*(X) = Y$. But such an X may be expressed in terms of S:
$$X = a_1 X^1 + a_2 X^2 + \cdots + a_n X^n.$$

It follows, using the properties of an isomorphism, that
$$\begin{aligned} Y = \text{Id}^*(X) &= a_1 \text{Id}^*(X^1) + a_2 \text{Id}^*(X^2) + \cdots + a_n \text{Id}^*(X^n) \\ &= a_1 \text{Id}(X^1) + a_2 \text{Id}(X^2) + \cdots + a_n \text{Id}(X^n) \\ &= \text{a member of } W^*. \end{aligned}$$

This shows that every vector in R is also in W^*.

2. Let Y be in W^*. Then we may write
$$\begin{aligned} Y &= a_1 \text{Id}(X^1) + a_2 \text{Id}(X^2) + \cdots + a_n \text{Id}(X^n) \\ &= \text{Id}(a_1 X^1 + a_2 X^2 + \cdots + a_n X^n) \\ &= \text{Id}(\text{a member of } V^*) \\ &= \text{Id}^*(\text{a member of } V^*) \\ &= \text{a member of } R. \end{aligned}$$

This shows that every vector in W^* is also in R. Combining this result with that of part 1 shows that $R = W^*$.

3. Suppose now that S is a basis for V^*; then S is linearly independent. We wish to show that T is linearly independent. Suppose that this is not true. Then there exists a nontrivial combination of T-vectors that equals Θ:
$$a_1 \text{Id}(X^1) + a_2 \text{Id}(X^2) + \cdots + a_n \text{Id}(X^n) = \Theta$$
or
$$\text{Id}(a_1 X^1 + a_2 X^2 + \cdots + a_n X^n) = \Theta.$$

But (recall Exercise 6 of Section 2.3) Id must identify the Θ of V^* with that of W^*. Thus,
$$a_1 X^1 + a_2 X^2 + \cdots + a_n X^n = \Theta.$$

This is impossible! (Why?)

Theorem 2.15. If V has dimension n, then V has a basis containing n vectors.

Proof. By hypothesis (recall the definition of dimension) there are isomorphisms that identify V with R^{n1}. Choose one. Suppose (so that the application of Theorem 2.14 that follows will be more clear) that we denote the inverse of this isomorphism by Id. Thus, Id is an isomorphism identifying each n by 1 column matrix with a vector in V:
$$\text{Id}\left(\begin{bmatrix} x_1 \\ x_2 \\ \vdots \\ x_3 \end{bmatrix}\right) = X.$$

Now we know that R^{n1} has a basis containing n vectors. (Recall Exercise 1a of Section 2.5.) For example, we could take $S = \{N^1, N^2, \ldots, N^n\}$ as a basis, where N^j is a column of zeros except for the jth entry, which is 1. It then follows, applying Theorem 2.14, that
$$T = \{\text{Id}(N^1), \text{Id}(N^2), \ldots, \text{Id}(N^n)\}$$
is a basis for V containing n vectors. (It follows, of course, that every basis for V has n vectors.) □

Theorem 2.15 is the converse of Theorem 2.10. This means that the same information about a vector space is imparted by saying either that it has dimension n or that it has a basis containing n vectors. To emphasize this point, we shall now give a new definition for the *dimension* of a vector space, one which depends on the notion of a basis. Also, this new definition will be broader than Definition 2.5, for it includes the case of vector spaces that are not n-dimensional.

84 VECTORS

Definition 2.12
1. A vector space having a single vector (recall Exercise 6 of Section 2.1) is said to have *dimension zero*.
2. A vector space having a basis consisting of n vectors is said to have *dimension n*.
3. All other vector spaces are said to be *infinite dimensional*.

EXERCISE SET 2.6

1 Find a maximal independent subset of

$$S = \left\{ \begin{bmatrix} 1 \\ 2 \\ 3 \end{bmatrix}, \begin{bmatrix} 4 \\ 7 \\ 11 \end{bmatrix}, \begin{bmatrix} 4 \\ 8 \\ 12 \end{bmatrix}, \begin{bmatrix} 3 \\ 5 \\ 8 \end{bmatrix} \right\}.$$

2 a) Why is it immediate that the set

$$T = \left\{ \begin{bmatrix} 1 \\ 0 \\ 0 \end{bmatrix}, \begin{bmatrix} 1 \\ 2 \\ 0 \end{bmatrix}, \begin{bmatrix} 1 \\ 2 \\ 3 \end{bmatrix}, \begin{bmatrix} 37 \\ 453 \\ -29 \end{bmatrix} \right\}$$

is linearly dependent?

b) Why is it immediate that the set

$$S = \left\{ \begin{bmatrix} 1 \\ 0 \\ 0 \\ 0 \end{bmatrix}, \begin{bmatrix} 1 \\ 2 \\ 0 \\ 0 \end{bmatrix}, \begin{bmatrix} 1 \\ 2 \\ 3 \\ 0 \end{bmatrix}, \begin{bmatrix} 37 \\ 453 \\ -29 \\ 0 \end{bmatrix} \right\}$$

is linearly dependent?

c) Find a maximal independent subset of S.
d) Find a basis for the space V spanned by S.
e) Find the dimension of V.

3 Given that the vectors $X^1 = [1, 2, 0, 3, 0]$, $X^2 = [2, 1, 0, 5, 0]$, and $X^3 = [3, 3, 0, -7, 0]$ are linearly independent, find additional vectors $X^4, X^5, \ldots$ such that the resulting set will be a basis for R^{15}.

4 Theorem 2.16. If V has dimension n, then every linearly independent set of n vectors is a basis for V.

Prove this theorem by completing the following argument: Suppose that $S = \{X^1, X^2, \ldots, X^n\}$ is an independent set that does *not* span V. Then a maximal independent subset of S can have at most ....

5 Theorem 2.17. If V has dimension n, then every spanning set having exactly n vectors is a basis for V.

Prove this theorem by completing the following argument: Suppose that $S = \{X^1, X^2, \ldots, X^n\}$ is a spanning set that is *not* linearly independent. Then at least one X^i must be dependent on ....

6 Let Id be the isomorphism from R^{14} onto P^3 defined by
$$\text{Id}([a, b, c, d]) = at^3 + bt^2 + ct + d.$$
Let V be the subspace of R^{14} spanned by the set
$$S = \{[1, 0, 2, 1], [1, 1, 0, 1], [0, -1, 2, 0]\}.$$
Then, according to Theorem 2.14, the isomorphism Id identifies V with some subspace of P^3. Call this latter space W.
a) Find a spanning set for W.
b) Find a basis for W.

7 a) Let V be the subspace of R^{13} spanned by $[1, 1, 0]$ and $[0, 1, 1]$. Let $X = [2, 3, 1]$. Show that X is in V and then find a second vector Y such that $\{X, Y\}$ will be a basis for V.
b) Let W be the subspace P^2 spanned by $t^2 + t$ and $t + 1$. Let
$$p(t) = 2t^2 + 3t + 1.$$
Show that p is in W and then find a second vector q such that $\{p, q\}$ will be a basis for W.

8 *A theorem about differential equations.* The set of functions satisfying a differential equation of the form
$$a_n y^{(n)} + a_{n-1} y^{(n-1)} + \cdots + a_0 y = 0,$$
where $a_n \neq 0$, forms a vector space. Moreover, the dimension of that space is exactly n.
a) Prove the first part of this theorem for the special case $n = 2$.

For parts (b) and (c) the reader may make use of the entire theorem.
b) Show that the functions e^t and e^{2t} both satisfy the equation $y'' - 3y' + 2y = 0$. Then prove that the complete solution of this equation is given by
$$y = \alpha e^t + \beta e^{2t}.$$
c) Show that each of the functions 1, t, and te^t satisfies the equation $y''' - 2y'' + y' = 0$. Then prove that the complete solution of this equation is given by
$$y = \alpha + \beta e^t + \gamma t e^t.$$

9 Our interest in this text is restricted primarily to finite-dimensional vector spaces. However, the reader should appreciate that our definitions allow for the possibility of infinite spanning sets and infinite bases. This will be made more clear on completing the following.
a) Let $S = \{1, t, t^2, t^3, \ldots\}$. Then S is an infinite subset of the space of all polynomial functions, P^∞. Prove that S is a spanning set for P^∞. Also, prove that P^∞ has no finite spanning sets.
b) Prove that S is linearly independent. [*Hint:* Use the fact that a polynomial of degree n can have the value 0 for at most n values of t.]

2.7 THE ROW SPACE OF A MATRIX

We devote this section to the following problem: Let $S = \{X^1, X^2, \ldots, X^n\}$ be a spanning set for a certain vector space V, but suppose that S is linearly dependent. How do we find a basis for V?

Let us take a direct, constructive approach. Pick any X^j as the first vector of the desired basis; relabel it as Y^1. From the remaining members of S, pick a second vector, say X^k, such that the set $\{Y^1, X^k\}$ is independent. Let X^k be the second vector in our basis; relabel it as Y^2. From the remaining vectors, pick a third vector, say X^m, such that the set $\{Y^1, Y^2, X^m\}$ is independent. Relabel X^m as Y^3, and so on. This process must come to a halt. That is, we eventually arrive at a set of k vectors, $S_1 = \{Y^1, Y^2, \ldots, Y^k\}$, that cannot be enlarged upon: we simply run out of vectors that are independent of those already chosen. Thus, S_1 is a maximal independent subset of a spanning set for V. By Theorem 2.12, S_1 is a basis for V.

To literally follow the plan outlined above would, in many cases, be very tedious. Suppose, for example, that we started with a spanning set containing ten vectors, and that V has dimension 4. Consider how many false selections might be made before isolating a linearly independent set of four vectors. Fortunately, there is a more efficient procedure, one that is dependent on the following theorem.

Theorem 2.18. Let A and B be m by n matrices that are row equivalent. Then the vector space of 1 by n matrices spanned by the rows of A is identical with that spanned by the rows of B.

Proof. Let $A = [a_{ij}]_{mn}$ and $B = [b_{ij}]_{mn}$. Also, let

$$A^i = i\text{th row of } A = [a_{i1}, a_{i2}, \ldots, a_{in}]$$

and

$$B^i = i\text{th row of } B = [b_{i1}, b_{i2}, \ldots, b_{in}].$$

Then what we have to prove is this: The subspace of R^{1n} spanned by $\{A^1, A^2, \ldots, A^n\}$ is identical with that spanned by $\{B^1, B^2, \ldots, B^n\}$.

Now, we are given that the matrix B can be derived from A by performing appropriate row operations upon A (Definition 1.13). Suppose then, as a special case, that B can be derived from A by performing just one row operation. There are three possibilities:

1. The row operation is a row interchange. Then the sets $\{A^1, A^2, \ldots, A^n\}$ and $\{B^1, B^2, \ldots, B^n\}$ are identical; consequently, they span identical spaces.

2. The row operation is a row multiplication. Then for some k we have

$$B^k = cA^k, \qquad c \neq 0,$$

and

$$B^i = A^i, \qquad i \neq k.$$

Suppose now that X is in the space spanned by the A-vectors. Then we write
$$X = a_1 A^1 + a_2 A^2 + \cdots + a_k A^k + \cdots + a_n A^n.$$
It follows that
$$X = a_1 B^1 + a_2 B^2 + \cdots + \frac{a_k}{c} B^k + \cdots + a_n B^n.$$
Thus, X is in the space spanned by the B-vectors. On the other hand, suppose that Y belongs to this latter space. Then
$$Y = b_1 B^1 + b_2 B^2 + \cdots + b_k B^k + \cdots + b_n B^n,$$
and, as before, it follows that
$$Y = b_1 A^1 + b_2 A^2 + \cdots + c b_k A^k + \cdots + b_n B^n.$$
This shows that Y belongs also to the space spanned by the A-vectors. Thus, these two spaces are identical.

3. Suppose that c times the jth row of A is added to its kth row. Then we have
$$B^k = A^k + c A^j$$
and
$$B^i = A^i, \quad i \neq k.$$
Then, starting from
$$X = a_1 A^1 + \cdots + a_j A^j + \cdots + a_k A^k + \cdots + a_n A^n,$$
we can pass to
$$X = a_1 B^1 + \cdots + (a_j - c a_k) B^j + \cdots + a_k B^k + \cdots + a_n B^n;$$
for
$$a_k A^k = a_k (B^k - c A^j) = a_k B^k - c a_k B^j.$$
This argument is reversible. Therefore, as before, the two spaces are identical.

Thus, we have proved the following: If B is obtainable from A by performing just one row operation, then the space spanned by $\{A^1, \ldots, A^n\}$ is identical with that spanned by $\{B^1, \ldots, B^n\}$. But what if p row operations are required to pass from A to B? Then we have only to apply the foregoing result p times. (What is really implied here is a very simple *proof by mathematical induction.* We shall discuss such proofs in detail in a later section.)

Definition 2.13. Given an m by n matrix A, let V be the subspace of R^{1n} spanned by the rows of A. Then V is called the *row space* of A. The dimension of V is called the *row rank* of A. (The *column space* of A and the *column rank* of A are defined similarly.)

Theorem 2.18 may now be restated in the following more compact form: *If A and B are row equivalent, then the row space of A is identical with that of B.* But how does this knowledge help us to solve the problem posed at the beginning of this section?

Example 1. Let V be the subspace of R^{15} spanned by $A^1 = [1, 0, 2, 1, 3]$, $A^2 = [2, 1, 0, 0, 1]$, $A^3 = [1, 2, 2, 1, 1]$, and $A^4 = [4, 1, 4, 2, 7]$. Find a basis for V.

Solution. By Definition 2.13, the space V is the row space of the matrix

$$A = \begin{bmatrix} 1 & 0 & 2 & 1 & 3 \\ 2 & 1 & 0 & 0 & 1 \\ 1 & 2 & 2 & 1 & 1 \\ 4 & 1 & 4 & 2 & 7 \end{bmatrix}.$$

Now, as the reader may verify, A is row equivalent to

$$B = \begin{bmatrix} 1 & 0 & 0 & 0 & 1 \\ 0 & 1 & 0 & 0 & -1 \\ 0 & 0 & 1 & \frac{1}{2} & 1 \\ 0 & 0 & 0 & 0 & 0 \end{bmatrix}.$$

It follows (Theorem 2.18) that V is also spanned by $B^1 = [1, 0, 0, 0, 1]$, $B^2 = [0, 1, 0, 0, -1]$, and $B^3 = [0, 0, 1, \frac{1}{2}, 1]$. But this spanning set is *obviously* independent. Thus, we may take $\{B^1, B^2, B^3\}$ as a basis for V. (It should be noted that our method does not rely on the reduction of A to row echelon form. Rather, our object is to reduce A to any B for which a maximal independent set of rows stands out.)

Example 2. Let W be the space of polynomials spanned by

$$p_1(t) = t^4 + 2t^2 + t + 3,$$
$$p_2(t) = 2t^4 + t^3 + 1,$$
$$p_3(t) = t^4 + 2t^3 + 2t^2 + t + 1,$$

and

$$p_4(t) = 4t^4 + t^3 + 4t^2 + 2t + 7.$$

Find a basis for W.

Solution. Evidently, W is a subspace of P^4. Let

$$\mathrm{Id}(at^4 + bt^3 + ct^2 + dt + e) = [a, b, c, d, e].$$

Then Id is an isomorphism from P^4 to R^{15}. But also, by this same isomorphism, W is isomorphic to some subspace of R^{15}, namely, the subspace spanned by

$$S = \{\text{Id}(p_1(t)), \text{Id}(p_2(t)), \text{Id}(p_3(t)), \text{Id}(p_4(t))\}.$$

Now it happens—it was planned this way—that $S = \{A^1, A^2, A^3, A^4\}$ where the "A rows" are the same as those of Example 1. Thus, the space of polynomials W is isomorphic to the space V of Example 1. But V has $\{B^1, B^2, B^3\}$ as a basis. It follows, using Theorem 2.14, that W has the basis

$$\{\text{Id}^{-1}(B^1), \text{Id}^{-1}(B^2), \text{Id}^{-1}(B^3)\} = \{t^4 + 1, t^3 - 1, t^2 + \tfrac{1}{2}t + 1\}.$$

EXERCISE SET 2.7

1 Let

$$A = \begin{bmatrix} 1 & 2 & 0 & 0 & 4 & 2 & 0 \\ 0 & 0 & 1 & 0 & 5 & 1 & 0 \\ 0 & 0 & 0 & 1 & 3 & 6 & 0 \\ 0 & 0 & 0 & 0 & 0 & 0 & 1 \\ 0 & 0 & 0 & 0 & 0 & 0 & 0 \end{bmatrix}.$$

a) What is the row rank of A?
b) Prove: If A is in row echelon form, then its row rank is equal to the number of its nonzero rows. [*Hint:* Let $A^1, A^2, \ldots, A^n$ be the nonzero rows of A and explain why no one of these rows is dependent on the others.]

2 Find a basis for the space spanned by the given set.
a) $\{[1, 2, 0, 3], [3, 6, 1, 9], [0, 1, 0, 2], [3, 4, 1, 5]\}$
b) $\{t^3 + 2t^2 + 3, t^2 + 2, 3t^3 + 4t^2 + t + 5, 3t^3 + 6t^2 + t + 9\}$
c) $\left\{ \begin{bmatrix} 1 & 3 \\ 2 & 0 \end{bmatrix}, \begin{bmatrix} 3 & 5 \\ 4 & 1 \end{bmatrix}, \begin{bmatrix} 3 & 9 \\ 6 & 1 \end{bmatrix}, \begin{bmatrix} 0 & 2 \\ 1 & 0 \end{bmatrix} \right\}$

3 Let V be the space spanned by $S = \{A^1, A^2, A^3, A^4, A^5\}$, where $A^1 = [1, 0, 2, 1, 5]$, $A^2 = [0, 0, 1, 1, 3]$, $A^3 = [1, 0, 3, 0, 4]$, $A^4 = [0, 0, 1, 2, 5]$, and $A^5 = [0, 0, 0, 1, 2]$.
a) Find the dimension of V.
b) Find a three-vector subset of S that is linearly dependent and one that is independent.

4 Student Q is given a certain subset of R^{1n}, say S, and is asked whether or not S is linearly independent. He proceeds as follows: He constructs a matrix A whose rows are the vectors in S, he reduces A to row echelon form, say B.

a) Noticing that B has a zero row, he concludes that S is dependent. Explain Q's reasoning. [*Hint:* A suitable explanation may be based on Theorems 2.13 and 2.18.]

b) Noticing that B has no zero rows, he concludes that S is independent. Explain Q's reasoning. [*Hint:* A suitable explanation may be based on Theorems 2.17 and 2.18.]

5 Use the method of Exercise 4 to determine whether the given set is linearly dependent or independent.

a) $\{[1, 0, 3, 2], [1, 0, 2, 1], [0, 1, 0, 1], [1, 3, 3, 5]\}$

b) $\left\{ \begin{bmatrix} 1 & 0 & 3 \\ 2 & 2 & 0 \end{bmatrix}, \begin{bmatrix} 1 & 0 & 2 \\ 1 & 3 & 4 \end{bmatrix}, \begin{bmatrix} 0 & 1 & 0 \\ 1 & 0 & 1 \end{bmatrix}, \begin{bmatrix} 1 & 3 & 3 \\ 5 & 0 & 0 \end{bmatrix} \right\}$

6 a) Complete Definition 2.13 by writing out in full detail the definitions for the *column space* of a matrix and the *column rank* of a matrix.

b) State a theorem analogous to Theorem 2.18 for the case where A and B are column equivalent.

c) Find the column rank of

$$A = \begin{bmatrix} 1 & 0 & 2 & 1 \\ 2 & 1 & 5 & 4 \\ 0 & 2 & 2 & 4 \end{bmatrix}.$$

[*Hint:* We may find instead the row rank of A'.]

7 Let

$$A = \begin{bmatrix} 1 & 0 & 2 & 1 & 5 \\ 1 & 0 & 3 & 0 & 4 \\ 0 & 0 & 1 & 1 & 3 \\ 0 & 0 & 0 & 1 & 2 \end{bmatrix}.$$

a) Find the row rank of A.

b) Find the column rank of A.

c) Make a conjecture and test it on other matrices.

8 Let V be the space of functions spanned by the set

$$\{t + e^t, 2t + e^{2t}, 3t - 1, e^t + 5e^{2t}, t - e^t + 1\}.$$

Find the dimension of V. [*Hint:* V is evidently a subspace of the space having the basis $\{1, t, e^t, e^{2t}\}$. This suggests a way of identifying the members of V with members of R^{14}.]

9 The set
$$S = \{[1, 0, 1, 0, 2], [0, 0, 2, 1, 3], [1, 0, 2, 2, 0]\}$$

is linearly independent. Find two other vectors which, along with S, will produce a basis for R^{15}. [*Hint:* First reduce $[S^1, S^2, S^3]'$ to row echelon form.]

2.8 CHANGE-OF-BASIS MATRIX

Thus far we have considered only matrices whose entries are real numbers. However, it is useful to adopt a more general point of view, permitting our rectangular arrays to be quite arbitrary. A rather simple generalization would be to consider matrices of complex numbers. An arithmetic of such matrices is immediate, for the operations defined for *real matrices* continue to make sense for *complex matrices*. Moreover, the various properties of matrix arithmetic that we have studied (associative laws, distributive laws, etc.) continue to hold for complex matrices.

We might go too far. For example, we might consider matrices whose entries are the words used in this book. But in this case no reasonable arithmetic exists: we find ourselves with a lot of rectangular arrays of words but with no suitable way to add or multiply these arrays. Thus, it is reasonable to restrict our attention to matrices whose entries can be added and multiplied by numbers.

Our immediate concern here is with a hybrid matrix arithmetic involving n by n matrices of real numbers and n by 1 matrices of vectors. For example, consider the product

$$\begin{bmatrix} 1 & 2 & 3 \\ 4 & 5 & 6 \\ 7 & 8 & 9 \end{bmatrix} \begin{bmatrix} X^1 \\ X^2 \\ X^3 \end{bmatrix},$$

where X^1, X^2, and X^3 are from some vector space V. The reader must at first protest: What kind of matrix arithmetic is this? Are we dealing with matrices of numbers? Or are we dealing with matrices of vectors? Well, we are dealing with both. In other words, we are contemplating a rather complicated algebraic structure in which there are two kinds of "matrix numbers."

We are not going to study this structure, cataloging its various arithmetic properties, for our use of it will be quite limited. Thus, we shall proceed rather informally. First, let us carry out the multiplication indicated above:

$$\begin{bmatrix} 1 & 2 & 3 \\ 4 & 5 & 6 \\ 7 & 8 & 9 \end{bmatrix} \begin{bmatrix} X^1 \\ X^2 \\ X^3 \end{bmatrix} = \begin{bmatrix} (X^1 + 2X^2 + 3X^3) \\ (4X^1 + 5X^2 + 6X^3) \\ (7X^1 + 8X^2 + 9X^3) \end{bmatrix} = \begin{bmatrix} Y^1 \\ Y^2 \\ Y^3 \end{bmatrix}.$$

Thus (this is evidently true in general) an n by n *real matrix* times an n by 1 *vector matrix* is an n by 1 vector matrix. Moreover, each Y^j is a linear combination of the X-vectors.

In addition to this fact, we shall need one other: If A and B are n by n real matrices, and $\mathcal{C}$ is an n by 1 matrix of vectors, then

$$A(B\mathcal{C}) = (AB)\mathcal{C}.$$

That is, we have an associative law for such hybrid matrix products. (Our earlier proof for the associativity of real matrix multiplication continues to be valid in the present context.)

We make one final observation before proceeding to the main result of this section. A basis is—although it has very special properties—a *set* of vectors. As such, there is no implied order given to the members of a basis. In other words, if $\{X^1, X^2, X^3\}$ is a basis for the space V, then $\{X^2, X^1, X^3\}$ is the same basis. In what follows, however, we shall require that our bases be *ordered*. Thus, we shall speak of an *ordered basis*, and, to exploit the hybrid matrix arithmetic described earlier, we will represent an ordered basis having n vectors by an n by 1 column matrix.

Theorem 2.19. *Change-of-basis matrix.* Let $\mathcal{B}$ be an ordered basis for an n-dimensional space V and let A be any nth-order, nonsingular matrix. Then $\mathcal{C}$, where $\mathcal{C} = A\mathcal{B}$, is also an ordered basis for V. Conversely, if $\mathcal{B}$ and $\mathcal{C}$ are both ordered bases for V, then there is a unique, nonsingular matrix A such that $\mathcal{C} = A\mathcal{B}$.

Proof

1. Let
$$\mathcal{B} = \begin{bmatrix} B^1 \\ \vdots \\ B^n \end{bmatrix} \quad \text{and} \quad \mathcal{C} = \begin{bmatrix} C^1 \\ \vdots \\ C^n \end{bmatrix}.$$

Then, using our hypothesis, we may write
$$\mathcal{B} = (A^{-1}A)\mathcal{B} = A^{-1}(A\mathcal{B}) = A^{-1}\mathcal{C}.$$

That is,
$$\begin{bmatrix} B^1 \\ \vdots \\ B^n \end{bmatrix} = A^{-1} \begin{bmatrix} C^1 \\ \vdots \\ C^n \end{bmatrix}.$$

It follows that each B^j may be expressed as a linear combination of the vectors in $\mathcal{C}$. Now suppose that X is any vector in V. Then X can be expressed as a linear combination of the vectors in $\mathcal{B}$, for $\mathcal{B}$ is a basis. As already noted, these vectors may be expressed in terms of the vectors of $\mathcal{C}$. It follows that X may be expressed as a linear combination of the members in $\mathcal{C}$. This shows that $\mathcal{C}$ is a spanning set for V. But $\mathcal{C}$ contains exactly n vectors; by Theorem 2.17, it is necessarily a basis for V.

2. If we are given that $\mathcal{B}$ and $\mathcal{C}$ are both ordered bases for V, then it follows that each vector in $\mathcal{C}$ may be expressed in terms of those in $\mathcal{B}$:

$$\begin{cases} C^1 = a_{11}B^1 + a_{12}B^2 + \cdots + a_{1n}B^n, \\ C^2 = a_{21}B^1 + a_{22}B^2 + \cdots + a_{2n}B^n, \\ \vdots \\ C^n = a_{n1}B^1 + a_{n2}B^2 + \cdots + a_{nn}B^n. \end{cases}$$

Equivalently,
$$\mathcal{C} = A\mathcal{B}, \quad \text{where } A = [a_{ij}]_{nn}.$$

Thus, there does exist a matrix A that transforms $\mathcal{B}$ into $\mathcal{C}$. Moreover, it is unique. For suppose that $\mathcal{C} = A_1\mathcal{B}$, where $A_1 \neq A$. It would then follow that at least one C^j had a different representation in terms of the B-vectors than that exhibited above. But, by Theorem 2.9, the representation of a vector as a linear combination of a set of basis vectors is unique.

Now by identical reasoning, there is a unique matrix A^* such that $\mathcal{B} = A^*\mathcal{C}$. It then follows that
$$\mathcal{C} = A\mathcal{B} = A(A^*\mathcal{C}) = (AA^*)\mathcal{C}.$$

Thus, using what we have already proved, AA^* is the matrix that transforms the ordered basis $\mathcal{C}$ into the ordered basis $\mathcal{C}$. But, evidently, $\mathcal{C} = I\mathcal{C}$. Therefore, $AA^* = I$. Similarly, from
$$\mathcal{B} = A^*\mathcal{C} = A^*(A\mathcal{B}) = (A^*A)\mathcal{B}$$
we may conclude that
$$A^*A = I.$$

It follows that A has A^* as its inverse; consequently, A is nonsingular.

Example 1. Let
$$\mathcal{B} = \begin{bmatrix} t^2 \\ t \\ 1 \end{bmatrix} \quad \text{and} \quad \mathcal{C} = \begin{bmatrix} t^2 + 1 \\ 2t - 3 \\ 2t^2 - 5 \end{bmatrix}.$$

Then $\mathcal{B}$ and $\mathcal{C}$ are both bases for P^2. (See Example 3, Section 2.5). Find the *change-of-basis matrix* A such that $\mathcal{C} = A\mathcal{B}$.

Solution. We follow the pattern laid out in part 2 of the proof of Theorem 2.19. The representation of the $\mathcal{C}$-vectors in terms of the $\mathcal{B}$-vectors is immediate in this case:
$$t^2 + 1 = (1)t^2 + (0)t + (\ 1\ )1,$$
$$2t - 3 = (0)t^2 + (2)t + (-3)1,$$
$$2t^2 - 5 = (2)t^2 + (0)t + (-5)1.$$

It follows that
$$\mathcal{C} = \begin{bmatrix} 1 & 0 & 1 \\ 0 & 2 & -3 \\ 2 & 0 & -5 \end{bmatrix} \mathcal{B}.$$

Example 2. Let

$$\mathcal{B} = \begin{bmatrix} \begin{bmatrix} 5 \\ 8 \end{bmatrix} \\ \begin{bmatrix} 2 \\ 2 \end{bmatrix} \end{bmatrix} \quad \text{and} \quad \mathcal{C} = \begin{bmatrix} \begin{bmatrix} 1 \\ 4 \end{bmatrix} \\ \begin{bmatrix} 0 \\ 6 \end{bmatrix} \end{bmatrix}.$$

Then $\mathcal{B}$ and $\mathcal{C}$ are both ordered bases for R^{21}. Find A such that $\mathcal{C} = A\mathcal{B}$.

Solution. We seek the numbers a_{11}, a_{12}, a_{21}, and a_{22} that satisfy the system

$$\begin{cases} C^1 = a_{11}B^1 + a_{12}B^2, \\ C^2 = a_{21}B^1 + a_{22}B^2. \end{cases}$$

We handle each equation separately:

$$\begin{bmatrix} 1 \\ 4 \end{bmatrix} = a_{11} \begin{bmatrix} 5 \\ 8 \end{bmatrix} + a_{12} \begin{bmatrix} 2 \\ 2 \end{bmatrix}.$$

There is exactly one solution: $a_{11} = 1$ and $a_{12} = -2$.

$$\begin{bmatrix} 0 \\ 6 \end{bmatrix} = a_{21} \begin{bmatrix} 5 \\ 8 \end{bmatrix} + a_{22} \begin{bmatrix} 2 \\ 2 \end{bmatrix}.$$

There is exactly one solution: $a_{21} = 2$ and $a_{22} = -5$. Thus,

$$\begin{bmatrix} C^1 \\ C^2 \end{bmatrix} = \begin{bmatrix} 1 & -2 \\ 2 & -5 \end{bmatrix} \begin{bmatrix} B^1 \\ B^2 \end{bmatrix}.$$

EXERCISE SET 2.8

For exercises 1 through 3, the reader has had no formal instructions. Just do what comes naturally.

1 Let

$$A = \begin{bmatrix} i & 3+i & 0 \\ 2+i & -i & 5i \end{bmatrix}, \quad B = \begin{bmatrix} 1 & 2-i \\ 3-i & i \\ 0 & -2 \end{bmatrix},$$

and

$$C = \begin{bmatrix} i & 1 \\ 2-i & 3i \end{bmatrix},$$

where i is the imaginary unit ($i^2 = -1$).

a) Find $AB + 3iC$.
b) Find $A + \overline{A}$. (Recall that $\overline{a+bi} = a - bi$.)

2 Let

$$A = \begin{bmatrix} 1 & t \\ \sin t & 0 \end{bmatrix}, \quad \text{and} \quad B = \begin{bmatrix} e^t & 0 \\ t & \cos t \end{bmatrix}.$$

a) Find $2A(A + B)$.
b) Find dA/dt.
c) Show that
$$\frac{d(AB)}{dt} = A\frac{dB}{dt} + \frac{dA}{dt}B.$$

3 *Matrices of matrices.* Find $A + B$ and AB, where

$$A = \begin{bmatrix} \begin{bmatrix} 1 & 2 \\ 3 & 4 \end{bmatrix} & \begin{bmatrix} 1 & 2 \\ -1 & 3 \end{bmatrix} \\ \begin{bmatrix} 2 & 0 \\ 1 & 2 \end{bmatrix} & \begin{bmatrix} 1 & 1 \\ 2 & 1 \end{bmatrix} \end{bmatrix} \quad \text{and} \quad B = A'.$$

4 Two ordered bases for R^{13} are given by

$$\mathcal{C} = \begin{bmatrix} [1, 2, 3] \\ [2, 3, 5] \\ [3, 5, 9] \end{bmatrix} \quad \text{and} \quad \mathcal{D} = \begin{bmatrix} [1, 1, 2] \\ [1, 2, 4] \\ [0, 0, 1] \end{bmatrix}.$$

Find independently matrices A and B such that $\mathcal{D} = A\mathcal{C}$ and $\mathcal{C} = B\mathcal{D}$. Then verify that A and B are inverses.

5 a) Let

$$\mathcal{C} = \begin{bmatrix} 1 \\ t \\ t^2 \end{bmatrix} \quad \text{and} \quad \mathcal{D} = \begin{bmatrix} t^2 \\ t \\ 1 \end{bmatrix}.$$

Find A such that $\mathcal{D} = A\mathcal{C}$.

b) Let

$$\mathcal{C} = \begin{bmatrix} t^2 + 2 \\ 3t^2 - t \\ t^2 + 5t \end{bmatrix} \quad \text{and} \quad \mathcal{D} = \begin{bmatrix} t^2 \\ t \\ 1 \end{bmatrix}.$$

Find A such that $\mathcal{D} = A\mathcal{C}$. [*Hint:* It may be more convenient to first find A^{-1}.]

6 a) Let $Y^1 = 1$, $Y^2 = t$, and $Y^3 = t + 1$. Show that

$$\begin{bmatrix} Y^1 \\ Y^2 \\ Y^3 \end{bmatrix} = \begin{bmatrix} 0 & -1 & 1 \\ -1 & 0 & 1 \\ 1 & 1 & 0 \end{bmatrix} \begin{bmatrix} Y^1 \\ Y^2 \\ Y^3 \end{bmatrix}.$$

b) Prove or disprove: Let $\mathcal{C}$ be an n by 1 matrix of vectors and let A be an nth-order matrix of numbers such that $\mathcal{C} = A\mathcal{C}$. Then $A = I$.

c) At the end of part 2 of the proof of Theorem 2.19 we concluded from $\mathcal{C} = (AA^*)\mathcal{C}$ that $AA^* = I$. Explain.

7 Let V be the set of all function matrices of the form

$$\begin{bmatrix} a + bt \\ ce^t \end{bmatrix},$$

with addition and numerical multiplication defined in the expected way. Then V is a vector space.

a) Let
$$X = \begin{bmatrix} 2+3t \\ 4e^t \end{bmatrix} \quad \text{and} \quad Y = \begin{bmatrix} 1-t \\ e^t \end{bmatrix}.$$
Find $2X + 3Y$.

b) Find a basis for V and thus determine the dimension of V.

8 Prove: The set $S = \{A^1, A^2, A^3, A^4\}$ is linearly independent if and only if
$$T = \{A^1, A^1 + A^2, A^1 + A^2 + A^3, A^1 + A^2 + A^3 + A^4\}$$
is linearly independent. [*Hint:* Consider a matrix calculation that would take you from the vectors of S to those of T.]

2.9 COORDINATE SYSTEMS

We begin this section with a short digression: we shall consider the notion of a *numeral system*. It is taken for granted here that the reader has some appreciation for the difference between numbers and the symbols used to denote numbers, the latter being called *numerals*. Also, it is assumed that the reader recognizes the fact that different numeral systems are possible for the same number system, and that a "good" numeral system is one which facilitates computation. (The decimal system is a good system; the Roman numeral system is a poor one.)

Now, our interest here will be in reviewing the way in which *base systems* are introduced. Our own decimal system is such a base system, the base being 10. However, so that we be as little prejudiced by past experience as possible, we shall consider a different base system.

Let N be the set of nonnegative integers, and let us assume that the only labeling scheme existing for N is that which uses the common word names for the numbers in N. Thus, we denote the numbers in N by

zero, one, two, three, ...

We now choose some number from N other than zero or one and declare it to be the *base* of our numeral system; let that number be three. The numbers less than three are now referred to as digits. Since the digits will play a special role in our numeral system, we invent more convenient symbols for them. For example, let us write—not being awfully inventive— 0 for zero, 1 for one, and 2 for two.

Next, we come to the essential idea behind every base system: each number in N can be expressed in the form
$$a_k(\text{three})^k + a_{k-1}(\text{three})^{k-1} + \cdots + a_1(\text{three}) + a_0,$$
where the a_j are digits. Moreover, except for the order of the terms, this

representation is unique. As illustrations, consider the following:

seven $= 2(\text{three}) + 1$,
fourteen $= 1(\text{three})^2 + 1(\text{three}) + 2$,
forty-six $= 1(\text{three})^{\text{three}} + 2(\text{three})^2 + 0(\text{three}) + 1$.

If we now agree that power-of-three representations are always to be written, as above, in the order of descending powers of three, then those representations may be abbreviated by simply listing the digit coefficients of any such representation in their order of appearance. Thus, we are led to introduce the following symbols (numerals) for seven, fourteen, and forty-six:

21 for seven,

112 for fourteen,

1201 for forty-six.

More generally, if

$$n = a_k(\text{three})^k + a_{k-1}(\text{three})^{k-1} + \cdots + a_1(\text{three}) + a_0,$$

we write for n

$$a_k a_{k-1} \ldots a_1 a_0.$$

This system of numerals based on the number three is commonly called the ternary system, and the k-tuple of digits corresponding to a given number is called the *ternary numeral* of that number. Algorithms for doing arithmetic—carrying, borrowing, patterns for multiplication and division, etc.—are analogous to those that the reader employs when he uses decimal numerals, and if the reader has never performed any calculations using other than the decimal system, he should spend an "hour" teaching himself how to use ternary numerals.

The reason for this seeming digression is this: There is an analogy between choosing a base b for the number system N and choosing an ordered basis $\mathcal{B}$ for a vector space V: the first choice establishes a numeral system for N; the second choice establishes a *coordinate system* for V.

For example, let V be the space of all polynomials of the form

$$X = at^2 + (a+b)t + b.$$

(Thus, V is a subspace of P^2.) By writing

$$X = a(t^2 + t) + b(t+1)$$

and noting that the set $\{t^2 + t, t + 1\}$ is linearly independent, we see that we may choose $\mathcal{B} = [t^2 + t, t + 1]$ as an ordered basis for V. (Prior to this we have represented an ordered basis by a column matrix; how-

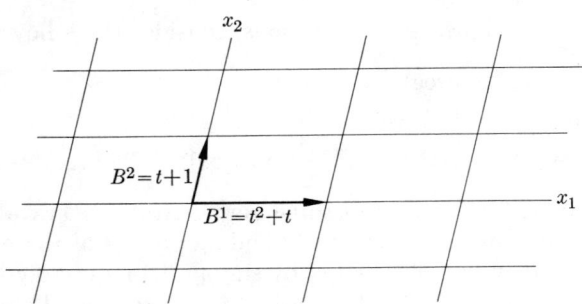

Figure 2.3

ever, a row matrix serves the same purpose of preserving the order of the basis vectors.) It follows that V has dimension 2 and, which is to say the same thing, V is isomorphic to R^{21}.

Now, we have seen earlier (Section 2.3, the discussion following Example 1) that there is a natural way of visualizing R^{21} as a "plane of arrows." It turns out that the same interpretation may be given to any vector space of dimension 2, in particular, to our polynomial space V. We reason as follows:

Let P be a plane of arrows emanating from a single point. We choose any two arrows that "span" our plane and simply decree that these arrows shall represent the basis polynomials $B^1 = t^2 + t$ and $B^2 = t + 1$ (see Fig. 2.3). Next, superimpose upon these arrows an x_1-axis and an x_2-axis, the lengths of the arrows establishing the scales on the axes and their heads determining the positive sides of the axes. In this way we have established a *coordinate system* for our polynomial space V. For let X be any polynomial in V. Then, because $\mathcal{B} = [t^2 + t, t + 1]$ is an ordered basis for V, we are assured that there exists a unique pair of numbers x_1 and x_2 such that

$$X = x_1(t^2 + t) + x_2(t + 1).$$

It is natural, then, to represent X by the arrow that is x_1 times the arrow representing $(t^2 + t)$ plus x_2 times the arrow representing $(t + 1)$. In other words, we represent X by the arrow whose *coordinates* are x_1 and x_2 (see Fig. 2.4).

The point is this: The acknowledged fact that V is isomorphic to R^{21} has now been given a vivid geometric interpretation, and we may—if we choose—picture the polynomials of V as a plane of arrows (or points) embedded in a certain coordinate system. It should be understood, however, that different pictures will result depending on which arrows are chosen to depict the basis vectors. For example, we might have

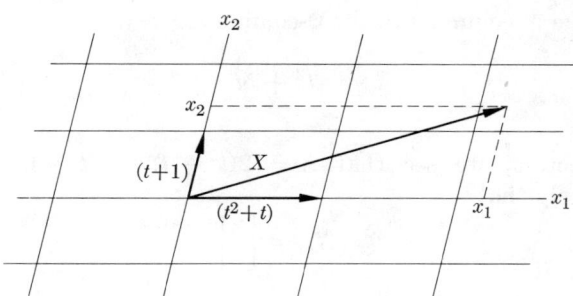

Figure 2.4

represented $(t^2 + t)$ and $(t + 1)$ by perpendicular arrows having the same length, thus producing a *rectangular coordinate system*. However, there is at this point no motivation for choosing such a special pair of arrows. (In Chapter 5, we shall see that there is a context in which polynomials—or any vectors—do have specified lengths and directions.)

Finally, we note that the visualization of a vector space as a space of arrows embedded in a coordinate system is possible only for vector spaces of dimension 1, 2, or 3. However, as is brought out in the definition that follows, the language of coordinate systems continues to find use in higher dimensional vector spaces.

Definition 2.14. Let V be an n-dimensional vector space, let $\mathcal{B}$ be an ordered basis for V, and let $\text{Id}_\mathcal{B}$ be the isomorphism from V to R^{n1} defined as follows: If

$$X = x_1 B^1 + x_2 B^2 + \cdots + x_n B^n,$$

then

$$\text{Id}_\mathcal{B}(X) = \begin{bmatrix} x_1 \\ x_2 \\ \vdots \\ x_n \end{bmatrix}.$$

Then the column space R^{n1} when viewed as the range of $\text{Id}_\mathcal{B}$ is called the *coordinate system for V based upon* $\mathcal{B}$. The column $\text{Id}_\mathcal{B}(X)$, which we usually denote by $X_\mathcal{B}$, is called the $\mathcal{B}$-*column for X;* and the jth entry of $X_\mathcal{B}$ is called the *jth $\mathcal{B}$-coordinate of X.*

Example 1. Let V be the subspace of P^2 consisting of all polynomials of the form $at^2 + (a + b)t + b$. (Recall the discussion preceding Definition 2.14.) We introduce two ordered bases for V:

$$\mathcal{B} = [t^2 + t, \, t + 1],$$
$$\mathcal{C} = [t^2 - 1, \, t^2 + 2t + 1].$$

Find both the $\mathcal{B}$-column and the $\mathcal{C}$-column for

$$X = 2t^2 + 5t + 3.$$

Solution

1. By inspection, we see that $X = 2(t^2 + t) + 3(t + 1)$. It follows immediately that

$$X_{\mathcal{B}} = \begin{bmatrix} 2 \\ 3 \end{bmatrix}.$$

2. Finding $X_{\mathcal{C}}$ is a little harder. We seek the numbers x_1 and x_2 such that

$$\begin{aligned} 2t^2 + 5t + 3 &= x_1 C^1 + x_2 C^2 \\ &= x_1(t^2 - 1) + x_2(t^2 + 2t + 1) \\ &= (x_1 + x_2)t^2 + (2x_2)t + (x_2 - x_1). \end{aligned}$$

Evidently, x_1 and x_2 must satisfy the system

$$\begin{cases} x_1 + x_2 = 2, \\ 2x_2 = 5, \\ -x_1 + x_2 = 3. \end{cases}$$

The solution is $x_1 = -\frac{1}{2}$ and $x_2 = \frac{5}{2}$. Thus,

$$X_{\mathcal{C}} = \begin{bmatrix} -\frac{1}{2} \\ \frac{5}{2} \end{bmatrix}. \quad \square$$

A remark on notation. Having introduced a coordinate system for a given vector space, it is often convenient to use the columns of that system as *labels* for the vectors. When this is done—so as to avoid ambiguity—it is customary to "tag" the column with the symbol designating the basis of the coordinate system being employed. For example, referring to Example 1, we might write

$$2t^2 + 5t + 3 = \begin{bmatrix} 2 \\ 3 \end{bmatrix}_{\mathcal{B}}$$

or

$$2t^2 + 5t + 3 = \begin{bmatrix} -\frac{1}{2} \\ \frac{5}{2} \end{bmatrix}_{\mathcal{C}}.$$

We do not mean, of course, that a certain polynomial is *equal* to a certain column matrix; rather, we are saying that the $\mathcal{B}$-column for $2t^2 + 5t + 3$ (or $\mathcal{C}$-column) will serve as a *label* for this polynomial. (In a similar way a subscript is often used to identify the base of a numeral system. For example, we might write—with obvious meaning—$14 = 112_3$.)

Example 2. Let V be the subspace of R^{41} having the ordered basis

$$\mathcal{B} = \left[\begin{bmatrix} 1 \\ 2 \\ 0 \\ 3 \end{bmatrix}, \begin{bmatrix} 0 \\ 1 \\ 2 \\ 5 \end{bmatrix}, \begin{bmatrix} 1 \\ -2 \\ 3 \\ -2 \end{bmatrix} \right].$$

Find the $\mathcal{B}$-column for the vector

$$X = \begin{bmatrix} 5 \\ -3 \\ 7 \\ -5 \end{bmatrix}.$$

Solution. We begin by setting

$$\begin{bmatrix} 5 \\ -3 \\ 7 \\ -5 \end{bmatrix} = x_1 \begin{bmatrix} 1 \\ 2 \\ 0 \\ 3 \end{bmatrix} + x_2 \begin{bmatrix} 0 \\ 1 \\ 2 \\ 5 \end{bmatrix} + x_3 \begin{bmatrix} 1 \\ -2 \\ 3 \\ -2 \end{bmatrix}.$$

Solving this equation, we obtain the single solution $x_1 = 2$, $x_2 = -1$, and $x_3 = 3$. Thus, we may write

$$X = \begin{bmatrix} 5 \\ -3 \\ 7 \\ -5 \end{bmatrix} = 2 \begin{bmatrix} 1 \\ 2 \\ 0 \\ 3 \end{bmatrix} + (-1) \begin{bmatrix} 0 \\ 1 \\ 2 \\ 5 \end{bmatrix} + 3 \begin{bmatrix} 1 \\ -2 \\ 3 \\ -2 \end{bmatrix}$$

$$= 2B^1 + (-1)B^2 + 3B^3.$$

It then follows that

$$X_\mathcal{B} = \begin{bmatrix} 2 \\ -1 \\ 3 \end{bmatrix}.$$

Definition 2.15. Let $\mathcal{N} = [N^1, N^2, \ldots, N^n]$, where N^j is the jth column of the n by n identity matrix. Then $\mathcal{N}$, an ordered basis for R^{n1}, is called the *natural basis* for R^{n1}. The corresponding coordinate system is called the *natural coordinate system* for R^{n1}.

Example 3. Let $\mathcal{N}$ be the natural basis for R^{31}. Find the $\mathcal{N}$-coordinates of

$$X = \begin{bmatrix} 2 \\ 3 \\ 4 \end{bmatrix}.$$

Solution. Evidently,

$$\begin{bmatrix} 2 \\ 3 \\ 4 \end{bmatrix} = 2N^1 + 3N^2 + 4N^3.$$

Thus,
$$X_{\mathfrak{N}} = \begin{bmatrix} 2 \\ 3 \\ 4 \end{bmatrix}.$$

It follows that the $\mathfrak{N}$-coordinates of X are

$$x_1 = 2, \qquad x_2 = 3, \qquad \text{and} \qquad x_3 = 4. \ \square$$

It should be appreciated by the reader (this is the point of Example 3) that the vectors of a given vector space do not automatically have coordinates: they have coordinates only after an ordered basis has been designated. However, in the case of a column space R^{n1} (or a row space) it is nevertheless the custom to think of these vectors as being inherently coordinatized, the jth entry of the column being the jth coordinate. This practice may be justified by accepting the following convention: When reference is made to the coordinates of a vector in R^{n1} without specifying a basis, then it is to be assumed that the speaker has in mind the natural coordinates of the vector.

EXERCISE SET 2.9

1 An ordered basis for P^2 is given by $\mathfrak{B} = [t^2, 2t, 1+t]$.
  a) $X = 2t^2 + 6t + 4$. Find $X_{\mathfrak{B}}$.
  b) $Y = t^2 + t - 9$. Find the second $\mathfrak{B}$-coordinate of Y.
  c) Suppose that we come upon the "equality"
  $$X = \begin{bmatrix} 2 \\ 3 \\ 4 \end{bmatrix}_{\mathfrak{B}}.$$
  What is X, really?

2 V is the subspace of R^{31} having the ordered basis
$$\mathfrak{B} = \begin{bmatrix} \begin{bmatrix} 1 \\ 1 \\ 1 \end{bmatrix}, \begin{bmatrix} 1 \\ 2 \\ 3 \end{bmatrix} \end{bmatrix}.$$

  a) $X = \begin{bmatrix} 4 \\ 2 \\ 0 \end{bmatrix}$. Find $X_{\mathfrak{B}}$.

  b) $Y_B = \begin{bmatrix} -1 \\ 2 \end{bmatrix}$. Find Y.

  c) $Z = 2X + 3Y$. Find Z, but do the arithmetic in V's $\mathfrak{B}$-coordinate system.

3 Let $\mathfrak{N}$ be the natural basis for R^{41}, let $\mathfrak{B} = [N^2, N^1, N^4, N^3]$, and let $X = [5, 2, 3, 7]'$.
  a) Find the third $\mathfrak{N}$-coordinate of X.
  b) Find the third $\mathfrak{B}$-coordinate of X.

4 Two ordered bases for R^{22} are given by

$$\mathfrak{B} = \left[\begin{bmatrix} 1 & 0 \\ 0 & 0 \end{bmatrix}, \begin{bmatrix} 0 & 1 \\ 0 & 0 \end{bmatrix}, \begin{bmatrix} 0 & 0 \\ 1 & 0 \end{bmatrix}, \begin{bmatrix} 0 & 0 \\ 0 & 1 \end{bmatrix} \right]$$

and

$$\mathfrak{C} = \left[\begin{bmatrix} 1 & 0 \\ 0 & 0 \end{bmatrix}, \begin{bmatrix} 1 & 1 \\ 0 & 0 \end{bmatrix}, \begin{bmatrix} 1 & 1 \\ 1 & 0 \end{bmatrix}, \begin{bmatrix} 1 & 1 \\ 1 & 1 \end{bmatrix} \right].$$

  a) $X_{\mathfrak{C}} = [1, 2, 3, 4]'$. Find $X_{\mathfrak{B}}$.
  b) $X_{\mathfrak{B}} = [-1, 0, -2, 3]'$. Find $X_{\mathfrak{C}}$.
  c) $X = \begin{bmatrix} a & b \\ c & d \end{bmatrix}$. Find $X_{\mathfrak{B}}$ and $X_{\mathfrak{C}}$.
  d) Find a matrix A such that $X_{\mathfrak{C}} = AX_{\mathfrak{B}}$ for any X in R^{22}.

5 Choose $\mathfrak{B} = [t, 1]$ as an ordered basis for P^1.
  a) Depicting B^1 and B^2 as equal-length arrows pointing to the "east" and "north," respectively, sketch a coordinate system for P^1. (Use a large scale; use a ruler.)
  b) Let $X = 2t + 1$. Place X on your sketch.
  c) Let $C^1 = t + 1$ and $C^2 = 2t - 1$. Place C^1 and C^2 on your sketch, and then superimpose the $\mathfrak{C}$-coordinate system on your $\mathfrak{B}$-coordinate system. (Let the axis determined by C^1 be the u_1-axis, and that determined by C^2, the u_2-axis.)
  d) By making suitable projections on the u_1- and u_2-axes, estimate the $\mathfrak{C}$-coordinates of X.
  e) Determine exactly the $\mathfrak{C}$-coordinates of X.

6 Two ordered bases for R^{12} are given by $\mathfrak{B} = [[2, 1], [1, 3]]$ and $\mathfrak{C} = [[1, 0], [0, 1]]$.
  a) Find $X_{\mathfrak{B}}$ and $X_{\mathfrak{C}}$, where $X = [1, -3]$.
  b) Find the $\mathfrak{B}$-columns for B^1 and B^2.
  c) Find the $\mathfrak{C}$-columns for C^1 and C^2.

7 Prove: If $\mathfrak{B} = [B^1, B^2, \ldots, B^n]$ is an ordered basis for some space V, then the $\mathfrak{B}$-column for each B^i is N^i.

8 a) Let V be the subspace of R^{14} consisting of all vectors of the form $[a, a + 2b, 2a + b, b]$. Given that

$$[a, a + 2b, 2a + b, b] = \begin{bmatrix} a \\ b \end{bmatrix}_{\mathfrak{B}},$$

find $\mathfrak{B}$.

b) Let V be as in part (a). Given that

$$[2, 8, 7, 3] = \begin{bmatrix} 1 \\ 1 \end{bmatrix}_e \quad \text{and} \quad [3, 11, 10, 4] = \begin{bmatrix} 2 \\ 1 \end{bmatrix}_e,$$

find $\mathfrak{C}$.

9 Let

$$\mathfrak{B} = \left[\begin{bmatrix} t^2 \\ 0 \end{bmatrix}, \begin{bmatrix} t \\ 0 \end{bmatrix}, \begin{bmatrix} 1 \\ 0 \end{bmatrix}, \begin{bmatrix} 0 \\ t \end{bmatrix}, \begin{bmatrix} 0 \\ 1 \end{bmatrix} \right].$$

Then $\mathfrak{B}$ is an ordered basis for the space V of all function matrices having the form

$$\begin{bmatrix} at^2 + bt + c \\ dt + e \end{bmatrix}.$$

a) $X = \begin{bmatrix} 2t^2 + 3t - 1 \\ 4t + 5 \end{bmatrix}$. Find $X_{\mathfrak{B}}$.

b) Find the second $\mathfrak{B}$-coordinate of X.

2.10 CHANGING COORDINATE SYSTEMS

Often in the application of mathematics we have occasion to deal with a *vector-to-number function*, that is, a function, say F, that has a vector space for its domain and a subset of the real-number system for its range. The use of such a function is often facilitated by introducing a coordinate system for its domain and then deriving a formula for $F(X)$ in terms of the coordinates of X. The idea is best illustrated by considering a specific function.

Let F be defined on the polynomial space P^1 by

$$F(p) = \left\{ \int_0^1 [2p(t)] \, dt \right\}^2.$$

For example, if $p(t) = 2(t - 5) + (t + 8)$, then

$$F(p) = \left\{ \int_0^1 2[2(t - 5) + (t + 8)] \, dt \right\}^2$$

$$= \left\{ \int_0^1 (6t - 4) \, dt \right\}^2$$

$$= \left\{ (3t^2 - 4t) \Big|_0^1 \right\}^2$$

$$= 1.$$

Now, suppose that we introduce a coordinate system into P^1 by choosing the basis $\mathfrak{B} = [t, 1]$. (From now on we shall invariably mean *ordered basis* when we write *basis*.) Then every member of P^1 may be expressed

in a unique way in terms of this basis:
$$p(t) = x_1 t + x_2.$$
It follows that
$$\begin{aligned}
F(p) &= \left\{\int_0^1 [2x_1 t + 2x_2]\, dt\right\}^2 \\
&= \left\{(x_1 t^2 + 2x_2 t)\Big|_0^1\right\}^2 \\
&= (x_1 + 2x_2)^2 \\
&= x_1^2 + 4x_1 x_2 + 4x_2^2.
\end{aligned}$$

The expression $x_1^2 + 4x_1 x_2 + 4x_2^2$ cannot in itself be regarded as a formula for computing $F(p)$, for p does not even show up in that expression. What do we have then? Well, we have a formula for a function that takes us from the $\mathcal{B}$-column for p to $F(p)$. Call this function f; its relation to F is probably best revealed by means of a diagram (see Fig. 2.5). In other words, we have two ways of passing from the polynomial p to the number w: We can go directly from p to w by using the function F, or we can first determine the $\mathcal{B}$-column for p and then employ f, the $\mathcal{B}$-*coordinate manifestation of F.*

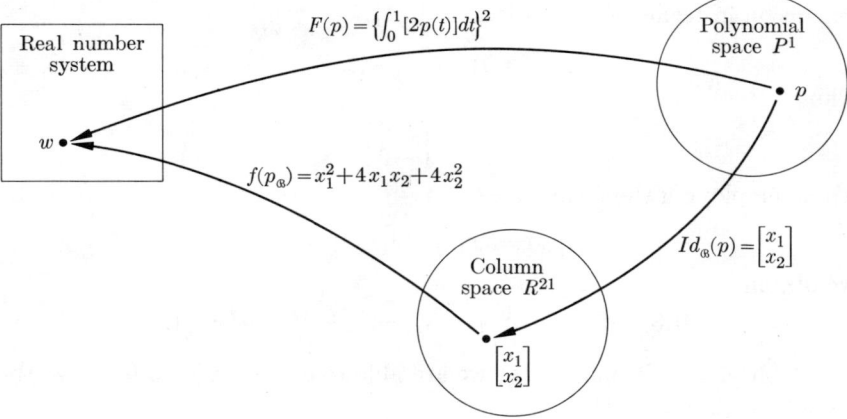

Figure 2.5

The distinction between a vector-to-number function and a particular coordinate manifestation of that function will be made more clear, perhaps, by displaying a second coordinate formula for F. Suppose, then, that we choose a new basis for P^1, say $\mathcal{C} = [t, -2t + 1]$. Then, as before, every member of P^1 has a unique representation in terms of this basis:
$$p(t) = u_1 t + u_2(-2t + 1).$$

(Having already used x_1 and x_2 for the $\mathcal{B}$-coordinates of p, we shift to u_1 and u_2 for its $\mathcal{C}$-coordinates.) It follows that

$$F(p) = \left\{\int_0^1 2[u_1 t + u_2(-2t + 1)]\, dt\right\}^2$$
$$= \left\{\int_0^1 [(2u_1 - 4u_2)t + 2u_2]\, dt\right\}^2$$
$$= \left\{[(u_1 - 2u_2)t^2 + (2u_2)t]\Big|_0^1\right\}^2$$
$$= (u_1 - 2u_2 + 2u_2)^2$$
$$= u_1^2.$$

We are thus led to a second coordinate function for F. Calling it g, we may write
$$F(p) = g(p_\mathcal{C}) = u_1^2.$$

Let us check our two coordinate formulas by using them to find $F(p)$, where $p(t) = 2(t - 5) + (t + 8)$. [We showed earlier—by a direct application of F—that $F(p) = 1$.]

By simplifying the given formula for $p(t)$ we are led directly to its expansion in terms of the basis $\mathcal{B} = [t, 1]$:
$$p(t) = 3t - 2.$$
Thus,
$$p_\mathcal{B} = \begin{bmatrix} 3 \\ -2 \end{bmatrix}.$$
Then, employing the formula
$$f(p_\mathcal{B}) = x_1^2 + 4x_1 x_2 + 4x_2^2,$$
we obtain
$$f([3, -2]') = 3^2 + 4(3)(-2) + 4(-2)^2 = 1.$$

With just a bit more effort we are able to expand $p(t)$ in terms of the basis $\mathcal{C} = [t, -2t + 1]$:
$$p(t) = -1(t) - 2(-2t + 1).$$
Thus,
$$p_\mathcal{C} = \begin{bmatrix} -1 \\ -2 \end{bmatrix}.$$
Then, employing the formula
$$g(p_\mathcal{C}) = u_1^2,$$
we obtain
$$g([-1, -2]') = (-1)^2 = 1.$$

We make an observation: A vector-to-number function may be "captured" in a variety of coordinate formulas, and the complexity of such formulas varies with the coordinate system chosen. We are thus led to the following question: How can we anticipate at the beginning a "good" coordinate system? Unfortunately, this question, in general, has no ready answer. Consider our function F. What was there about this function, as originally defined by means of an integral, to suggest that the coordinate system based on $\mathcal{C} = [t, -2t + 1]$ would lead us to the very simple formula u_1^2? Apparently nothing; the choice of $\mathcal{C}$ would seem almost miraculous.

Often, in practice, a good coordinate formula can be achieved in the following way: We begin by choosing any convenient basis and then derive the corresponding coordinate formula. Next, that formula is studied in search of clues that indicate a superior coordinate system, one which permits a simpler formula. Finally, that simpler formula is derived. These are admittedly rather vague instructions; let us see how they work with respect to the function F.

Suppose, as before, that we have chosen the basis $\mathcal{B} = [t, 1]$ and have derived the coordinate formula

$$f(p_\mathcal{B}) = x_1^2 + 4x_1 x_2 + 4x_2^2.$$

Now,

$$x_1^2 + 4x_1 x_2 + 4x_2^2 = (x_1 + 2x_2)^2.$$

From this formula it is almost evident that the substitutions

$$x_1 = u_1 - 2u_2 \quad \text{and} \quad x_2 = u_2$$

will lead us to the very simple formula

$$(x_1 + 2x_2)^2 = (u_1 - 2u_2 + 2u_2)^2 = u_1^2.$$

What do these substitutions mean? Well, whenever we deal with two coordinate systems simultaneously, there will always be formulas that relate a vector's coordinates in one system to those in the other. In other words, when we made our substitutions for x_1 and x_2, we might have said: Let us choose a new basis $\mathcal{C}$ for P^1 in such a way that the $\mathcal{C}$-coordinates of a given p, say u_1 and u_2, are related to its $\mathcal{B}$-coordinates, say x_1 and x_2, as follows:

$$x_1 = u_1 - 2u_2 \quad \text{and} \quad x_2 = u_2.$$

Then, with respect to the coordinate system based on $\mathcal{C}$ we have the coordinate formula

$$g(p_\mathcal{C}) = u_1^2.$$

That there actually is such a basis $\mathcal{C}$ is easily proved by finding it. We begin by writing
$$x_1 B^1 + x_2 B^2 = p(t) = u_1 C^1 + u_2 C^2,$$
which is nothing more than an acknowledgement that the x_j and the u_j are different sets of coordinates for the same polynomial. Using matrix notation, we may write
$$[x_1, x_2] \begin{bmatrix} B^1 \\ B^2 \end{bmatrix} = [u_1, u_2] \begin{bmatrix} C^1 \\ C^2 \end{bmatrix}.$$
We may also write our coordinate substitutions in matrix form:
$$[x_1, x_2] = [u_1, u_2] \begin{bmatrix} 1 & 0 \\ -2 & 1 \end{bmatrix}.$$
Making this substitution—and using associativity—we obtain
$$[u_1, u_2] \begin{bmatrix} 1 & 0 \\ -2 & 1 \end{bmatrix} \begin{bmatrix} B^1 \\ B^2 \end{bmatrix} = [u_1, u_2] \begin{bmatrix} C^1 \\ C^2 \end{bmatrix}.$$
And, since the C-coordinate representation is unique, we may conclude that
$$\begin{bmatrix} C^1 \\ C^2 \end{bmatrix} = \begin{bmatrix} 1 & 0 \\ -2 & 1 \end{bmatrix} \begin{bmatrix} B^1 \\ B^2 \end{bmatrix}.$$
That is, $C^1 = B^1$ and $C^2 = -2B^1 + B^2$. Thus, we are led anew to the "miraculous basis"
$$[C^1, C^2] = [t, -2t + 1].$$

Let us briefly summarize our long discussion. The study of a vector-to-number function is often facilitated by introducing a coordinate system and subsequently deriving a coordinate formula for the function in question. However, despite having established one coordinate system, we may have good reason to introduce a second one. This is done, most often, not by choosing a new basis directly, but by writing formulas that relate the old coordinates to the new ones.

If the reader has by now grown receptive to the idea that we often have occasion to use more than one coordinate system for the same vector space, then he should appreciate the following theorem.

Theorem 2.20. *The change-of-coordinates formula.* Let $\mathcal{B}$ and $\mathcal{C}$ be ordered bases for an n-dimensional vector space V, and suppose that A is the change-of-basis matrix such that $\mathcal{C} = A\mathcal{B}$. Then for any X in V we have
$$X_{\mathcal{C}} = (A^{-1})' X_{\mathcal{B}}.$$
Moreover, the converse proposition is also true; that is, if $X_{\mathcal{C}} = AX_{\mathcal{B}}$,

then $\mathcal{C} = (A^{-1})'\mathcal{B}$. [Or, if A transforms the basis $\mathcal{B}$ into $\mathcal{C}$, then $(A^{-1})'$ transforms the $\mathcal{B}$-column of X into its $\mathcal{C}$-column. And conversely.]

Proof

1. Let
$$X_\mathcal{B} = \begin{bmatrix} b_1 \\ b_2 \\ \vdots \\ b_n \end{bmatrix} \quad \text{and} \quad X_\mathcal{C} = \begin{bmatrix} c_1 \\ c_2 \\ \vdots \\ c_n \end{bmatrix}.$$

This means that
$$c_1 C^1 + c_2 C^2 + \cdots + c_n C^n = X = b_1 B^1 + b_2 B^2 + \cdots + b_n B^n.$$

Or, in matrix form,
$$[c_1, c_2, \ldots, c_n]\mathcal{C} = [b_1, b_2, \ldots, b_n]\mathcal{B}.$$

Or, even more compactly,
$$X'_\mathcal{C}\mathcal{C} = X'_\mathcal{B}\mathcal{B}.$$

But, $\mathcal{C} = A\mathcal{B}$. Thus, we may write
$$X'_\mathcal{C} A \mathcal{B} = X'_\mathcal{B} \mathcal{B}.$$

It then follows (why?) that
$$X'_\mathcal{C} A = X'_\mathcal{B}.$$

Solving for $X_\mathcal{C}$ gives
$$X'_\mathcal{C} = X'_\mathcal{B} A^{-1}.$$

Finally, taking the transpose of each side yields
$$X_\mathcal{C} = (A^{-1})' X_\mathcal{B}.$$

2. The converse proposition is easily proved by simply reversing the steps in part 1.

Example 1. Let $\mathcal{B} = [t^2, t, 1]$ and $\mathcal{C} = [1, 2t+1, 3t^2+2t+1]$. Then $\mathcal{B}$ and $\mathcal{C}$ are bases for P^2, and as such they determine coordinate systems in that space. Find the change-of-coordinates formula that transforms $\mathcal{B}$-columns into $\mathcal{C}$-columns.

Solution. We observe first that
$$\begin{bmatrix} 1 \\ 2t+1 \\ 3t^2+2t+1 \end{bmatrix} = \begin{bmatrix} 0 & 0 & 1 \\ 0 & 2 & 1 \\ 3 & 2 & 1 \end{bmatrix} \begin{bmatrix} t^2 \\ t \\ 1 \end{bmatrix}.$$

Thus,
$$A = \begin{bmatrix} 0 & 0 & 1 \\ 0 & 2 & 1 \\ 3 & 2 & 1 \end{bmatrix}$$

is the change-of-basis matrix for passing from $\mathcal{B}$ to $\mathcal{C}$. But then, by Theorem 2.20, $(A^{-1})'$ is the change-of-coordinates matrix that takes $\mathcal{B}$-columns into $\mathcal{C}$-columns. A direct computation produces

$$(A^{-1})' = \begin{bmatrix} 0 & -\frac{1}{2} & 1 \\ -\frac{1}{3} & \frac{1}{2} & 0 \\ \frac{1}{3} & 0 & 0 \end{bmatrix}.$$

Thus, the desired formula is

$$X_{\mathcal{C}} = \begin{bmatrix} 0 & -\frac{1}{2} & 1 \\ -\frac{1}{3} & \frac{1}{2} & 0 \\ \frac{1}{3} & 0 & 0 \end{bmatrix} X_{\mathcal{B}}.$$

Example 2. Let $\mathcal{B} = [[1, 0, 0], [1, 1, 0], [1, 1, 1]]'$. Then $\mathcal{B}$ is a basis for R^{13}, and as such it determines a coordinate system for R^{13}. Now suppose that a new coordinate system is introduced by means of the following change-of-coordinates formula:

$$X_{\mathcal{C}} = \begin{bmatrix} 1 & 1 & 1 \\ 2 & 1 & 2 \\ 1 & 0 & 2 \end{bmatrix} X_{\mathcal{B}}.$$

Find the basis $\mathcal{C}$.

Solution. The change-of-coordinates matrix is given by

$$A = \begin{bmatrix} 1 & 1 & 1 \\ 2 & 1 & 2 \\ 1 & 0 & 2 \end{bmatrix}.$$

By appealing to the converse part of Theorem 2.20, we are assured that the corresponding change-of-basis matrix is given by

$$(A^{-1})' = \begin{bmatrix} -2 & 2 & 1 \\ 2 & -1 & -1 \\ -1 & 0 & 1 \end{bmatrix}.$$

Thus,

$$\mathcal{C} = \begin{bmatrix} -2 & 2 & 1 \\ 2 & -1 & -1 \\ -1 & 0 & 1 \end{bmatrix} \begin{bmatrix} [1, 0, 0] \\ [1, 1, 0] \\ [1, 1, 1] \end{bmatrix} = \begin{bmatrix} [1, & 3, & 1] \\ [0, & -2, & -1] \\ [0, & 1, & 1] \end{bmatrix}.$$

EXERCISE SET 2.10

1 The space R^{21} has

$$\mathcal{B} = \left[\begin{bmatrix} 2 \\ 1 \end{bmatrix}, \begin{bmatrix} 3 \\ 4 \end{bmatrix} \right]'$$

as an ordered basis. A second ordered basis is introduced by the formula

$$\mathcal{C} = \begin{bmatrix} 3 & 2 \\ 1 & 1 \end{bmatrix} \mathcal{B}.$$

a) Find $\mathcal{C}$.
b) Let $X = [1, 3]'$. Find $X_{\mathcal{C}}$ in a direct manner, i.e., by exhibiting X as a linear combination of C^1 and C^2.
c) Find the change-of-coordinates matrix that transforms $\mathcal{B}$-columns into $\mathcal{C}$-columns.
d) Find $X_{\mathcal{C}}$ by first finding $X_{\mathcal{B}}$ and then using the matrix found in part (c).

2 Let R^{31} be *coordinatized* by

$$\mathcal{B} = \left[\begin{bmatrix} 0 \\ 0 \\ 1 \end{bmatrix}, \begin{bmatrix} 0 \\ 1 \\ 1 \end{bmatrix}, \begin{bmatrix} 1 \\ 1 \\ 1 \end{bmatrix} \right].$$

A new coordinate system is introduced by

$$X_{\mathcal{C}} = \begin{bmatrix} 1 & 0 & -1 \\ -1 & 1 & 1 \\ 1 & 0 & 0 \end{bmatrix} X_{\mathcal{B}}.$$

a) Let $X = [3, 2, 1]'$. Find $X_{\mathcal{C}}$.
b) Find $\mathcal{C}$.
c) Check the result obtained under part (a).

3 Let V be the space having the basis $\mathcal{B} = [e^t, e^{2t}]$. Define G on V by the formula $G(f) = [f'(0)]^2$.

a) Find the $\mathcal{B}$-coordinate manifestation of G.
b) Find the basis $\mathcal{C}$ such that the $\mathcal{C}$-coordinate manifestation of G is given by $g(f_{\mathcal{C}}) = u_1^2$, where u_1 and u_2 are the $\mathcal{C}$-coordinates of f. [*Hint:* Recall the function F considered in the text.]

4 Let f be defined on R^{21} by

$$f\left(\begin{bmatrix} x_1 \\ x_2 \end{bmatrix} \right) = x_1^2 + 4x_1 x_2 + 8x_2^2.$$

a) Interpreting x_1 and x_2 as the natural coordinates for a typical vector X in R^{21}, find a matrix A such that $f(X_{\mathcal{R}}) = X_{\mathcal{R}}' A X_{\mathcal{R}}$.
b) Find a matrix B such that $B'AB = I$.

c) Introduce a new coordinate system for R^{21} by means of the formula $X_{\mathcal{H}} = BX_{\mathcal{C}}$. Now find the $\mathcal{C}$-coordinate formula for f. (Call this function g; take u_1 and u_2 as the $\mathcal{C}$-coordinates of X.)

d) The function f may be visualized (in an obvious way) as a surface extending over a plane. Relative to this observation, why do you suppose that mathematicians have labeled the relation existing between A and I as a "congruence relation"?

5 Let F be defined on R^{22} by

$$F\left(\begin{bmatrix} a & b \\ c & d \end{bmatrix}\right) = 2(ad - bc).$$

a) Let $\mathcal{B}$ be the basis of Exercise 4, Section 2.9. Find the $\mathcal{B}$-coordinate formula for F. (Take x_1, x_2, x_3, and x_4 as the $\mathcal{B}$-coordinates of a typical X in R^{22}.)

b) There is a basis $\mathcal{C}$ such that the $\mathcal{C}$-coordinate formula for F is

$$g(X_{\mathcal{C}}) = 2u_1^2 - 2u_2^2 + \tfrac{1}{2}u_3^2 - \tfrac{1}{2}u_4^2.$$

Find the $\mathcal{C}$-column for

$$\begin{bmatrix} a & b \\ c & d \end{bmatrix}$$

and then use the foregoing formula to find

$$F\left(\begin{bmatrix} a & b \\ c & d \end{bmatrix}\right).$$

6 Let f be defined on R^{21} by

$$f(X) = 4x_1^2 + 4x_1x_2 + 3x_2^2.$$

Interpreting x_1 and x_2 as the natural coordinates of X, let a new coordinate system be introduced by

$$X_{\mathcal{H}} = \begin{bmatrix} 1 & -\tfrac{1}{2} \\ 0 & 1 \end{bmatrix} X_{\mathcal{B}}.$$

a) Taking u_1 and u_2 as the $\mathcal{B}$-coordinates of X, find the $\mathcal{B}$-coordinate formula for f.

b) Make a sketch of the natural coordinate system wherein N^1 and N^2 are depicted by equal-length arrows pointing to the "east" and "north," respectively. Locate B^1 and B^2 on this coordinate system, and then superimpose upon B^1 and B^2 a u_1-axis and a u_2-axis.

c) Interpreting X as a point rather than an arrow, the set of all X's satisfying $f(X) = 4$ may be interpreted as a plane curve. Place this curve on the sketch obtained in part (b). [*Hint:* It will be easier to use the coordinate system based on $\mathcal{B}$.]

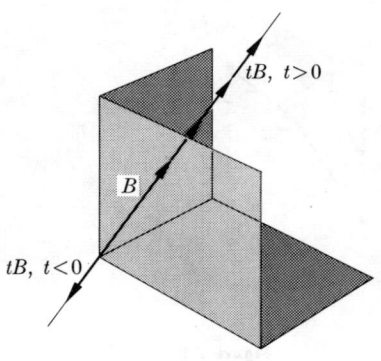 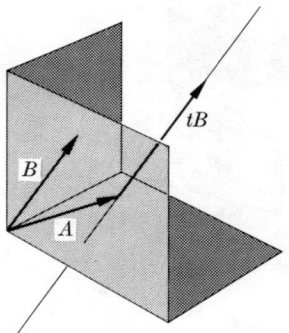

Figure 2.6   Figure 2.7

2.11 LINES AND PLANES

We close this chapter with what is essentially a digression. The remarks in this section will not lead us more deeply into the study of vector spaces; rather, they will serve to give us greater insight into the nature of a vector space.

Our discussion will be informal: we shall motivate our discussion by referring to the space of arrows emanating from a point, and we shall make free use of *pictorial reasoning*. We begin by talking about lines.

Let B be any arrow emanating from the origin of our arrow space. Numerical multiples of B will produce arrows in the same direction or in the opposite direction, depending on whether the multiplier is positive or negative (see Fig. 2.6). The collection of all such arrows evidently "points out" every point on the line through the origin that contains B. Such a line, then, is "captured" by the formula

$$X = tB, \quad -\infty < t < \infty.$$

More formally, our line may be interpreted as the range of this number-to-vector function that assigns the vector tB to the number t.

Now suppose that this line is translated as in Fig. 2.7. And suppose that A is any vector that reaches out to the translated line. Then every point on this new line is pointed out by a vector of the form $A + tB$. That is, the translated line is captured by the formula

$$X = A + tB, \quad -\infty < t < \infty.$$

We turn next to planes. Again we appeal to pictures. In Fig. 2.8, we show arrows B and C emanating from the origin, and in different directions. (This implies that C is not a multiple of B, that B and C, in the

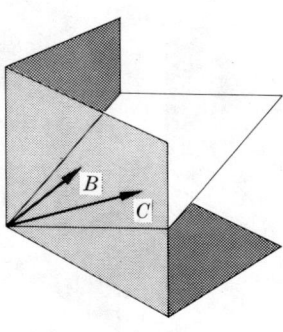 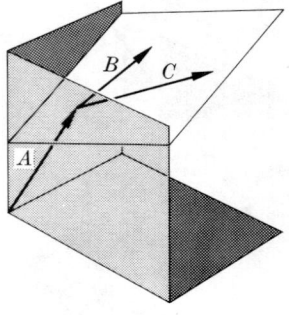

Figure 2.8 Figure 2.9

language of vector spaces, are linearly independent.) Now, there is one plane through the origin that contains $\mathfrak{B}$ and $\mathfrak{C}$. Moreover, every point in that plane can be reached by taking an appropriate linear combination of B and C. Thus, this plane may be identified with the range of the number-to-vector function

$$X = sB + tC, \quad -\infty < s < \infty, \; -\infty < t < \infty.$$

As with the line, if this plane is now translated away from the origin (see Fig. 2.9) and A is any vector reaching from the origin to the plane, then we simply add A to the above formula:

$$X = A + sB + tC, \quad -\infty < s < \infty, \; -\infty < t < \infty.$$

We are thus led to the following definition.

Definition 2.16. Let A, B^1, B^2, ..., B^k be vectors chosen from an n-dimensional vector space V. Suppose also that the set $\{B^1, B^2, \ldots, B^k\}$ is linearly independent. Then the range of the function

$$X = A + t_1 B^1 + t_2 B^2 + \cdots + t_k B^k,$$

where $-\infty < t_j < \infty$ for $j = 1, 2, \ldots, k$, is said to be a *k-plane in V*. A 1-plane in V is given a special name: it is called a *line*.

The reader may now quite properly ask: Is the usage of these geometrical terms justifiable? That is, do the sets of vectors that have here been called lines and planes actually "behave," somehow, like ordinary lines and planes? They do. We shall not prove this in a formal manner—this would take us too deeply into the study of geometry—but rather, we shall consider a number of specific examples that lend credence to our answer. (In later sections we shall obtain further evidence, so that in the end the reader will certainly find justifiable the mathematician's practice of thinking and speaking geometrically in quite arbitrary vector spaces.)

Example 1. Find a vector formula for the line in R^{41} that passes through the points [1, 2, 3, 4]' and [2, 4, 4, 3]', that is, through the "tips" of these vectors.

Solution. Let
$$A = \begin{bmatrix} 1 \\ 2 \\ 3 \\ 4 \end{bmatrix} \quad \text{and} \quad B = \begin{bmatrix} 2 \\ 4 \\ 4 \\ 3 \end{bmatrix}.$$

Then take
$$C = B - A = \begin{bmatrix} 1 \\ 2 \\ 1 \\ -1 \end{bmatrix}.$$

Interpreting C as we would in an arrow space (as the arrow joining the tips of A and B), we arrive at the following formula for the line in question:

$$X = A + tC = \begin{bmatrix} 1 \\ 2 \\ 3 \\ 4 \end{bmatrix} + t \begin{bmatrix} 1 \\ 2 \\ 1 \\ -1 \end{bmatrix},$$

or
$$X = \begin{bmatrix} t+1 \\ 2t+2 \\ t+3 \\ -t+4 \end{bmatrix}.$$

Example 2. Find the 2-plane in R^{41} that passes through the points [2, 3, 6, 1]', [4, 4, 6, 4]', and [5, 6, 7, 3]'.

Solution. Let
$$A = \begin{bmatrix} 2 \\ 3 \\ 6 \\ 1 \end{bmatrix}, \quad B = \begin{bmatrix} 4 \\ 4 \\ 6 \\ 4 \end{bmatrix}, \quad \text{and} \quad C = \begin{bmatrix} 5 \\ 6 \\ 7 \\ 3 \end{bmatrix}.$$

Then take
$$D = B - A = \begin{bmatrix} 2 \\ 1 \\ 0 \\ 3 \end{bmatrix} \quad \text{and} \quad E = C - A = \begin{bmatrix} 3 \\ 3 \\ 1 \\ 2 \end{bmatrix}.$$

Interpreting D and E as we would in an arrow space, we arrive at the following formula for the 2-plane in question:

$$X = A + sD + tE = \begin{bmatrix} 2 \\ 3 \\ 6 \\ 1 \end{bmatrix} + s \begin{bmatrix} 2 \\ 1 \\ 0 \\ 3 \end{bmatrix} + t \begin{bmatrix} 3 \\ 3 \\ 1 \\ 2 \end{bmatrix},$$

or

$$X = \begin{bmatrix} 2s + 3t + 2 \\ s + 3t + 3 \\ t + 6 \\ 3s + 2t + 1 \end{bmatrix}.$$

Example 3. Given a line and plane in ordinary 3-space, it makes sense to ask whether or not they intersect, and if so, where? It is constructive to consider the same question in R^4[1]: Do the line of Example 1 and the plane of Example 2 intersect? If so, what is the point (vector) of intersection?

Solution. Let L and P denote the line and plane under consideration. We have available the following vector formulas for L and P:

$$L: \quad X = \begin{bmatrix} t+1 \\ 2t+2 \\ t+3 \\ -t+4 \end{bmatrix}; \quad P: \quad X = \begin{bmatrix} 2s+3t+2 \\ s+3t+3 \\ t+6 \\ 3s+2t+1 \end{bmatrix}.$$

In terms of these formulas the first question takes the following form: Is there a vector

$$\begin{bmatrix} a \\ b \\ c \\ d \end{bmatrix}$$

that arises from each of these formulas?

It would be natural, perhaps, to resolve this question by examining for solutions the equation

$$\begin{bmatrix} t+1 \\ 2t+2 \\ t+3 \\ -t+4 \end{bmatrix} = \begin{bmatrix} 2s+3t+2 \\ s+3t+3 \\ t+6 \\ 3s+2t+1 \end{bmatrix}.$$

But this would be wrong. To insist that there were such an s and t (for there to be an intersection point) would be to insist that a point of intersection had to arise from both formulas for the very same choice of t. This is too restrictive.

2.11 LINES AND PLANES

A correct procedure is as follows: Suppose that A^* is an intersection point. Suppose further that A^* arises from the line's formula for the choice $t = \alpha$, and from the plane's formula for the choice $s = \beta$ and $t = \gamma$. That is, we have

$$A^* = \begin{bmatrix} \alpha + 1 \\ 2\alpha + 2 \\ \alpha + 3 \\ -\alpha + 4 \end{bmatrix} \quad \text{and} \quad A^* = \begin{bmatrix} 2\beta + 3\gamma + 2 \\ \beta + 3\gamma + 2 \\ \gamma + 6 \\ 3\beta + 2\gamma + 1 \end{bmatrix}.$$

It would then follow that

$$\begin{bmatrix} \alpha + 1 \\ 2\alpha + 2 \\ \alpha + 3 \\ -\alpha + 4 \end{bmatrix} = \begin{bmatrix} 2\beta + 3\gamma + 2 \\ \beta + 3\gamma + 3 \\ \gamma + 6 \\ 3\beta + 2\gamma + 1 \end{bmatrix}.$$

But this argument is reversible. Thus, any triple (α, β, γ) that satisfies this last equation will lead us to an intersection point. (Either substitute α in the formula for L, or β and γ in the formula for P.)

To find values for α, β, and γ, we first put our equation into a more familiar form:

$$\begin{bmatrix} 1 & -2 & -3 \\ 2 & -1 & -3 \\ 1 & 0 & -1 \\ -1 & -3 & -2 \end{bmatrix} \begin{bmatrix} \alpha \\ \beta \\ \gamma \end{bmatrix} = \begin{bmatrix} 1 \\ 1 \\ 3 \\ -3 \end{bmatrix}.$$

Our next move would be to reduce the augmented matrix to row echelon form. On the way, however, we discover that

$$\begin{bmatrix} 1 & -2 & -3 & 1 \\ 2 & -1 & -3 & 1 \\ 1 & 0 & -1 & 3 \\ -1 & -3 & -2 & -3 \end{bmatrix} \stackrel{\text{(row)}}{=} \begin{bmatrix} 1 & -2 & -3 & 1 \\ 0 & 1 & 1 & -\frac{1}{3} \\ 0 & 0 & 0 & \frac{8}{3} \\ 0 & -5 & -5 & -2 \end{bmatrix}.$$

It follows—look at the third row—that our system has no solution. Thus, L and P do not intersect. □

Does the foregoing result suggest that the lines and 2-planes of R^{41} do not have properties analogous to ordinary lines and planes? Have we gone too far in introducing this geometric language into higher dimensional vector spaces? Not at all. For even in ordinary 3-space a line and plane do not have to intersect; the line could be parallel to the plane. This is precisely what has happened to us here; our line L is parallel, in some sense, to the plane P.

To understand this we must first decide what it means for a line in R^{41} to be parallel to a 2-plane in R^{41}. As always, we look at a picture.

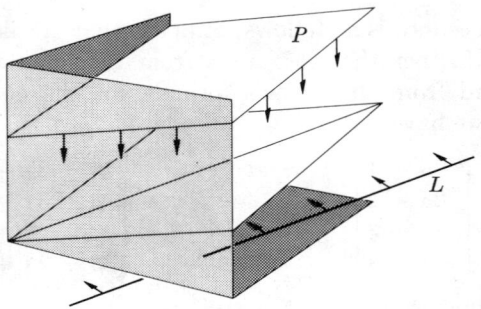

Figure 2.10

Consider Fig. 2.10, and suppose that L is parallel to P. Then, by an appropriate translation, the line L can be made to lie in the plane P. And if both plane and line are translated appropriately, we can have line and plane both passing through the origin. In this favorable position, the line may be viewed as a vector subspace of the plane. We are thus led to the following analytic description of parallelism: If the vector C is in the space spanned by D and E, then the line

$$X = A + tC$$

is parallel to the 2-plane

$$X = B + sD + tE.$$

(This definition does not preclude the possibility of the parallel line lying in the 2-plane.)

Now recall the vectors C, D, and E of Examples 1 and 2. We have

$$C = \begin{bmatrix} 1 \\ 2 \\ 1 \\ -1 \end{bmatrix}, \quad D = \begin{bmatrix} 2 \\ 1 \\ 0 \\ 3 \end{bmatrix}, \quad \text{and} \quad E = \begin{bmatrix} 3 \\ 3 \\ 1 \\ 2 \end{bmatrix}.$$

It is easy to see that $C = E - D$. It follows that every multiple of C lies in the space spanned by D and E. Thus, our line and our plane are parallel.

Let us review our discussion. It was shown that the line L fails to intersect the 2-plane P. It was then shown—and this fact seemed to justify the lack of an intersection point—that L is parallel to P. Now to stop here might lead the reader to a wrong conclusion: he might conclude that another line, one not parallel to P, would be obliged to intersect P. This would be incorrect! To see that this is so, recall the system in α, β, and γ upon which the existence of an intersection point depended. This system has four equations but only three unknowns. It is an unusual occurrence for such a system to have a solution. Thus, given at random a line and a 2-plane in R^4[1] it is not to be expected that they should intersect.

2.11   LINES AND PLANES

The situation described above is much like that involving two lines in ordinary 3-space. They can fail to intersect because they are parallel, and they can also fail to intersect because they are *skew*. It happens that in R^{41} (or any 4-dimensional vector space) a line can be parallel to a 2-plane and it can also be *skew* to a 2-plane. The anticipated relationship between a line and a plane arises in R^{41} only when we consider a line along with a 3-plane. It then turns out—the intersection-determining system now has four equations in four unknowns—that the line will either intersect the 3-plane in a single point or be parallel to it (see Exercise 2).

EXERCISE SET 2.11

1. Let L_1 be the line in R^{41} that passes through $[-3, -4, 1, -4]'$ and $[0, 2, 10, 8]'$. Let L_2 be the line that passes through $[1, 2, 3, 4]'$ and $[2, 4, 4, 2]'$.
   a) Find a vector formula for L_1. Show that it is not parallel to the 2-plane P of Example 2, and show also that it intersects that 2-plane in a single point. Find that point.
   b) Repeat part (a) for the line L_2, only this time show that L_2 does not intersect P.

2. Let P be the 3-plane in R^{41} described by the vector formula

$$X = \begin{bmatrix} 1 \\ 2 \\ 3 \\ 4 \end{bmatrix} + s \begin{bmatrix} 1 \\ 2 \\ 1 \\ 0 \end{bmatrix} + t \begin{bmatrix} 2 \\ 2 \\ 1 \\ 1 \end{bmatrix} + u \begin{bmatrix} 3 \\ 4 \\ 5 \\ 2 \end{bmatrix}.$$

   a) Let L_1 be the line described by

$$X^1 = \begin{bmatrix} 2 \\ 0 \\ 2 \\ 4 \end{bmatrix} + r \begin{bmatrix} 1 \\ 4 \\ 2 \\ 1 \end{bmatrix}.$$

   Show that L intersects P and find the intersection point.
   b) Let L_2 be the line described by

$$X^2 = \begin{bmatrix} 1 \\ 2 \\ 3 \\ 3 \end{bmatrix} + r \begin{bmatrix} 3 \\ 6 \\ 6 \\ 1 \end{bmatrix}.$$

   Show that L does not intersect P.
   c) Let L_3 be the line described by

$$X^3 = \begin{bmatrix} 1 \\ 2 \\ 3 \\ 4 \end{bmatrix} + r \begin{bmatrix} 3 \\ 6 \\ 6 \\ 1 \end{bmatrix}.$$

   Show that X lies in P.

d) Define the following: The line L described by $X = A + tB$ is *parallel* to the 3-plane P described by $X = C + sD + tE + uF$. Then show that L_2 and L_3 are both parallel to the 3-plane of this exercise.

3 Let P_1 be the 2-plane in R^{31} defined by

$$X^1 = \begin{bmatrix} 1 \\ 2 \\ 3 \end{bmatrix} + s \begin{bmatrix} 1 \\ 2 \\ 0 \end{bmatrix} + t \begin{bmatrix} 1 \\ 1 \\ 2 \end{bmatrix}.$$

Let P_2 be defined by

$$X^2 = \begin{bmatrix} 2 \\ 2 \\ 7 \end{bmatrix} + s \begin{bmatrix} 2 \\ 3 \\ 2 \end{bmatrix} + t \begin{bmatrix} 1 \\ 4 \\ -4 \end{bmatrix}.$$

Show that P_1 and P_2 are identical.

4 The discussion in the text is restricted to the representation of k-planes by vector formulas; it is possible, by introducing a coordinate system, to represent a k-plane by a *coordinate equation*.

a) Let S be the 2-plane in R^{31} defined by

$$X = \begin{bmatrix} 1 \\ 2 \\ 3 \end{bmatrix} + s \begin{bmatrix} 1 \\ 0 \\ 2 \end{bmatrix} + t \begin{bmatrix} 0 \\ 1 \\ 1 \end{bmatrix}.$$

Show that the natural coordinates of every X in S satisfy a certain equation of the form $ax_1 + bx_2 + cx_3 = d$. [*Hint:* Let $[\alpha, \beta, \gamma]'$ be on S. It follows that the equation $X = [\alpha, \beta, \gamma]'$ has a solution. Consider the corresponding augmented matrix.]

b) Let S be the set of vectors in R^{31} whose natural coordinates satisfy the equation

$$2x_1 + 5x_2 + 3x_3 = 4.$$

Show that S is a 2-plane. [*Hint:* Let $x_1 = s$, $x_2 = t$, and solve for x_3. Then determine A, B, and C such that $X = A + sB + tC$.]

5 Three lines in R^{41} are defined by

$$X = t\begin{bmatrix} 1 \\ 2 \\ 1 \\ 2 \end{bmatrix} + \begin{bmatrix} 2 \\ 0 \\ -1 \\ 2 \end{bmatrix}, \quad X = t\begin{bmatrix} 7 \\ 0 \\ 0 \\ 7 \end{bmatrix} + \begin{bmatrix} 1 \\ -2 \\ -2 \\ 0 \end{bmatrix}, \quad X = t\begin{bmatrix} 1 \\ 1 \\ 2 \\ 5 \end{bmatrix} + \begin{bmatrix} 5 \\ 2 \\ 6 \\ 20 \end{bmatrix}.$$

Show that all three lines intersect in a single point. Find that point.

6 a) Consider the vector equation

$$t_1 A + t_2 B + t_3 C + D = t_4 E + F,$$

where A, B, C, D, E, and F are given vectors in R^{41}. Also, it is given that the set $\{A, B, C\}$ is linearly independent. Now, what condition must be imposed upon E to prevent this equation from having a unique solution $[t_1, t_2, t_3, t_4]$? [*Hint:* The reader may use the as yet unproven fact that the row rank of a matrix is equal to its column rank.]

b) Prove: If L is a line in R^{41} and P is a 3-plane in R^{41}, then either L intersects P in a single point or L is parallel to P.

CHAPTER 3

LINEAR TRANSFORMATIONS

3.1 LINEAR TRANSFORMATIONS

Corresponding to each real number c, there is a real function f defined by $f(x) = cx$. Graphically, f is a line through the origin having slope c. (We are assuming, of course, that the graph of f is displayed on a plane that has been coordinatized in the natural manner.) Analytically, f is distinguished from other continuous functions by the following properties: For any two numbers a and b, we have both

$$f(a + b) = f(a) + f(b)$$

and

$$f(ab) = af(b).$$

That is, if we restrict our attention to continuous functions, then the "multiples of x" are the only ones having these properties. (We shall not pause to prove this.)

This observation is made especially interesting by the realization that many students of high school mathematics are eager to assume that any newly introduced function has these properties. Consider the following list of "false identities" and recall past sins:

$$\text{``}(x + y)^2 = x^2 + y^2\text{,''}$$
$$\text{``}\sin (x + y) = \sin x + \sin y\text{,''}$$
$$\text{``}\sin 2x = 2 \sin x\text{,''}$$
$$\text{``}\log (x + y) = \log x + \log y\text{.''}$$

On the other hand, if we widen our outlook, considering functions that assign vectors to vectors, then we do come upon a variety of functions that have the special properties of $f(x) = cx$.

Example 1. Let V be the set of all real functions $f(t)$ that are differentiable for all t. Then V, along with the usual operations for adding functions and multiplying functions by numbers, is a vector space. Now let D be the function defined on V by the following rule: If f belongs to V, then D assigns to f its derivative. For example,

$$D(t^2) = 2t,$$
$$D(\sin t) = \cos t,$$
$$D(e^{3t}) = 3e^{3t}.$$

Then, for any two members of V and for any number c, we may write

$$D(f + g) = D(f) + D(g)$$

and

$$D(cf) = cD(f).$$

These are simply the familiar differentiation rules as expressed in terms of the function D.

It should be observed that the range of V, call it W, is not identical with V. In particular, $D(t^{4/3}) = \frac{4}{3}t^{1/3}$, and this latter function is not differentiable for all t. Thus, the function having the form $t^{4/3}$ is in V while that assigned to it by D is not.

Now although the range of D is not identical with the vector space V, it is nevertheless a vector space in its own right. For suppose that f and g both belong to W. This means that there are functions F and G in V such that $D(F) = f$ and $D(G) = g$. Since V is a vector space, we are assured that the functions $F + G$ and cF are also in V. Thus, it makes sense to talk about $D(F + G)$ and $D(cF)$. Now, by the properties of D noted earlier—those special properties of $f(x) = cx$—we may write

$$D(F + G) = f + g \quad \text{and} \quad D(cF) = cf.$$

It follows that the functions $f + g$ and cf belong to W. This shows that W is a subspace of W^*, where W^* is the set of real functions that are defined for all t along with the usual function operations.

Before leaving this example, it should be noted that the function D has no inverse; or, which is to say the same thing, the function D is not one-to-one. This follows, in particular, from the observation that

$$D(t^2) = 2t \quad \text{and} \quad D(t^2 + 39) = 2t.$$

Example 2. Let L be defined on R^{22} by

$$L\left(\begin{bmatrix} \alpha & \beta \\ \gamma & \delta \end{bmatrix}\right) = \begin{bmatrix} 1 & 3 \\ 2 & 7 \end{bmatrix} \begin{bmatrix} \alpha & \beta \\ \gamma & \delta \end{bmatrix}.$$

3.1 LINEAR TRANSFORMATIONS

For example,
$$L\left(\begin{bmatrix} 3 & -2 \\ 1 & 4 \end{bmatrix}\right) = \begin{bmatrix} 1 & 3 \\ 2 & 7 \end{bmatrix}\begin{bmatrix} 3 & -2 \\ 1 & 4 \end{bmatrix} = \begin{bmatrix} 6 & 10 \\ 13 & 24 \end{bmatrix}.$$

Now, let A and B be any two vectors in R^{22}, and let c be any number. Then, using only certain properties of matrix arithmetic, we may write

$$L(A + B) = \begin{bmatrix} 1 & 3 \\ 2 & 7 \end{bmatrix}(A + B) = \begin{bmatrix} 1 & 3 \\ 2 & 7 \end{bmatrix}A + \begin{bmatrix} 1 & 3 \\ 2 & 7 \end{bmatrix}B = L(A) + L(B)$$

and

$$L(cA) = \begin{bmatrix} 1 & 3 \\ 2 & 7 \end{bmatrix}(cA) = c\left(\begin{bmatrix} 1 & 3 \\ 2 & 7 \end{bmatrix}A\right) = cL(A).$$

Thus, L also has the special properties of $f(x) = cx$.

In contrast to the function D of Example 1, the function L is one-to-one, and, consequently, it has an inverse L^{-1}. This follows from the fact that
$$\begin{bmatrix} 1 & 3 \\ 2 & 7 \end{bmatrix}$$
is nonsingular; indeed,
$$\begin{bmatrix} 1 & 3 \\ 2 & 7 \end{bmatrix}^{-1} = \begin{bmatrix} 7 & -3 \\ -2 & 1 \end{bmatrix}.$$

For suppose that $L(A) = L(B)$, where $A \neq B$. That is, we have
$$\begin{bmatrix} 1 & 3 \\ 2 & 7 \end{bmatrix}A = \begin{bmatrix} 1 & 3 \\ 2 & 7 \end{bmatrix}B.$$

Then premultiplying both sides by
$$\begin{bmatrix} 7 & -3 \\ -2 & 1 \end{bmatrix}$$
produces $A = B$: a contradiction. Thus, L cannot assign the same vector to different members of R^{22}. Finally, and the reader may verify this, a formula for L^{-1} is given by
$$L^{-1}(Y) = \begin{bmatrix} 7 & -3 \\ -2 & 1 \end{bmatrix}Y.$$

Definition 3.1. Let V and W both be vector spaces. Let L be a function that assigns to each vector in V some vector in W. (Not all of W need be involved.) Suppose also that L has the following properties: For any two vectors X and Y in V, and for any scalar c, we have
1. $L(X + Y) = L(X) + L(Y)$.
2. $L(cX) = cL(X)$.

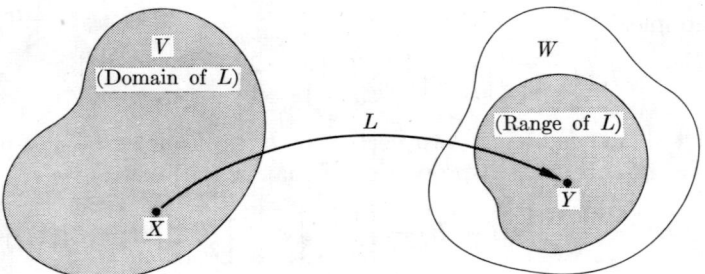

Figure 3.1

Then L is said to be a *linear transformation from V into W* (see Fig. 3.1). (If it is clear that every vector in W is used, that is, that the range of L is exactly W, then we replace the word "into" by "onto.")

Theorem 3.1. The range of a linear transformation is a vector space.

Proof. This is proved in Example 1 for the linear transformation D of that example. The proof given there did not in any way depend on the special nature of D, but only on those properties that make D a linear transformation.)

Theorem 3.2. If a linear transformation L has an inverse, then that inverse is also a linear transformation.

Proof. Let X and Y belong to the range space of L. Then $X + Y$ also belongs. Thus, we may write

$$L(L^{-1}(X + Y)) = X + Y,$$

and, because L is linear,

$$L(L^{-1}(X) + L^{-1}(Y)) = L(L^{-1}(X)) + L(L^{-1}(Y))$$
$$= X + Y.$$

It follows, because L assigns the vector $X + Y$ only once, that

$$L^{-1}(X + Y) = L^{-1}(X) + L^{-1}(Y).$$

Similarly, the reader may show that

$$L^{-1}(cX) = cL^{-1}(X).$$

Thus, L^{-1} is a linear transformation.

Theorem 3.3. Let L be a linear transformation defined in a vector space V, where V has the basis $\mathcal{B} = [B^1, B^2, \ldots, B^n]$. If we know $L(B^1)$, $L(B^2), \ldots, L(B^n)$, then we can determine $L(X)$ for any X in V. (Briefly: A linear transformation is completely determined by its action on a set of basis vectors.)

Proof. Let X be in V. Then $X = b_1 B^1 + b_2 B^2 + \cdots + b_n B^n$, and
$$L(X) = L(b_1 B^1 + b_2 B^2 + \cdots + b_n B^n)$$
$$= L(b_1 B^1) + L(b_2 B^2) + \cdots + L(b_n B^n)$$
$$= b_1 L(B^1) + b_2 L(B^2) + \cdots + b_n L(B^n).$$

(Give reasons for the last two steps.) It follows that we know $L(X)$ if we know $L(B^j)$ for $j = 1, 2, \ldots, n$.

EXERCISE SET 3.1

1 Let L be defined on R^{21} by
$$L\left(\begin{bmatrix} x_1 \\ x_2 \end{bmatrix}\right) = \begin{bmatrix} 2x_1 + 3x_2 \\ -x_1 + 5x_2 \end{bmatrix}.$$
a) Find $L([1, 4]')$.
b) Find a matrix A such that $L(X) = AX$ for all X in R^{21}.
c) Prove that L is a linear transformation.

2 Let F be defined on R^{21} by
$$F\left(\begin{bmatrix} x_1 \\ x_2 \end{bmatrix}\right) = \begin{bmatrix} 2x_1 + 3x_2 \\ 5x_1 x_2 \end{bmatrix}.$$
Prove that F is not a linear transformation. [*Hint:* Find a vector X and a number c such that $F(cX) \neq cF(X)$.]

3 Let L be the linear transformation defined on R^{31} by
$$L(X) = \begin{bmatrix} 3 & 2 & 1 \\ 6 & 4 & 2 \end{bmatrix} X.$$
a) Find $L([a, b, c]')$.
b) Find a spanning set for the range of L.
c) What is the dimension of the range of L?

4 Let D be the differentiation transformation of Example 1. Let L be defined on P^3 by
$$L[p(t)] = t^2 D[p(t)].$$
a) Find $L(t^3 + 2t^2 + 7)$.
b) Prove that L is a linear transformation.
c) Find the dimension of the range space of L. [*Hint:* Express $L[p(t)]$ as a linear combination of vectors in P^4.]

5 Let L be a linear transformation with domain R^{22}. Suppose that only the following is known about L:

$$L\left(\begin{bmatrix} 1 & 1 \\ 1 & 1 \end{bmatrix}\right) = \begin{bmatrix} 2 \\ 3 \\ 4 \end{bmatrix}, \quad L\left(\begin{bmatrix} 0 & 1 \\ 1 & 1 \end{bmatrix}\right) = \begin{bmatrix} 0 \\ 0 \\ 0 \end{bmatrix},$$

$$L\left(\begin{bmatrix} 0 & 0 \\ 1 & 1 \end{bmatrix}\right) = \begin{bmatrix} 1 \\ -1 \\ -2 \end{bmatrix}, \quad \text{and} \quad L\left(\begin{bmatrix} 0 & 0 \\ 0 & 1 \end{bmatrix}\right) = \begin{bmatrix} 3 \\ 2 \\ 2 \end{bmatrix}.$$

a) Find $L\left(\begin{bmatrix} 5 & 6 \\ 3 & 7 \end{bmatrix}\right)$.

b) Find a basis for the range space of L.

6 a) Describe the relation, if any, between the notion of an isomorphism between vector spaces and that of a linear transformation

b) Prove: If L is a linear transformation whose range dimension is less than its domain dimension, then L is not one-to-one.

7 Prove the following: If L is a linear transformation from V into W, then

a) $L(\Theta_V) = \Theta_W$. b) $L(X - Y) = L(X) - L(Y)$.

8 a) Prove: If $[B^1, B^2, \ldots, B^n]$ is a basis for V, then $[L(B^1), L(B^2), \ldots, L(B^n)]$ is a spanning set for the range of V.

b) Use the foregoing theorem to find a spanning set for the range of the transformation L found in Exercise 4.

c) Prove: The dimension of the range of a linear transformation never exceeds that of its domain.

9 Let L be the linear transformation defined on R^{12} by

$$L(X) = X \begin{bmatrix} 1 & 2 \\ 1 & 3 \end{bmatrix}.$$

a) Find $L([2, 7])$.

b) Let $Y = L(X)$. Find a matrix A such that $X = YA$.

c) Prove that L is one-to-one. [*Hint:* Start with the assertion $L(X) = L(Y)$.]

d) Write a formula for $L^{-1}(Y)$ that is analogous to that given for $L(X)$.

10 Let V be the vector space consisting of all real functions that are continuous on the closed interval [0, 1]. Let L be defined on V by

$$L[f(t)] = f(1) \cdot t + \int_0^1 f(t)\, dt.$$

a) Find $L(1/(t+1))$. b) Prove that L is linear.

c) What are the dimensions of the domain and range of L?

11 Let L be defined on P^2 by $L[p(t)] = (t^2 + 1)p(t)$.

a) Prove that L is linear. b) Find the dimension of its range space.

c) Prove that L is one-to-one and then find a formula for its inverse.

12 Prove that the two conditions listed under Definition 3.1 are equivalent to the single condition $L(aX + bY) = aL(X) + bL(Y)$.

3.2 THE KERNEL OF A LINEAR TRANSFORMATION

A function f is one-to-one if no range element is assumed more than once by f. Thus, a proof of one-to-oneness may be carried out as follows: Let b be any element in the range of f. Then show that the equation $f(x) = b$ has exactly one solution, say a. This shows that b arises in the range of f just once, i.e., as the mate of a.

For linear transformations—because of their special properties—a simpler proof is available. To show that L is one-to-one, we have only to show that the equation $L(X) = \Theta_W$ has a single solution, that solution necessarily being Θ_V (recall Exercise 7, Section 3.1). We reason as follows: Let L have domain V and range W, and suppose that $L(X) = \Theta_W$ has only Θ_V as a solution. Now take B as any vector in W. We wish to show that the equation $L(X) = B$ has only one solution. Well, suppose that there are two (distinct) solutions, say A_1 and A_2. Then we could write

$$L(A_1 - A_2) = L(A_1) - L(A_2) = B - B = \Theta_W.$$

However, only $L(\Theta_V) = \Theta_W$. Therefore, $A_1 - A_2 = \Theta_V$. But $A_1 \neq A_2$!

Definition 3.2. Let L be a linear transformation with domain V and range W. An equation of the form $L(X) = B$ is said to be *homogeneous* or *nonhomogeneous* according as $B = \Theta_W$ or $B \neq \Theta_W$. By the *kernel* of L we mean the solution set of the homogeneous equation $L(X) = \Theta_W$.

Theorem 3.4. If the kernel of L contains only the zero-vector, then L is one-to-one. And conversely.

Proof. The first part of this theorem has already been proved in the discussion preceding Definition 3.2. The converse part is completely trivial (Explain.)

Example 1. Let V be the subspace of P^3 spanned by t^3 and t^2. Thus, the polynomials in V are of the form $at^3 + bt^2$. Define L on V by the formula $L[p(t)] = p'(t)$. Now prove that L is a one-to-one transformation, and obtain a formula for L^{-1}.

Solution. Let $p(t) = at^3 + bt^2$. Then $L[p(t)] = 3at^2 + 2bt$. It follows that $L[p(t)] = \Theta$ only if $a = 0$ and $b = 0$. Thus, the zero polynomial is the only member of V that satisfies the equation $L(X) = \Theta$. In other words, the kernel of L contains only the zero vector. By Theorem 3.4, the transformation L is one-to-one.

Let $q(t)$ be in the range of L. Then $L^{-1}[q(t)]$ is that polynomial in V whose derivative is $q(t)$. This leads us to the following "verbal formula" for L^{-1}: $L^{-1}[q(t)]$ is that antiderivative of $q(t)$ obtained by choosing 0 as

the so-called constant of integration. A more elegant formula is the following:
$$L^{-1}[q(t)] = \int_0^t q(u)\,du.$$
For example,
$$L(2t^3 - 7t^2) = 6t^2 - 14t$$
and
$$L^{-1}(6t^2 - 14t) = \int_0^t (6u^2 - 14u)\,du$$
$$= (2u^3 - 7u^2)\big|_0^t$$
$$= 2t^3 - 7t^2.$$

Example 2. Let L be defined on R^{31} by
$$L(X) = \begin{bmatrix} 1 & 1 & 3 \\ 1 & 2 & 4 \\ 2 & 3 & 7 \end{bmatrix} X.$$

Show that K, the kernel of L, is a subspace of R^{31}, and find its dimension.

Solution. By definition, K is the solution set of the homogeneous equation
$$\begin{bmatrix} 1 & 1 & 3 \\ 1 & 2 & 4 \\ 2 & 3 & 7 \end{bmatrix} X = \begin{bmatrix} 0 \\ 0 \\ 0 \end{bmatrix}.$$

The reader may verify that
$$\begin{bmatrix} 1 & 1 & 3 & 0 \\ 1 & 2 & 4 & 0 \\ 2 & 3 & 7 & 0 \end{bmatrix} \overset{(\text{row})}{=} \begin{bmatrix} 1 & 0 & 2 & 0 \\ 0 & 1 & 1 & 0 \\ 0 & 0 & 0 & 0 \end{bmatrix}.$$

It follows that
$$X = \begin{bmatrix} -2t \\ -t \\ t \end{bmatrix} = t \begin{bmatrix} -2 \\ -1 \\ 1 \end{bmatrix}.$$

Thus, K is an infinite set of vectors. (It follows that L has no inverse.) Moreover, because K is the set of all multiples of a single vector, K is a subspace of the domain of L. Since K has the basis $\{[-2, -1, 1]'\}$, the dimension of K is 1. (That K is a subspace of the domain of L is not accidental.)

Theorem 3.5. The kernel of L is a subspace of the domain of L.

Proof. See Exercise 5. □

The domain of a linear transformation is a vector space (by Definition 3.1); the range of a linear transformation is a vector space (by Theorem

3.1); and now, the kernel of a linear transformation is a vector space. If these spaces are all of finite dimension, then it turns out that there is an interesting numerical relationship between the dimensions of these spaces. We show this relationship first for a specific transformation.

Example 3. Let L be the transformation of Example 2. Find the dimension of its range space.

Solution. For arbitrary X in R^{31}, we have

$$L(X) = \begin{bmatrix} 1 & 1 & 3 \\ 1 & 2 & 4 \\ 2 & 3 & 7 \end{bmatrix} \begin{bmatrix} x_1 \\ x_2 \\ x_3 \end{bmatrix} = \begin{bmatrix} (x_1 + x_2 + 3x_3) \\ (x_1 + 2x_2 + 4x_3) \\ (2x_1 + 3x_2 + 7x_3) \end{bmatrix}$$

$$= x_1 \begin{bmatrix} 1 \\ 1 \\ 2 \end{bmatrix} + x_2 \begin{bmatrix} 1 \\ 2 \\ 3 \end{bmatrix} + x_3 \begin{bmatrix} 3 \\ 4 \\ 7 \end{bmatrix}.$$

It follows that the range of L is the space spanned by

$$S = \left\{ \begin{bmatrix} 1 \\ 1 \\ 2 \end{bmatrix}, \begin{bmatrix} 1 \\ 2 \\ 3 \end{bmatrix}, \begin{bmatrix} 3 \\ 4 \\ 7 \end{bmatrix} \right\}.$$

Now S is linearly dependent: $S^3 = 2S^1 + S^2$. But

$$R = \left\{ \begin{bmatrix} 1 \\ 1 \\ 2 \end{bmatrix}, \begin{bmatrix} 1 \\ 2 \\ 3 \end{bmatrix} \right\}$$

is linearly independent. Thus, R is a basis for the range of L (by Theorem 2.12). It follows that the range of L has dimension 2.

Finally, recalling the result of Example 2, we note the following numerical relationship:

Domain dimension = range dimension + kernel dimension.

Theorem 3.6. *The dimension theorem.* Let L be a linear transformation with domain V and range W, spaces of dimensions n and m, respectively. Let K, the kernel of L, have dimension k. Then

$$n = m + k.$$

Proof. Let $\{X^1, X^2, \ldots, X^k\}$ be a basis for K and let $\{X^1, \ldots, X^k, X^{k+1}, \ldots, X^n\}$ be a basis for V. (That such a basis for V exists is easily justified. Let S be the union of the given basis for K with any basis for V. Then S surely spans V. Now select a maximal independent subset of S, starting with the vectors $X^1, X^2, \ldots, X^k$.) The theorem is proved

by showing that the set
$$S^* = \{L(X^{k+1}), L(X^{k+2}), \ldots, L(X^n)\}$$
is a basis for the range of L. (Why?)

1. S^* *is a spanning set for* W. Let Y be any vector in W. Then $Y = L(X)$ for some X in V. It follows that we may write
$$X = a_1 X^1 + \cdots + a_k X^k + a_{k+1} X^{k+1} + \cdots + a_n X^n.$$
But then
$$\begin{aligned} Y = L(X) &= L(a_1 X^1 + \cdots + a_k X^k) + L(a_{k+1} X^{k+1} + \cdots + a_n X^n) \\ &= \Theta_W + L(a_{k+1} X^{k+1} + \cdots + a_n X^n) \\ &= a_{k+1} L(X^{k+1}) + \cdots + a_n L(X^n). \end{aligned}$$
Thus, each Y in W is expressible as a linear combination of the vectors in S^*.

2. S^* *is linearly independent.* Suppose that scalars $a_{k+1}, a_{k+2}, \ldots, a_n$ are chosen so that
$$a_{k+1} L(X^{k+1}) + \cdots + a_n L(X^n) = \Theta_W.$$
Then, equivalently, we have
$$L(a_{k+1} X^{k+1} + \cdots + a_n X^n) = \Theta_W.$$
This means that the vector $Y = a_{k+1} X^{k+1} + \cdots + a_n X^n$ is in K, the kernel of L. As such, it must be expressible as a linear combination of the vectors X^1 through X^k. Thus, there are scalars $a_1, a_2, \ldots, a_k$ such that
$$Y = a_1 X^1 + \cdots + a_k X^k.$$
But then, exploiting our two representations for Y, we may write
$$\Theta_V = Y - Y = a_1 X^1 + \cdots + a_k X^k - a_{k+1} X^{k+1} - \cdots - a_n X^n.$$
Since the vectors X^1 through X^n constitute a linearly independent set, we may conclude that each $a_j = 0$. In particular,
$$a_{k+1} = a_{k+2} = \cdots = a_n = 0.$$
It follows, that S^* is linearly independent.

Example 4. Let L be defined on R^{31} by
$$L\left(\begin{bmatrix} x_1 \\ x_2 \\ x_3 \end{bmatrix}\right) = \begin{bmatrix} x_1 \\ x_2 \\ 0 \end{bmatrix}.$$

3.2 THE KERNEL OF A LINEAR TRANSFORMATION

Since
$$\begin{bmatrix} x_1 \\ x_2 \\ 0 \end{bmatrix} = x_1 \begin{bmatrix} 1 \\ 0 \\ 0 \end{bmatrix} + x_2 \begin{bmatrix} 0 \\ 1 \\ 0 \end{bmatrix},$$

it follows immediately that the range space of L has dimension 2.

Next, we set $L(X) = \Theta$, that is,
$$\begin{bmatrix} x_1 \\ x_2 \\ 0 \end{bmatrix} = \begin{bmatrix} 0 \\ 0 \\ 0 \end{bmatrix}.$$

It follows that X is in the kernel of L if and only if its first two entries are zeros; in other words, X must be of the form
$$\begin{bmatrix} 0 \\ 0 \\ x_3 \end{bmatrix} = x_3 \begin{bmatrix} 0 \\ 0 \\ 1 \end{bmatrix}.$$

Thus, the kernel of L has dimension 1.

The vector spaces here associated with L have a nice arrow interpretation. Taking N^1, N^2, and N^3 as in Fig. 3.2, the arrow $L(X)$ may be viewed as the "shadow" of X when projected upon the x_1x_2-plane by a light source "above" that plane. The range and kernel of L assume the form of a plane and a line, respectively; their dimensions add up to three.

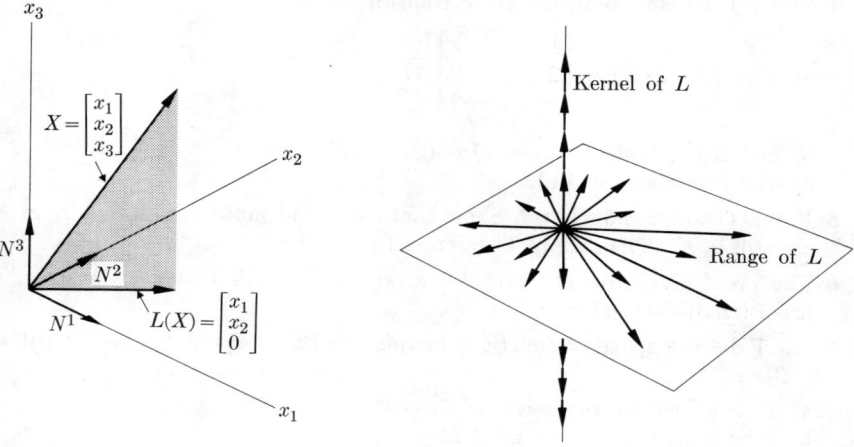

Figure 3.2

EXERCISE SET 3.2

1. Solve the equation $L(X) = \theta$ and thence determine whether or not L is one-to-one.
   a) $L(X) = [1, 2]X$, X in R^{21}.
   b) $L(X) = 77X$, X in R^{22}.
   c) $L[p(t)] = p'(t)$, p in P^{∞}.

2. For each of the linear transformations given below, make arrow sketches analogous to those of Fig. 3.2.

   a) $L\left(\begin{bmatrix} x_1 \\ x_2 \end{bmatrix}\right) = \begin{bmatrix} x_1 \\ 0 \end{bmatrix}$.

   b) $L\left(\begin{bmatrix} x_1 \\ x_2 \end{bmatrix}\right) = \frac{1}{2}\left(\begin{bmatrix} x_1 \\ x_2 \end{bmatrix} + \begin{bmatrix} x_2 \\ x_1 \end{bmatrix}\right)$.

   c) $L\left(\begin{bmatrix} x_1 \\ x_2 \\ x_3 \end{bmatrix}\right) = \begin{bmatrix} 0 \\ 0 \\ x_3 \end{bmatrix}$.

3. Let V be the space of functions having the basis [sin t, cos t].
   a) Let L be defined on V by
   $$L[f(t)] = \int_0^{\pi} f(t)\, dt.$$
   Find the kernel of L.
   b) $L[f(t)] = \int_0^{2\pi} f(t)\, dt$. Find the kernel of L.
   c) $L[f(t)] = f'(t)$. Prove that L is one-to-one and find a formula for L^{-1}.

4. Verify Theorem 3.6 for the given transformation.

   a) X in R^{31}, $L(X) = \begin{bmatrix} 1 & 0 & 2 \\ 2 & 1 & 2 \\ 0 & 1 & -2 \end{bmatrix} X$.

   b) $p(t)$ in P^3, $L[p(t)] = (t - 1)p'''(t)$.
   c) The L of Exercise 3(a).

5. Prove Theorem 3.5. [*Hint:* Show that sums and multiples of vectors in K are still in K. Then apply Theorem 2.1.]

6. The proof given for Theorem 3.6 makes no sense if $k = n$. Construct a proof for this special case.

7. Let V be the space of functions having the basis $[e^t, e^{-t}, t]$. Let $L[f(t)] = f''(t) - f(t)$.
   a) Find a basis for the kernel of L.
   b) Find a basis for the range of L.
   c) Verify Theorem 3.6.

8 Same as Exercise 7, where L is defined on R^{22} by

$$L(X) = \begin{bmatrix} 1 & 2 \\ 2 & 4 \end{bmatrix} X.$$

9 In R^{13} a *cross product* of vectors is defined by

$$X \times Y = [x_2 y_3 - x_3 y_2, \; x_3 y_1 - x_1 y_3, \; x_1 y_2 - x_2 y_1].$$

Now, let L be defined over R^{13} by the formula $L(X) = X \times A$, where A is some nonzero vector in V.
a) Verify that L is a linear transformation.
b) Prove that the kernel of L is the space spanned by A.

10 Let V be the subspace of R^{31} that satisfies the equation $x_3 = 2x_1 + 5x_2$. Let L be defined on V by

$$L(X) = \begin{bmatrix} 1 & 0 & 2 \\ 2 & 1 & 1 \end{bmatrix} X.$$

a) What is the range of L?
b) Solve the equation $L(X) = [2, 0]'$
c) Prove that L is one-to-one.
d) Find a matrix A such that $L^{-1}(Y) = AY$.
e) Verify that $L^{-1}(L(X)) = X$ and $L(L^{-1}(Y)) = Y$.

3.3 LINEAR EQUATIONS

In the last section, under Definition 3.2, we introduced the terms *homogeneous equation* and *nonhomogeneous equation*. The reason for having such a case distinction is provided by the following theorem.

Theorem 3.7. Let L be a linear transformation from V into W, and suppose that A is a solution of the equation $L(X) = B$. Then every solution of this equation is obtained by adding to A the vectors belonging to the kernel of L. (In other words, all solutions of a nonhomogeneous equation may be found by finding just one solution and then adding to it all of the solutions of the corresponding homogeneous equation.)

Proof. There are two things that must be proved:
1. Let C be any vector belonging to the kernel of L; thus, $L(C) = \Theta$. Then, exploiting the linearity of L, we may write

$$L(A + C) = L(A) + L(C) = B + \Theta = B.$$

This shows that adding a kernel vector to A does indeed produce a new (provided that $C \neq \Theta$) solution for $L(X) = B$.

2. Let A^* be any solution of $L(X) = B$ other than A. Then we may write
$$\Theta = B - B = L(A^*) - L(A) = L(A^* - A).$$

It follows that $(A^* - A)$ is in the kernel of L. Denoting this vector by C^*, we have
$$A^* - A = C^*,$$
or
$$A^* = A + C^*.$$

This shows that every solution of $L(X) = B$ may be obtained by adding to A the appropriate kernel vector.

Example 1. Let V be the space of real functions that are *twice differentiable* on the interval $(-\infty, \infty)$, and let L be defined on V by
$$L[f(t)] = f''(t) - 3f'(t) + 2f(t).$$

Solve the nonhomogeneous equation
$$L[f(t)] = 6t^2 - 20t + 13.$$

Solution. Assuming no prior knowledge about equations of this sort, we have no idea as to whether or not the given equation has solutions. Assuming that it does, however, we do have (because of Theorem 3.7) a program for finding all such solutions:

1. We seek first a single solution of the given equation. We begin by asking a question: What sort of function would—on being combined with its first and second derivatives in the manner prescribed by L—be transformed into $6t^2 - 20t + 13$? Certainly, functions like $\sin t$, $\ln t$, and e^t seem to be unlikely candidates; rather, it would seem more reasonable to anticipate a polynomial. Thus, we make an educated guess: we try a function of the form $f(t) = at^2 + bt + c$. That is, we substitute this polynomial into the equation and then determine whether or not the coefficients a, b, and c can be chosen in such a way as to make $f(t)$ a solution. Several computations are required:

$$2f(t) = 2(at^2 + bt + c),$$
$$-3f'(t) = -3(2at + b),$$
$$f''(t) = 2a.$$

Thus,
$$L[f(t)] = f''(t) - 3f'(t) + 2f(t)$$
$$= (2a)t^2 + (2b - 6a)t + (2c - 3b + 2a).$$

It follows that $f(t)$ will be a solution if

$$\begin{cases} 2a & = 6, \\ -6a + 2b & = -20, \\ 2a - 3b + 2c & = 13. \end{cases}$$

The reader may verify that this system has exactly one solution: $a = 3$, $b = -1$, and $c = 2$. Thus,

$$f(t) = 3t^2 - t + 2$$

is one solution of the given nonhomogeneous equation.

2. We seek next the kernel of L, that is, the solution space of the homogeneous equation $L[f(t)] = \Theta$:

$$f''(t) - 3f'(t) + 2f(t) = \Theta.$$

Again we ask a leading question: What sort of function would in this combination with its first and second derivatives "cancel itself"? Answer: An exponential function, for derivatives of exponential functions are functions of the same species. Thus, we try $f(t) = e^{at}$:

$$2f(t) = 2(e^{at}),$$
$$-3f'(t) = -3(ae^{at}),$$

and

$$f''(t) = a^2 e^{at}.$$

It follows that

$$\begin{aligned} L[f(t)] &= 2e^{at} - 3ae^{at} + a^2 e^{at} \\ &= (a^2 - 3a + 2)e^{at}. \end{aligned}$$

Evidently, $f(t)$ will be a solution if

$$a^2 - 3a + 2 = 0,$$

that is, if $a = 1$ or $a = 2$.

We have thus shown that the functions e^t and e^{2t} are in the kernel of L. Now, it can be proved that the kernel of L, say K, has dimension 2. (This is not a trivial matter; it follows from the fact that the second derivative is the highest derivative involved in computing $L[f(t)]$.) Assuming this, and noting that e^t and e^{2t} are linearly independent, we may take $\mathcal{B} = [e^t, e^{2t}]$ as a basis for K.

3. Finally, appealing to Theorem 3.7, we see that all solutions of the nonhomogeneous equation are given by

$$f(t) = ae^t + be^{2t} + (3t^2 - t + 2).$$

Example 2. Let L be defined on R^{31} by

$$L(X) = \begin{bmatrix} 1 & 3 & 4 \\ 1 & 2 & 5 \end{bmatrix} X.$$

Solve the nonhomogeneous equation

$$L(X) = \begin{bmatrix} 3 \\ 5 \end{bmatrix}.$$

Solution

1. As in Example 1, we seek first a single solution of the given equation. Noting that the given equation is equivalent to the system

$$\begin{cases} x_1 + 3x_2 + 4x_3 = 3, \\ x_1 + 2x_2 + 5x_3 = 5, \end{cases}$$

we are led to the augmented matrix

$$\begin{bmatrix} 1 & 3 & 4 & 3 \\ 1 & 2 & 5 & 5 \end{bmatrix}.$$

Leaving the computations to the reader, we have

$$\begin{bmatrix} 1 & 3 & 4 & 3 \\ 1 & 2 & 5 & 5 \end{bmatrix} \stackrel{(\text{row})}{=} \begin{bmatrix} 1 & 0 & 7 & 9 \\ 0 & 1 & -1 & -2 \end{bmatrix}.$$

It follows that the original system is equivalent to

$$\begin{cases} x_1 = 9 - 7x_3, \\ x_2 = -2 + x_3. \end{cases}$$

Then, taking $x_3 = 0$ (or any convenient choice), we obtain the solution $X = [9, -2, 0]'$.

2. Proceeding as in part 1, we first determine the augmented matrix corresponding to the homogeneous equation $L(X) = \Theta$ and then reduce it to row echelon form:

$$\begin{bmatrix} 1 & 3 & 4 & 0 \\ 1 & 2 & 5 & 0 \end{bmatrix} \stackrel{(\text{row})}{=} \begin{bmatrix} 1 & 0 & 7 & 0 \\ 0 & 1 & -1 & 0 \end{bmatrix}.$$

The resulting system is

$$\begin{cases} x_1 + 7x_3 = 0, \\ x_2 - x_3 = 0. \end{cases}$$

Taking $x_3 = s$, we may describe all solutions of this system—all kernel vectors—by the formula

$$X = \begin{bmatrix} -7s \\ s \\ s \end{bmatrix} = s \begin{bmatrix} -7 \\ 1 \\ 1 \end{bmatrix}.$$

3. Finally, appealing to Theorem 3.7, we see that all solutions of the nonhomogeneous equation are given by

$$X = s \begin{bmatrix} -7 \\ 1 \\ 1 \end{bmatrix} + \begin{bmatrix} 9 \\ -2 \\ 0 \end{bmatrix}. \quad \square$$

It is instructive to compare Examples 1 and 2. In the case of Example 1, the application of Theorem 3.7 proved most useful. Indeed, this is the way such equations are solved in practice. On the other hand, the nonhomogeneous equation of Example 2 and the corresponding homogeneous one are both readily solved by reduction of the associated augmented matrices, the nonhomogeneous case being no more difficult than the homogeneous one. Thus, in this case the 2-step program based on Theorem 3.7 is really a misguided one.

However, and this is a very important point, we have shown how two very different kinds of equations can be solved by the same mode of attack. Why is this important? Well, suppose in the future that we come upon a new kind of equation. Then, trying to profit by our present experience, we might reasonably be expected to try to interpret that equation in the language of vector spaces and linear transformations. Suppose that such an interpretation is possible, and suppose further that the given equation is nonhomogeneous. Then, as in Example 1, the 2-step method of solution based on Theorem 3.7 may well be an effective one. Furthermore, in recognizing such a course of action, we demonstrate an appreciation for the mathematician's art of abstracting: for it is only by thinking in terms of abstract vector spaces and abstract linear transformations that we can "see" particular instances of these abstractions in a variety of contexts.

We now close this section with an application of the *dimension theorem;* we use it to prove a theorem which may have already been anticipated by the reader.

Theorem 3.8. The row space of a matrix has the same dimension as its column space. (Or, the row rank of a matrix equals its column rank.)

Proof. Let A be an m by n matrix, let L be the linear transformation from R^{n1} into R^{m1} defined by $L(X) = AX$, and suppose that the row rank of A is r. Then we may argue as follows:

1. If the row rank of A is r, then the row echelon reduction of A has r nonzero rows. It follows that the homogeneous equation $AX = \Theta$ is solved by solving for r coordinates of X in terms of the remaining $(n - r)$ coordinates. This means—since a solution of $AX = \Theta$ is a kernel vector—that a typical kernel vector may be expressed as a linear combination of $(n - r)$ columns from R^{n1}; moreover, this set

of columns is independent because each column has an entry 1 where the others have the entry 0. Thus, the dimension of the kernel is $(n - r)$.

2. The column rank of A = the dimension of the space spanned by the columns of A

$\qquad\qquad\qquad\quad$ = the dimension of the range of A
$\qquad\qquad\qquad\quad$ (see Exercise 10)
$\qquad\qquad\qquad\quad$ = $n - $ (dimension of the kernel of A)
$\qquad\qquad\qquad\quad$ = $n - (n - r)$
$\qquad\qquad\qquad\quad$ = r
$\qquad\qquad\qquad\quad$ = the row rank of A.

Definition 3.3. By the *rank* of a matrix we mean either its row rank or its column rank.

EXERCISE SET 3.3

1 Let V be the space of real functions that are differentiable for all t. Let L be defined on V by $L[f(t)] = f'(t)$. Solve the equation $L[f(t)] = 2t + \cos t$ using the 2-step method based on Theorem 3.7.

2 Let V be as in Exercise 1. Let $L[f(t)] = f'(t) - f(t)$. Solve the equation $L[f(t)] = 3t - 5$. [*Hint:* Use Example 1 as a model.]

3 Let V be the space of functions having the basis [sin t, cos t, 1]. Let L be defined on V by

$$L[f(t)] = \int_0^\pi f(t)\, dt.$$

a) Find (by inspection) a single solution of $L[f(t)] = 100$. [*Hint:* Think about areas under curves.]
b) Solve the homogeneous equation $L[f(t)] = 0$.
c) Solve the nonhomogeneous equation $L[f(t)] = 100$.

4 Let L be defined on R^{22} by

$$L(X) = \begin{bmatrix} 1 & 1 \\ 2 & 2 \end{bmatrix} X.$$

Consider the nonhomogeneous equation

$$L(X) = \begin{bmatrix} 25 & 31 \\ 50 & 62 \end{bmatrix}.$$

a) Verify that

$$\begin{bmatrix} 25 & 31 \\ 0 & 0 \end{bmatrix}$$

is a solution.

b) Verify that
$$\left\{ \begin{bmatrix} 1 & 0 \\ -1 & 0 \end{bmatrix}, \begin{bmatrix} 0 & 1 \\ 0 & -1 \end{bmatrix} \right\}$$
is a basis for the kernel of L.
c) Find all solutions of the given equation.

5 Verify Theorem 3.8 for the given matrix.

a) $\begin{bmatrix} 1 & 0 & 1 & -1 \\ 3 & 1 & 2 & -1 \\ 3 & 2 & 1 & 1 \end{bmatrix}$ b) $\begin{bmatrix} 1 & 3 & 5 & 7 \\ 2 & 4 & 6 & 8 \end{bmatrix}$ c) $[1, 2, 3, 4]$

6 a) Let V be the space spanned by the set
$$S = \left\{ \begin{bmatrix} 1 \\ 2 \\ 3 \\ \sqrt{17} \end{bmatrix}, \begin{bmatrix} 1 \\ 2 \\ 3 \\ 259 \end{bmatrix}, \begin{bmatrix} 1 \\ 2 \\ 3 \\ \pi \end{bmatrix}, \begin{bmatrix} 1 \\ 2 \\ 3 \\ 4 \end{bmatrix} \right\}.$$

Just by "looking at" S, make a guess as to the dimension of V.
b) Use Theorem 3.8 to determine the dimension of V.

7 Consider the *nonhomogeneous system*
$$\begin{cases} x_1 + x_2 + x_3 = 6, \\ x_1 + 2x_2 - x_3 = 2, \\ 2x_1 + 3x_2 = 8. \end{cases}$$

Student α discovers that $A = [1, 2, 3]'$ is a solution; student β discovers that $B = [4, 0, 2]'$ is a solution. Both students discover that the solution of the corresponding *homogeneous system* is given by
$$X = t \begin{bmatrix} -3 \\ 2 \\ 1 \end{bmatrix}.$$

Applying Theorem 3.7, our two students turn in the following solutions for the given system:
$$\alpha: X = t \begin{bmatrix} -3 \\ 2 \\ 1 \end{bmatrix} + \begin{bmatrix} 1 \\ 2 \\ 3 \end{bmatrix},$$
$$\beta: X = t \begin{bmatrix} -3 \\ 2 \\ 1 \end{bmatrix} + \begin{bmatrix} 4 \\ 0 \\ 2 \end{bmatrix}.$$

Show directly—without appealing to Theorem 3.7—that these two solutions are identical.

8 Let L be defined on a certain vector space V, and suppose that the kernel of L is k-dimensional, where $k > 0$. Also, suppose that B is in the range of L. Now, prove that the solutions of the equation $L(X) = B$ form a k-plane in V.

9 Let V be as in Example 1. Let L be defined on V by $L[f(t)] = f''(t) - f(t)$. Solve the equation $L[f(t)] = t^3$.

10 Let L be defined on R^{n1} by the formula $L(X) = AX$, where A is m by n.
   a) Prove that the dimension of the range of L is equal to the column rank of A. [*Hint:* Use the notation $A = [C^1, C^2, \ldots, C^n]$ and recall Exercise 8(a) of Section 3.1.]
   b) Asked to find the dimension of the range of L, a certain student proceeds to reduce A to row echelon form, and then gives as his answer the number of nonzero rows found in the reduced matrix. Is his answer correct? Explain.

11 Let L be defined on R^{31} by $L(X) = AX$, where A is 3 by 3. Find an A for which the kernel of L is: (a) zero-dimensional, (b) one-dimensional, (c) two-dimensional, (d) three-dimensional.

3.4 MATRIX REPRESENTATIONS

Let A be an m by n matrix and let L be a function defined on R^{n1} by

$$L(X) = AX.$$

Then, of course, L is a linear transformation. But this is not the only way in which linear transformations arise. We have seen, for example, that when the domain is a space of functions, then the operations of differentiation and integration also lead to linear transformations. Now it is interesting, and also most important from both a theoretical and a computational point of view, that any linear transformation between finite dimensional vector spaces can be *represented by a matrix*. To give meaning to this announcement we first consider an example.

Example 1. Let $D(at^3 + bt^2 + ct + d) = 3at^2 + 2bt + c$. Thus, D is the *differentiation transformation*, restricted to the polynomial space P^3. Its range is P^2. Now, choose $\mathcal{B} = [t^3, t^2, t, 1]$ and $\mathcal{C} = [t^2, t, 1]$ as ordered bases for the domain and range of D. It then follows that

$$at^3 + bt^2 + ct + d = \begin{bmatrix} a \\ b \\ c \\ d \end{bmatrix}_\mathcal{B} \quad \text{and} \quad 3at^2 + 2bt + c = \begin{bmatrix} 3a \\ 2b \\ c \end{bmatrix}_\mathcal{C}.$$

The problem is this: We wish to find a matrix A such that

$$\begin{bmatrix} 3a \\ 2b \\ c \end{bmatrix}_\mathcal{C} = A \begin{bmatrix} a \\ b \\ c \\ d \end{bmatrix}_\mathcal{B}.$$

3.4 MATRIX REPRESENTATIONS

To find such a matrix, we have only to answer the question: How can we express $3a$, $2b$, and c as linear combinations of a, b, c, and d? The answer is

$$\begin{cases} 3a = 3 \cdot a + 0 \cdot b + 0 \cdot c + 0 \cdot d, \\ 2b = 0 \cdot a + 2 \cdot b + 0 \cdot c + 0 \cdot d, \\ c = 0 \cdot a + 0 \cdot b + 1 \cdot c + 0 \cdot d. \end{cases}$$

It follows that

$$\begin{bmatrix} 3a \\ 2b \\ c \end{bmatrix}_e = \begin{bmatrix} 3 & 0 & 0 & 0 \\ 0 & 2 & 0 & 0 \\ 0 & 0 & 1 & 0 \end{bmatrix} \begin{bmatrix} a \\ b \\ c \\ d \end{bmatrix}_\mathcal{B}.$$

The significance of the matrix A is simply this: We can now differentiate polynomials in P^3 by using matrix multiplication. For example, suppose that we wish to find $D(2t^3 - 7t^2 + 6t - 2)$. Then we might proceed as suggested by the following diagram:

$$2t^3 - 7t^2 + 6t - 2 \xrightarrow{\text{Id}_\mathcal{B}} \begin{bmatrix} 2 \\ -7 \\ 6 \\ -2 \end{bmatrix}_\mathcal{B}$$

(multiplication by A)

$$6t^2 - 14t + 6 \xleftarrow{\text{Id}_e^{-1}} \begin{bmatrix} 6 \\ -14 \\ 6 \end{bmatrix}_e$$

It is in this sense that D is *represented* by the matrix A. (We are not suggesting here that this is a good, or even a reasonable, way of differentiating a polynomial. Rather, this is just one illustration of the way in which the operation of quite arbitrary linear transformations can be duplicated by matrix multipliers.)

It must be emphasized that the entries of A depend not only on the transformation D but also on the ordered bases chosen for the domain and range of D. Indeed, even if we keep the same basis vectors, changing only their order, the matrix representing D will change. For example, let

$$\mathcal{B}^* = [t^3, t, t^2, 1] \quad \text{and} \quad \mathcal{C}^* = [1, t, t^2].$$

Then the matrix that "does the work" of D is given by

$$A^* = \begin{bmatrix} 0 & 1 & 0 & 0 \\ 0 & 0 & 2 & 0 \\ 3 & 0 & 0 & 0 \end{bmatrix}.$$

For with respect to these ordered bases we have

$$at^3 + bt^2 + ct + d = \begin{bmatrix} a \\ c \\ b \\ d \end{bmatrix}_{\mathcal{B}^*} \quad \text{and} \quad 3at^2 + 2bt + c = \begin{bmatrix} c \\ 2b \\ 3a \end{bmatrix}_{\mathcal{C}^*}.$$

It is easily verified that

$$\begin{bmatrix} c \\ 2b \\ 3a \end{bmatrix}_{\mathcal{C}^*} = \begin{bmatrix} 0 & 1 & 0 & 0 \\ 0 & 0 & 2 & 0 \\ 3 & 0 & 0 & 0 \end{bmatrix} \begin{bmatrix} a \\ c \\ b \\ d \end{bmatrix}_{\mathcal{B}^*}.$$

Finally, a close look at A reveals the following: The entries in the jth column of A are the $\mathcal{C}$-coordinates of $D(B^j)$. This is verified below:

$$D(B^1) = D(t^3) = 3t^2 = \begin{bmatrix} 3 \\ 0 \\ 0 \end{bmatrix}_e.$$

$$D(B^2) = D(t^2) = 2t = \begin{bmatrix} 0 \\ 2 \\ 0 \end{bmatrix}_e.$$

$$D(B^3) = D(t) = 1 = \begin{bmatrix} 0 \\ 0 \\ 1 \end{bmatrix}_e.$$

$$D(B^4) = D(1) = 0 = \begin{bmatrix} 0 \\ 0 \\ 0 \end{bmatrix}_e.$$

An analogous observation may be made about A^*.

Theorem 3.9. Let L be a linear transformation from V into W, the dimensions of these spaces being $n \neq 0$ and $m \neq 0$, respectively. Next, let $\mathcal{B}$ and $\mathcal{C}$ be ordered bases for V and W, respectively, and let A be the m by n matrix defined as follows: The jth column of A is the $\mathcal{C}$-column of $L(B^j)$. Finally, let $Y = L(X)$, where X is any vector in V. Then

$$Y_{\mathcal{C}} = AX_{\mathcal{B}}.$$

(That is, the $\mathcal{C}$-column of Y can be found by multiplying the $\mathcal{B}$-column of X by A.) Moreover, there is only one such matrix A.

3.4 MATRIX REPRESENTATIONS

Proof. Suppose that

$$X_\mathcal{B} = \begin{bmatrix} x_1 \\ x_2 \\ \vdots \\ x_n \end{bmatrix} \quad \text{and} \quad Y_\mathcal{C} = \begin{bmatrix} y_1 \\ y_2 \\ \vdots \\ y_m \end{bmatrix}.$$

Then we may write

$$X = x_1 B^1 + x_2 B^2 + \cdots + x_n B^n.$$

Now, exploiting the linearity of L, we have

$$Y = L(X) = x_1 L(B^1) + x_2 L(B^2) + \cdots + x_n L(B^n).$$

But each vector $L(B^j)$, being in the range space W, has a $\mathcal{C}$-label. Thus, suppose that

$$L(B^j) = \begin{bmatrix} a_{1j} \\ a_{2j} \\ \vdots \\ a_{mj} \end{bmatrix}_\mathcal{C}.$$

Equivalently,

$$L(B^j) = \sum_{k=1}^{m} a_{kj} C^k.$$

Returning to Y, we may write

$$Y = x_1 \left(\sum_{k=1}^{m} a_{k1} C^k \right) + x_2 \left(\sum_{k=1}^{m} a_{k2} C^k \right) + \cdots + x_n \left(\sum_{k=1}^{m} a_{kn} C^k \right)$$

$$= \sum_{j=1}^{n} x_j \left(\sum_{k=1}^{m} a_{kj} C^k \right)$$

$$= \sum_{j=1}^{n} \left(\sum_{k=1}^{m} x_j a_{kj} C^k \right)$$

$$= \sum_{k=1}^{m} \left(\sum_{j=1}^{n} x_j a_{kj} C^k \right)$$

$$= \sum_{k=1}^{m} \left(\sum_{j=1}^{n} x_j a_{kj} \right) C^k.$$

Thus, we have expressed Y as a linear combination of the basis vectors in $\mathcal{C}$. It follows that the kth $\mathcal{C}$-coordinate of Y is given by

$$y_k = \sum_{j=1}^{n} a_{kj} x_j.$$

That is,
$$y_1 = \sum_{j=1}^{n} a_{1j}x_j = a_{11}x_1 + a_{12}x_2 + \cdots + a_{1n}x_n,$$
$$y_2 = \sum_{j=1}^{n} a_{2j}x_j = a_{21}x_1 + a_{22}x_2 + \cdots + a_{2n}x_n,$$
$$\vdots$$
$$y_m = \sum_{j=1}^{n} a_{mj}x_j = a_{m1}x_1 + a_{m2}x_2 + \cdots + a_{mn}x_n.$$

Thus, $Y_\mathcal{C} = AX_\mathcal{B}$. Moreover, the fact that all of our steps are reversible implies that there is just one such matrix A.

Definition 3.4. The matrix A of Theorem 3.9 is called the $\mathcal{CB}$-*representation of L*. The formula
$$Y_\mathcal{C} = AX_\mathcal{B}$$
is called the $\mathcal{CB}$-*formula for L*. (If the $\mathcal{CB}$-representation of L is denoted by $L_{\mathcal{CB}}$, then the corresponding $\mathcal{CB}$-formula assumes a notationally elegant form: $Y_\mathcal{C} = L_{\mathcal{CB}} X_\mathcal{B}$.)

Theorem 3.9 may now be paraphrased as follows: *The jth column in the $\mathcal{CB}$-representation of L is the $\mathcal{C}$-label of $L(B_j)$.*

Example 2. Let L, $\mathcal{B}$, $\mathcal{C}$, $\mathcal{B}^*$, and $\mathcal{C}^*$ be as in Example 1. Then the $\mathcal{CB}$-representation of L is
$$\begin{bmatrix} 3 & 0 & 0 & 0 \\ 0 & 2 & 0 & 0 \\ 0 & 0 & 1 & 0 \end{bmatrix}.$$

The $\mathcal{C}^*\mathcal{B}^*$-representation of L is
$$\begin{bmatrix} 0 & 1 & 0 & 0 \\ 0 & 0 & 2 & 0 \\ 3 & 0 & 0 & 0 \end{bmatrix}.$$

(These results were established in Example 1.)

Example 3. Let L be defined on R^{22} by
$$L(X) = \begin{bmatrix} 1 & 2 \\ 3 & 4 \end{bmatrix} X.$$

Then the range of L is some subspace of R^{22}. Now, taking
$$\mathcal{B} = \mathcal{C} = \left[\begin{bmatrix} 1 & 0 \\ 0 & 0 \end{bmatrix}, \begin{bmatrix} 0 & 1 \\ 0 & 0 \end{bmatrix}, \begin{bmatrix} 0 & 0 \\ 1 & 0 \end{bmatrix}, \begin{bmatrix} 0 & 0 \\ 0 & 1 \end{bmatrix} \right],$$

find the $\mathcal{CB}$-formula for L.

Solution. We find the $\mathcal{C}$-label of $L(B^j)$, for $j = 1, 2, 3, 4$:

$$L(B^1) = \begin{bmatrix} 1 & 2 \\ 3 & 4 \end{bmatrix} \begin{bmatrix} 1 & 0 \\ 0 & 0 \end{bmatrix} = \begin{bmatrix} 1 & 0 \\ 3 & 0 \end{bmatrix} = \begin{bmatrix} 1 \\ 0 \\ 3 \\ 0 \end{bmatrix}_e,$$

$$L(B^2) = \begin{bmatrix} 1 & 2 \\ 3 & 4 \end{bmatrix} \begin{bmatrix} 0 & 1 \\ 0 & 0 \end{bmatrix} = \begin{bmatrix} 0 & 1 \\ 0 & 3 \end{bmatrix} = \begin{bmatrix} 0 \\ 1 \\ 0 \\ 3 \end{bmatrix}_e,$$

$$L(B^3) = \begin{bmatrix} 1 & 2 \\ 3 & 4 \end{bmatrix} \begin{bmatrix} 0 & 0 \\ 1 & 0 \end{bmatrix} = \begin{bmatrix} 2 & 0 \\ 4 & 0 \end{bmatrix} = \begin{bmatrix} 2 \\ 0 \\ 4 \\ 0 \end{bmatrix}_e,$$

$$L(B^4) = \begin{bmatrix} 1 & 2 \\ 3 & 4 \end{bmatrix} \begin{bmatrix} 0 & 0 \\ 0 & 1 \end{bmatrix} = \begin{bmatrix} 0 & 2 \\ 0 & 4 \end{bmatrix} = \begin{bmatrix} 0 \\ 2 \\ 0 \\ 4 \end{bmatrix}_e.$$

Thus, by Theorem 3.9,

$$\begin{bmatrix} y_1 \\ y_2 \\ y_3 \\ y_4 \end{bmatrix}_e = \begin{bmatrix} 1 & 0 & 2 & 0 \\ 0 & 1 & 0 & 2 \\ 3 & 0 & 4 & 0 \\ 0 & 3 & 0 & 4 \end{bmatrix} \begin{bmatrix} x_1 \\ x_2 \\ x_3 \\ x_4 \end{bmatrix}_{\mathcal{B}}.$$

As a spot check, we have

$$L\left(\begin{bmatrix} 2 & 5 \\ -3 & 1 \end{bmatrix}\right) = \begin{bmatrix} 1 & 2 \\ 3 & 4 \end{bmatrix} \begin{bmatrix} 2 & 5 \\ -3 & 1 \end{bmatrix} = \begin{bmatrix} -4 & 7 \\ -6 & 19 \end{bmatrix}$$

and

$$L\left(\begin{bmatrix} 2 & 5 \\ -3 & 1 \end{bmatrix}\right) = \begin{bmatrix} 1 & 0 & 2 & 0 \\ 0 & 1 & 0 & 2 \\ 3 & 0 & 4 & 0 \\ 0 & 3 & 0 & 4 \end{bmatrix} \begin{bmatrix} 2 \\ 5 \\ -3 \\ 1 \end{bmatrix}_{\mathcal{B}} = \begin{bmatrix} -4 \\ 7 \\ -6 \\ 19 \end{bmatrix}_e = \begin{bmatrix} -4 & 7 \\ -6 & 19 \end{bmatrix}.$$

(Although the transformation L is defined by a matrix formula, it must be understood that the matrix of that formula,

$$\begin{bmatrix} 1 & 2 \\ 3 & 4 \end{bmatrix},$$

is not a *matrix representation* for L.)

146 LINEAR TRANSFORMATIONS   3.4

EXERCISE SET 3.4

1 Let L be the linear transformation from P^2 into P^1 defined by

$$L[p(t)] = t\int_0^1 p(t)\,dt + \int_0^1 p(t-1)\,dt.$$

Let $\mathcal{B} = [t^2, t, 1]$ and $\mathcal{C} = [t, 1]$.

a) Find the $\mathcal{CB}$-formula for L.
b) Use each of the two formulas now available to find $L(2t^2 - 3t + 4)$.

2 Let V be the space of functions having the basis $\mathcal{B} = [\sin t, \cos t]$, and let L be defined on V by $L[f(t)] = 2f(t) - 3f'(t)$. It is easily verified that V is also the range of L. Thus, we may take $\mathcal{B}$ as both the *domain basis* and the *range basis*.

a) Find the $\mathcal{BB}$-formula for L.
b) Use each of the two formulas now available to find $L(2\sin t + 3\cos t)$.

3 Let L be defined on R^{13} by

$$L(X) = X\begin{bmatrix} 1 & 2 \\ 3 & 0 \\ 1 & 3 \end{bmatrix}.$$

Let $\mathcal{B} = [[1, 1, 1], [0, 1, 1], [0, 0, 1]]$ and $\mathcal{C} = [[1, 1], [0, 1]]$.

a) Find the $\mathcal{CB}$-formula for L.
b) Use each of the formulas now available to find $L([1, 2, 3])$.

4 Let $\mathfrak{N}$ be the natural basis for R^{21}; let

$$\mathcal{B} = \left[\begin{bmatrix} 0 \\ 1 \end{bmatrix}, \begin{bmatrix} 1 \\ 2 \end{bmatrix}\right];$$

and let L be defined on R^{21} by

$$L(X) = \begin{bmatrix} 1 & 2 \\ 4 & 7 \end{bmatrix} X.$$

Find the specified matrix representation for L.

a) $L_{\mathfrak{N}\mathfrak{N}}$ b) $L_{\mathcal{B}\mathfrak{N}}$
c) $L_{\mathfrak{N}\mathcal{B}}$ d) $L_{\mathcal{B}\mathcal{B}}$

5 Let $\mathcal{B} = [t^2, t, 1]$; let $\mathcal{C} = [t^2 + 2, t - 3, 2t^2 + t]$; and let L be the transformation from P^2 into P^2 whose $\mathcal{CB}$-representation is

$$A = \begin{bmatrix} 1 & -1 & 0 \\ 2 & 0 & 4 \\ 1 & 2 & 3 \end{bmatrix}.$$

a) Find $L(t^2 + 3t - 1)$.
b) Find $L[L(t^2 + 3t - 1)]$.

6 Let
$$\mathfrak{B} = \left[\begin{bmatrix} 0 \\ 1 \end{bmatrix}, \begin{bmatrix} 1 \\ 2 \end{bmatrix} \right] \quad \text{and} \quad \mathfrak{C} = \left[\begin{bmatrix} 2 \\ 0 \end{bmatrix}, \begin{bmatrix} 0 \\ 3 \end{bmatrix} \right];$$

and let L be the linear transformation from R^{21} into R^{21} about which the following is known: $L(B^1) = C^1$ and $L(B^2) = C^2$.

a) Find the $\mathfrak{C}\mathfrak{B}$-formula for L.
b) Find the $\mathfrak{N}\mathfrak{N}$-formula for L.
c) Use each of the formulas now available to find $L([2, 5]')$.

7 Let $\mathfrak{B}$ and $\mathfrak{C}$ be as in Exercise 5.

a) Using the method of Example 1, Section 2.10, find the matrix A such that $X_\mathfrak{C} = AX_\mathfrak{B}$, where X is a vector in P^2.
b) Let L be defined on P^2 by $L(X) = X$, the *identity transformation*. Find the $\mathfrak{C}\mathfrak{B}$-formula for L. Then replace $Y_\mathfrak{C}$ in that formula by $X_\mathfrak{C}$—this is permissible since $Y = X$—and compare the resulting formula with that obtained under part (a).
c) Prove: Let B and C both be bases for a vector space V, and let L be the identity transformation on V. Then the $\mathfrak{C}\mathfrak{B}$-representation for L is the matrix that transforms $\mathfrak{B}$-columns into $\mathfrak{C}$-columns.

8 Recall Example 1. We may equally well regard D as a linear transformation from P^3 into P^3. Under this interpretation, what is the $\mathfrak{B}\mathfrak{B}$-representation for D?

9 Let V be the space of real functions that are continuous on the interval $[0, 1]$. Let L be defined on V by
$$L[f(t)] = f(0)t + \int_0^1 f(t)\, dt.$$

a) Find $L(\sin t + e^t + 1)$.
b) Explain why L has no matrix representation.

10 Let F be defined on P^2 by
$$F[p(t)] = p(t) \int_0^1 p(t)\, dt.$$

a) Find $F(3t^2 + 2t + 1)$.
b) Explain why L has no matrix representation.

11 Let L be defined on P^2 by
$$L[p(t)] = p'(0) + p(0) + \int_0^1 p(t)\, dt.$$

Let $\mathfrak{B} = [1, t^2 + t, t + 1]$ and let $\mathfrak{C} = [13]$. Find the $\mathfrak{C}\mathfrak{B}$-representation for L.

12 Let V be the space of real functions having the basis $\mathfrak{B} = [e^t, te^t, t^2 e^t]$. Let D be the differentiation transformation. Find the $\mathfrak{B}\mathfrak{B}$-representation for D.

13 Read Appendix B.

3.5 ARITHMETIC OPERATIONS FOR LINEAR TRANSFORMATIONS

Given the sets D and R, let S be the set of all functions from D into R. Now, if R has an arithmetic structure, then arithmetic operations in S can be defined in terms of those in R. For example, suppose that $D = R =$ the real number system and that f and g are in S. Then the following definitions are possible:

1. Let h be defined by $h(t) = f(t) + g(t)$. Then h is the *sum* of f and g.
2. Let h be defined by $h(t) = cf(t)$. Then h is the *c-multiple* of f.

We could also define a *difference* function and a *product* function. On the other hand, a *division* in S cannot, in general, be defined: to introduce h by the formula $h(t) = f(t)/g(t)$ is not possible in the case where g is a function that takes on the value zero. (We are requiring here that our functions be defined for all real numbers.)

Another way to construct new functions out of old—and one which does not make use of the arithmetic structure of R—is to *compose* them. For example, given any functions f and g, we might introduce the function h defined by

$$h(t) = f(g(t)).$$

However, for this formula to make sense, the range of g must be a subset of the domain of f. If such compositions are to be possible for all pairs of functions chosen from a set S—and in either order—then the sets D and R must be identical.

Now suppose that this is the case. Could we then liken this "composing" to an arithmetic operation? That is, does it in any way resemble the operations that we have previously called additions or multiplications? Well, it is an associative operation. This simple fact is made clear, perhaps, only by introducing additional notation. Suppose, for example, that we denote the composition operation by $\circ$; thus, the function defined by $f(g(t))$ will be denoted by $f \circ g$. Then the associativity of $\circ$ is established by showing that

$$f \circ (g \circ h) = (f \circ g) \circ h.$$

This is done as follows:

Let $F = g \circ h$ and let $F^* = f \circ F$. Then for any t (in the domain of our functions) we may write $F^*(t) = f(F(t))$. But $F(t) = g(h(t))$. Thus,

$$F^*(t) = f(g(h(t))).$$

Similarly, let $H = f \circ g$ and let $H^* = H \circ h$. Then $H^*(t) = H(h(t))$. But for any x, we have $H(x) = f(g(x))$. Thus, letting $x = h(t)$, we have

$$H^*(t) = f(g(h(t))).$$

It follows that $F^* = H^*$; that is, $f \circ (g \circ h) = (f \circ g) \circ h$. On the other hand, the composition operation is not, in general, commutative. Consider for example, the real functions $\sin t$ and t^2. Surely, the functions defined by $\sin(t^2)$ and $(\sin t)^2$ are distinct.

Our primary interest here is with the combining operations that make sense for linear transformations. Because of the limited arithmetic structure of a vector space, there are just three such operations.

Definition 3.5. (See Exercise 7.)
1. The linear transformation L^* defined by $L^*(X) = L_1(X) + L_2(X)$ is called the *sum of L_1 and L_2*. We write $L^* = L_1 \oplus L_2$.
2. The linear transformation L^* defined by $L^*(X) = cL(X)$ is called the *c-multiple of L*. We write $L^* = c \odot L$.
3. The linear transformation L^* defined by $L^*(X) = L_1(L_2(X))$ is called the *composition of L_1 and L_2*. We write $L^* = L_1 \circ L_2$.

We now make an observation: there are three operations for linear transformations; there are three operations for matrices; and there is—for transformations between finite dimensional spaces—a scheme for identifying linear transformations with matrices. A conjecture follows: perhaps the linear transformation operations "parallel" the matrix operations. That is, perhaps linear transformations can be combined by performing corresponding operations upon their matrix representations. As we shall see, they can.

In this section we restrict our attention to the arithmetic structure involving the first two operations. Suppose, then, that T is the set of all linear transformations from V into W. Then T along with $\oplus$ and $\odot$ is a vector space. This follows from the fact that the linear transformation operations "inherit" the properties of the vector space operations in W. For example, consider the required commutativity of addition. Let

$$L = L_1 \oplus L_2 \quad \text{and} \quad L^* = L_2 \oplus L_1.$$

Then

$$L(X) = L_1(X) + L_2(X) = L_2(X) + L_1(X) = L^*(X)$$

Thus, $L = L^*$.

The other vector space axioms are verified in a similar manner. More interesting, however, is the following result.

Theorem 3.10. Let T be the vector space of linear transformations from V into W, where V and W are vector spaces of dimensions $n \neq 0$ and $m \neq 0$, respectively. Then T is isomorphic to R^{mn} and, consequently, T has dimension mn.

Proof. Let $\mathcal{B}$ and $\mathcal{C}$ be ordered bases for V and W, respectively, and let Mat be the function that assigns to each L in T its $\mathcal{CB}$-representation. The existence of such a function is guaranteed by Theorem 3.9. Moreover, it is clear that two different transformations cannot be assigned the same matrix; for if two transformations are different, then their $\mathcal{CB}$-formulas must be different. Thus, Mat is a one-to-one function from V into W. To show that it is an isomorphism, we must show that it has the two linear transformation properties:

$$\text{Mat}(c \odot L) = c \cdot \text{Mat}(L)$$

and

$$\text{Mat}(L_1 \oplus L_2) = \text{Mat}(L_1) + \text{Mat}(L_2).$$

We prove only the first, leaving the second for the reader to prove.

1. Let $Y = L(X)$, where X is any vector in V. Identifying X and Y with their $\mathcal{B}$- and $\mathcal{C}$-columns, respectively, we may write

$$Y_\mathcal{C} = \text{Mat}(L) X_\mathcal{B}.$$

It follows, employing the scalar multiplication of matrix arithmetic, that

$$c Y_\mathcal{C} = \{c \, \text{Mat}(L)\} X_\mathcal{B}.$$

2. Next, let $L^* = c \odot L$, and let $Y^* = L^*(X)$. This means that $Y^* = cL(X) = cY$. Since the function that assigns $\mathcal{C}$-labels to vectors in W is an isomorphism, it follows that

$$Y^*_\mathcal{C} = c Y_\mathcal{C}.$$

Substituting into the result obtained under part 1, we have

$$Y^*_\mathcal{C} = \{c \text{Mat}(L)\} X_\mathcal{B}.$$

3. On the other hand, since $Y^* = L^*(X)$, we may write

$$Y^*_\mathcal{C} = \text{Mat}(L^*) X_\mathcal{B}.$$

Comparing these last two results, and on considering the fact that the $\mathcal{CB}$-formula for a given transformation is unique, we conclude that

$$\text{Mat}(L^*) = c \cdot \text{Mat}(L).$$

That is,

$$\text{Mat}(c \odot L) = c \cdot \text{Mat}(L).$$

Example 1. Let T be the space of linear transformations from P^2 into P^2. Let L_1 and L_2 be the members of T defined by

$$L_1[p(t)] = p(0) \cdot t^2 + p'(t), \qquad L_2[p(t)] = p''(t) \cdot t^2 + p(0).$$

Finally, let $L = L_1 \oplus (2 \odot L_2)$.

3.5   ARITHMETIC OPERATIONS FOR LINEAR TRANSFORMATIONS   151

1. We first find $L(at^2 + bt + c)$ in a direct manner:
$$L_1(at^2 + bt + c) = ct^2 + 2at + b,$$
$$L_2(at^2 + bt + c) = 2at^2 + c.$$
Thus,
$$L(at^2 + bt + c) = (ct^2 + 2at + b) + 2(2at^2 + c)$$
$$= (4a + c)t^2 + (2a)t + (b + 2c).$$

2. We next find $L(at^2 + bt + c)$ by using the fact that T is isomorphic to R^{33}. Let $\mathcal{B} = [t^2, t, 1]$ and let Mat be the isomorphism that assigns to each L in T its $\mathcal{BB}$-representation. Then, appealing directly to the given formulas for L_1 and L_2, we construct $\text{Mat}(L_1)$ and $\text{Mat}(L_2)$:

$$L_1(t^2) = 0 \cdot t^2 + 2t = 2t = \begin{bmatrix} 0 \\ 2 \\ 0 \end{bmatrix}_\mathcal{B}, \quad L_2(t^2) = 2t^2 + 0 = 2t^2 = \begin{bmatrix} 2 \\ 0 \\ 0 \end{bmatrix}_\mathcal{B},$$

$$L_1(t) = 0 \cdot t + 1 = 1 = \begin{bmatrix} 0 \\ 0 \\ 1 \end{bmatrix}_\mathcal{B}, \quad L_2(t) = 0 \cdot t^2 + 0 = 0 = \begin{bmatrix} 0 \\ 0 \\ 0 \end{bmatrix}_\mathcal{B},$$

$$L_1(1) = 1 \cdot t^2 + 0 = t = \begin{bmatrix} 1 \\ 0 \\ 0 \end{bmatrix}_\mathcal{B}. \quad L_2(1) = 0 \cdot t^2 + 1 = 1 = \begin{bmatrix} 0 \\ 0 \\ 1 \end{bmatrix}_\mathcal{B}.$$

Therefore, $\text{Mat}(L_1) = \begin{bmatrix} 0 & 0 & 1 \\ 2 & 0 & 0 \\ 0 & 1 & 0 \end{bmatrix}$. Therefore, $\text{Mat}(L_2) = \begin{bmatrix} 2 & 0 & 0 \\ 0 & 0 & 0 \\ 0 & 0 & 1 \end{bmatrix}$.

It follows that
$$\text{Mat}(L) = \text{Mat}(L_1 \oplus (2 \odot L_2)) = \text{Mat}(L_1) + 2\,\text{Mat}(L_2)$$
$$= \begin{bmatrix} 4 & 0 & 1 \\ 2 & 0 & 0 \\ 0 & 1 & 2 \end{bmatrix},$$

and this means that
$$L(at^2 + bt + c) = \begin{bmatrix} 4 & 0 & 1 \\ 2 & 0 & 0 \\ 0 & 1 & 2 \end{bmatrix} \begin{bmatrix} a \\ b \\ c \end{bmatrix}_\mathcal{B} = \begin{bmatrix} (4a + c) \\ 2a \\ (b + 2c) \end{bmatrix}_\mathcal{B}$$
$$= (4a + c)t^2 + (2a)t + (b + 2c).$$

Example 2. Let T be the space of linear transformations from P^1 into P^1. Find a basis for T.

Solution. Let $\mathcal{B} = [t, 1]$. Let Mat assign to each L in T its $\mathcal{BB}$-representation. Then Mat is an isomorphism from T onto R^{22}. Now R^{22} has

the basis
$$\left\{\begin{bmatrix}1 & 0\\ 0 & 0\end{bmatrix}, \begin{bmatrix}0 & 1\\ 0 & 0\end{bmatrix}, \begin{bmatrix}0 & 0\\ 1 & 0\end{bmatrix}, \begin{bmatrix}0 & 0\\ 0 & 1\end{bmatrix}\right\}.$$

It follows (Theorem 2.14) that the set
$$\left\{\mathrm{Mat}^{-1}\begin{bmatrix}1 & 0\\ 0 & 0\end{bmatrix}, \mathrm{Mat}^{-1}\begin{bmatrix}0 & 1\\ 0 & 0\end{bmatrix}, \mathrm{Mat}^{-1}\begin{bmatrix}0 & 0\\ 1 & 0\end{bmatrix}, \mathrm{Mat}^{-1}\begin{bmatrix}0 & 0\\ 0 & 1\end{bmatrix}\right\}$$

is a basis for T. However, we have not really "found" a basis for T until we "exhibit" the transformations of the basis. This is easy to do. Let
$$L_1 = \mathrm{Mat}^{-1}\begin{bmatrix}1 & 0\\ 0 & 0\end{bmatrix}.$$

This means that L_1 is the transformation whose $\mathcal{B}\mathcal{B}$-representation is
$$\begin{bmatrix}1 & 0\\ 0 & 0\end{bmatrix}.$$

It follows that
$$L_1(at+b) = \begin{bmatrix}1 & 0\\ 0 & 0\end{bmatrix}\begin{bmatrix}a\\ b\end{bmatrix}_\mathcal{B} = \begin{bmatrix}a\\ 0\end{bmatrix}_\mathcal{B} = at.$$

Similarly,
$$L_2(at+b) = \begin{bmatrix}0 & 1\\ 0 & 0\end{bmatrix}\begin{bmatrix}a\\ b\end{bmatrix}_\mathcal{B} = \begin{bmatrix}b\\ 0\end{bmatrix}_\mathcal{B} = bt,$$

$$L_3(at+b) = \begin{bmatrix}0 & 0\\ 1 & 0\end{bmatrix}\begin{bmatrix}a\\ b\end{bmatrix}_\mathcal{B} = \begin{bmatrix}0\\ a\end{bmatrix}_\mathcal{B} = a,$$

and
$$L_4(at+b) = \begin{bmatrix}0 & 0\\ 0 & 1\end{bmatrix}\begin{bmatrix}a\\ b\end{bmatrix}_\mathcal{B} = \begin{bmatrix}0\\ b\end{bmatrix}_\mathcal{B} = b.$$

Thus, as a final answer to the problem posed we might write: A basis for T is given by the set

$$\{L_1(at+b) = at,\ L_2(at+b) = bt,\ L_3(at+b) = a,\ L_4(at+b) = b\},$$

where it is tacitly understood that this set is not what it appears to be: it is not a set of formulas; rather, it is the set of transformations defined by the listed formulas.

A more colorful answer is the following: The space of all linear transformations from P^1 into P^1 has as a basis the following set of transformations:

1. "Drop the constant."
2. "Interchange coefficients and then drop the constant."
3. "Differentiate."
4. "Evaluate at 0."

EXERCISE SET 3.5

1 Let L_1 and L_2 be the linear transformations from R^{21} into R^{21} whose $\mathfrak{MN}$-representations are, respectively,

$$A_1 = \begin{bmatrix} 1 & 2 \\ 3 & 4 \end{bmatrix} \quad \text{and} \quad A_2 = \begin{bmatrix} 1 & 1 \\ 2 & 2 \end{bmatrix}.$$

Find the $\mathfrak{MN}$-representation for L, where

$$L = \{3 \odot (L_1 \oplus L_2)\} \oplus \{(2 \odot L_2) \oplus L_1\}.$$

2 Let L_1 and L_2 be the linear transformations from P^2 to P^2 defined by

$$L_1[p(t)] = p(t+2), \qquad L_2[p(t)] = p(0)t^2 + \int_0^1 p(t)\, dt.$$

a) Let $L = L_1 \oplus L_2$. Find $L(at^2 + bt + c)$ without using matrices.
b) Use matrices to find $L(at^2 + bt + c)$. (Recall Example 1.)

3 Let T be the space of all linear transformations from P^2 into the one-dimensional space of real numbers. Verify that $\mathfrak{B} = [L_1, L_2, L_3]$ is a basis for T, where $L_1[p(t)] = p(0)$, $L_2[p(t)] = p'(0)$, and $L_3[p(t)] = \frac{1}{2}p''(0)$.

4 Let V be the vector space having the basis $\mathfrak{B} = [\cos t, \sin t]$. Let T be the space of linear transformations from V into V.

a) Find an ordered basis for T.
b) Let D be the transformation in T that differentiates. Express D as a linear combination of your basis vectors.

5 Complete the proof of Theorem 3.10 by showing that

$$\text{Mat}(L_1 \oplus L_2) = \text{Mat}(L_1) + \text{Mat}(L_2).$$

6 Recall the discussion preceding Theorem 3.10. Verify that the structure $(T, R, \oplus, \odot)$ satisfies Axioms 3, 5, 6, and 8 of Definition 2.1.

7 In Definition 3.5, it is tacitly assumed that the transformations $L_1 \oplus L_2$, $c \odot L_1$, and $L_1 \circ L_2$ are all linear transformations provided that L_1 and L_2 are. Prove this.

8 Recall Example 2. Let $\mathfrak{B} = [L_1, L_2, L_3, L_4]$. Find the $\mathfrak{B}$-label for L, where $L[p(t)] = p(t-2) + p'(t)$.

3.6 LINEAR ALGEBRAS

We now direct our attention to the third of the three linear transformation operations: the composition operation. We have in mind that the composition of transformations is "mirrored" in the multiplication of their matrix representations. Making this last statement precise is, perhaps, more difficult than its proof.

154 LINEAR TRANSFORMATIONS 3.6

Theorem 3.11. Let U, V, and W be vector spaces (of finite dimension) having the ordered bases $\mathcal{B}$, $\mathcal{C}$, and $\mathcal{D}$, respectively. Let L_1 be a linear transformation from U into V, and let L_2 be a linear transformation from V into W. Finally, let Mat be the function that assigns to transformations between these spaces their matrix representations with respect to the given bases. Then

$$\text{Mat}(L_2 \circ L_1) = \text{Mat}(L_2) \times \text{Mat}(L_1).$$

Proof. Taking X as any vector in U, we let

$$Y = L_1(X) \quad \text{and} \quad Z = L_2(Y).$$

It follows that

$$Y_\mathcal{C} = \text{Mat}(L_1) X_\mathcal{B} \quad \text{and} \quad Z_\mathcal{D} = \text{Mat}(L_2) Y_\mathcal{C}.$$

Combining these equations, exploiting the associativity of matrix multiplication, we have

$$Z_\mathcal{D} = \{\text{Mat}(L_2) \times \text{Mat}(L_1)\} X_\mathcal{B}.$$

On the other hand, recalling our definition of Z, we have

$$Z = L_2(L_1(X)),$$

or, letting $L = L_2 \circ L_1$,

$$Z = L(X).$$

But then we may write

$$Z_\mathcal{D} = \text{Mat}(L) X_\mathcal{B}.$$

It follows (since by Theorem 3.9 the $\mathcal{B}\mathcal{B}$-representation of L is unique) that

$$\text{Mat}(L) = \text{Mat}(L_2) \times \text{Mat}(L_1).$$

That is,

$$\text{Mat}(L_2 \circ L_1) = \text{Mat}(L_2) \times \text{Mat}(L_1).$$

Example 1. Let L_1 and L_2 be the linear transformations of Example 1, Section 3.5. Thus,

$$L_1(at^2 + bt + c) = ct^2 + 2at + b$$

and

$$L_2(at^2 + bt + c) = 2at^2 + c.$$

Now, let $L = L_2 \circ L_1$.

1. We first find a polynomial formula for L without using matrices:

$$L_2(L_1(at^2 + bt + c)) = L_2(ct^2 + 2at + b) = 2ct^2 + b.$$

2. We find the same formula using matrix calculations: First, we choose $\mathcal{B} = [t^2, t, 1]$. Then, borrowing our earlier results, we obtain

$$\text{Mat}(L_2 \circ L_1) = \text{Mat}(L_2) \times \text{Mat}(L_1)$$

$$= \begin{bmatrix} 2 & 0 & 0 \\ 0 & 0 & 0 \\ 0 & 0 & 1 \end{bmatrix} \times \begin{bmatrix} 0 & 0 & 1 \\ 2 & 0 & 0 \\ 0 & 1 & 0 \end{bmatrix}$$

$$= \begin{bmatrix} 0 & 0 & 2 \\ 0 & 0 & 0 \\ 0 & 1 & 0 \end{bmatrix}.$$

Thus,

$$L = L_2 \circ L_1 = \text{Mat}^{-1}\left(\begin{bmatrix} 0 & 0 & 2 \\ 0 & 0 & 0 \\ 0 & 1 & 0 \end{bmatrix} \right).$$

It follows that

$$L(at^2 + bt + c) = \begin{bmatrix} 0 & 0 & 2 \\ 0 & 0 & 0 \\ 0 & 1 & 0 \end{bmatrix} \begin{bmatrix} a \\ b \\ c \end{bmatrix}_\mathcal{B} = \begin{bmatrix} 2c \\ 0 \\ b \end{bmatrix}_\mathcal{B}$$

$$= 2ct^2 + b. \quad \square$$

Next, let T be the space of all linear transformations from V into V, where V is any vector space having dimension n. Then, according to Theorem 3.10, the structure $(T, R, \oplus, \odot)$ is a vector space of dimension n^2. But now, since the range of each member of T is a subspace of V (the common domain of all members of T), the operation of composition may be carried out for any two transformations in T—and in either order. Thus, we are prompted to consider the more complicated structure $(T, R, \oplus, \odot, \circ)$.

We follow the same program as before. Let $\mathcal{B}$ be an ordered basis for V, and let Mat be the function that assigns to each L in T its $\mathcal{B}\mathcal{B}$-representation. Then, appealing to both Theorem 3.10 and Theorem 3.11, we have

1. $\text{Mat}(L_1 \oplus L_2) = \text{Mat}(L_1) + \text{Mat}(L_2)$,
2. $\text{Mat}(c \odot L) = c \cdot \text{Mat}(L)$,
3. $\text{Mat}(L_1 \circ L_2) = \text{Mat}(L_1) \times \text{Mat}(L_2)$.

It follows that the arithmetic of T is "mirrored" in that of the set of all n by n matrices. More precisely, we would like to say that these two structures are *isomorphic*. However, this term has been defined (in this text) only for vector spaces. To extend its meaning we must first have a formal definition for the kind of structure that we are now studying.

Definition 3.6. Let $\mathcal{L}$ be any set of objects, let R be the real number system, and let there be three operations: an addition of objects in $\mathcal{L}$, denoted by $\oplus$; a multiplication of objects in $\mathcal{L}$ by members of R, denoted by $\odot$; and a multiplication of objects in $\mathcal{L}$, denoted by $\circ$. Suppose that the structure $(\mathcal{L}, R, \oplus, \odot, \circ)$ has the following characteristics:

1. The structure $(\mathcal{L}, R, \oplus, \odot)$ is a vector space.
2. We have the following axioms involving $\circ$:
    a) $X \circ (Y \circ Z) = (X \circ Y) \circ Z$.
    b) $X \circ (c \odot Y) = (c \odot X) \circ Y = c \odot (X \circ Y)$.
    c) $X \circ (Y \oplus Z) = (X \circ Y) \oplus (X \circ Z)$.
    d) $(X \oplus Y) \circ Z = (X \circ Z) \oplus (Y \circ Z)$.

Then the structure $(\mathcal{L}, R, \oplus, \odot, \circ)$ is called a *linear algebra*. The members of $\mathcal{L}$ are still called *vectors*, and those of R, *scalars*. (Actually, we have here a *linear algebra over the real number system*. It is possible to consider linear algebras over other number systems: the complex number system, for example.)

Definition 3.7. Given two linear algebras $(\mathcal{L}_1, R, \oplus, \odot, \circ)$ and $(\mathcal{L}_2, R, \oplus', \odot', \circ')$, suppose that there is a one-to-one function Id with domain $\mathcal{L}_1$ and range $\mathcal{L}_2$ having the following properties:

1. $\mathrm{Id}(X \oplus Y) = \mathrm{Id}(X) \oplus' \mathrm{Id}(Y)$.
2. $\mathrm{Id}(c \odot X) = c \odot' \mathrm{Id}(X)$.
3. $\mathrm{Id}(X \circ Y) = \mathrm{Id}(X) \circ' \mathrm{Id}(Y)$.

Then the linear algebras are said to be *isomorphic*.

It is not within the province of this text to make a detailed study of linear algebras in general. Our only interest here is in two linear algebras: (1) the set T of linear transformations from V into V (n-dimensional) along with the three linear transformation operations, and (2) the set M of all n by n matrices along with the three matrix operations. (That these structures are linear algebras is easily verified by checking out the axioms of Definition 3.6.) About these two linear algebras we have the following theorem.

Theorem 3.12. The linear algebras described above are isomorphic.

Proof. Theorem 3.10 "plus" Theorem 3.11. ∎

We conclude this section—and this chapter—with one final observation: the study of a given linear transformation (between finite dimensional vector spaces) may be effected by carrying out an analogous study of one of its matrix representations. That study may be simple or difficult depending on the nature of the matrix. Thus, it becomes important that we have some facility in passing from one matrix representation of a transformation to another. Now although such a find-a-new-representation problem may be handled by appealing directly to Theorem 3.9, it is more instructive to approach the problem from a change-of-coordinates point of view. An example will make this clear.

Example 2. Let $\mathcal{B} = [t^2, t, 1]$ and let $\mathcal{C} = [t, 1]$. Let L be the linear from P^2 into P^1 whose $\mathcal{CB}$-formula is

$$Y_\mathcal{C} = \begin{bmatrix} 1 & 0 & 2 \\ 3 & 1 & 1 \end{bmatrix} X_\mathcal{B}.$$

Now suppose that new coordinate systems are introduced into P^2 and P^1, respectively, by the following change-of-coordinates formulas:

$$X_{\mathcal{B}*} = \begin{bmatrix} 1 & 0 & 1 \\ 0 & 1 & -2 \\ 1 & 0 & 2 \end{bmatrix} X_\mathcal{B} \quad \text{and} \quad Y_{\mathcal{C}*} = \begin{bmatrix} 1 & 2 \\ 1 & 3 \end{bmatrix} Y_\mathcal{C}.$$

Equivalently,

$$X_\mathcal{B} = \begin{bmatrix} 1 & 0 & 1 \\ 0 & 1 & -2 \\ 1 & 0 & 2 \end{bmatrix}^{-1} X_{\mathcal{B}*} \quad \text{and} \quad Y_\mathcal{C} = \begin{bmatrix} 1 & 2 \\ 1 & 3 \end{bmatrix}^{-1} Y_{\mathcal{C}*}.$$

Then, by making the appropriate substitutions, the $\mathcal{CB}$-formula for L can be transformed into its $\mathcal{C}*\mathcal{B}*$-formula:

$$\begin{bmatrix} 1 & 2 \\ 1 & 3 \end{bmatrix}^{-1} Y_{\mathcal{C}*} = \begin{bmatrix} 1 & 0 & 2 \\ 3 & 1 & 1 \end{bmatrix} \left\{ \begin{bmatrix} 1 & 0 & 1 \\ 0 & 1 & -2 \\ 1 & 0 & 2 \end{bmatrix}^{-1} X_{\mathcal{B}*} \right\}.$$

or

$$Y_{\mathcal{C}*} = \left\{ \begin{bmatrix} 1 & 2 \\ 1 & 3 \end{bmatrix} \begin{bmatrix} 1 & 0 & 2 \\ 3 & 1 & 1 \end{bmatrix} \begin{bmatrix} 1 & 0 & 1 \\ 0 & 1 & -2 \\ 1 & 0 & 2 \end{bmatrix}^{-1} \right\} X_{\mathcal{B}*}.$$

Finally, the reader may verify the matrix calculations, we have

$$Y_{\mathcal{C}*} = \begin{bmatrix} 6 & 2 & 1 \\ 9 & 3 & 1 \end{bmatrix} X_{\mathcal{B}*}.$$

EXERCISE SET 3.6

1 Let L_1 be the transformation from R^{31} into R^{21} defined by

$$L_1(X) = \begin{bmatrix} 1 & 0 & 2 \\ 2 & 1 & -1 \end{bmatrix} X,$$

and let L_2 be the transformation from R^{21} into R^{21} defined by

$$L_2(X) = \begin{bmatrix} 1 & 0 \\ 1 & 2 \end{bmatrix} X.$$

Find an analogous matrix formula for the transformation L, where

a) $L = \{2 \odot L_1\} \oplus \{L_2 \circ L_1\}$.
b) $L = 3 \odot \{L_2 \circ L_2 \circ L_1\}$.

2 Let L_1 and L_2 be linear transformations from P^2 into P^2 defined, respectively, as follows:

$$L_1[p(t)] = \frac{d}{dt}[tp(t)] \quad \text{and} \quad L_2[p(t)] = p(t-1).$$

Find the $\mathcal{B}\mathcal{B}$-representation for L, where $\mathcal{B} = [t^2, t, 1]$.

a) $L = L_1 \circ \{3 \odot L_2\}$.
b) $L = L_1 \circ \{L_1 \oplus L_2\}$.
c) $L = \{L_1 \oplus L_2\} \circ L_1$.

3 Recall Example 2.

a) Starting with its $\mathcal{C}\mathcal{B}$-formula, find a polynomial formula for L, that is, find a formula of the form $L(at^2 + bt + c) = \alpha t + \beta$.
b) Starting with its $\mathcal{C}^*\mathcal{B}^*$-formula, find a polynomial formula for L.

4 Prove the following corollary to Theorem 3.11: Let L be a one-to-one linear transformation from U onto V (both of finite dimension), and let $\mathcal{B}$ and $\mathcal{C}$ be bases for U and V, respectively. Then the $\mathcal{C}\mathcal{B}$-representation for L and the $\mathcal{B}\mathcal{C}$-representation for L^{-1} are inverse matrices. [*Hint:* In Theorem 3.11, let $W = U$, $\mathfrak{D} = \mathcal{B}$, $L_1 = L$, and $L_2 = L^{-1}$.]

5 Let L be the linear transformation from P^2 into P^2 defined by $L[p(t)] = p(t-3)$, and let $\mathcal{B} = [t^2, t, 1]$.

a) Find $L_{\mathcal{B}\mathcal{B}}$, the $\mathcal{B}\mathcal{B}$-representation of L.
b) Find the inverse of $L_{\mathcal{B}\mathcal{B}}$, that is, find $(L_{\mathcal{B}\mathcal{B}})^{-1}$.
c) Find the $\mathcal{B}\mathcal{B}$-representation of L^{-1}, that is, find $(L^{-1})_{\mathcal{B}\mathcal{B}}$.
d) Use the result of part (c) to verify that $L^{-1}[q(t)] = q(t+3)$.

6 *A problem written in the language of physicists.* Suppose that the variables z_1 and z_2 are dependent on the variables y_1 and y_2 in the following manner:

$$\begin{cases} z_1 = 2y_1 + 3y_2, \\ z_2 = -5y_1 + 7y_2. \end{cases}$$

Suppose further that the y's are dependent on the variables x_1, x_2, and x_3 in the following manner:

$$\begin{cases} y_1 = 3x_1 - 2x_2 + x_3, \\ y_2 = 2x_1 + 7x_2. \end{cases}$$

Show, using two methods, how the z's depend on the x's.

7 Let L be the linear transformation from R^{31} into R^{31} defined by

$$L(X) = \begin{bmatrix} 1 & 0 & 2 \\ 3 & 1 & 0 \\ 2 & 1 & 2 \end{bmatrix} X.$$

a) Find the $\mathfrak{N}\mathfrak{N}$-formula for L.
b) Given the change-of-coordinates formula

$$X_\mathfrak{B} = \begin{bmatrix} 1 & 0 & 1 \\ 0 & 1 & 1 \\ 1 & 2 & 4 \end{bmatrix} X_\mathfrak{N},$$

find the $\mathfrak{B}\mathfrak{B}$-formula for L.

8 Let T be the set of linear transformations from R^{81} to R^{21}; let M be the set of 4 by 4 matrices. Explain the following statement: T and M are isomorphic as vector spaces but not as linear algebras.

9 Suppose that a multiplication of vectors in R^{31} is introduced by

$$\begin{bmatrix} x_1 \\ x_2 \\ x_3 \end{bmatrix} \times \begin{bmatrix} y_1 \\ y_2 \\ y_3 \end{bmatrix} = \begin{bmatrix} x_2 y_3 - x_3 y_2 \\ x_3 y_1 - x_1 y_3 \\ x_1 y_2 - x_2 y_1 \end{bmatrix}.$$

Show that the structure consisting of the vector space R^{31} along with this multiplication is not a linear algebra. [*Hint:* Look for a counterexample that denies 2(a) of Definition 3.6.]

10 Prove: Let $\mathfrak{B}$ and $\mathfrak{C}$ both be bases for V, and let L be a linear transformation from V into V. Then there exists a nonsingular matrix P such that

$$L_{\mathfrak{c}\mathfrak{c}} = P^{-1} L_{\mathfrak{B}\mathfrak{B}} P.$$

11 Let L be defined on P^3 by the formula $L[p(t)] = p'(t)$, and let $\mathfrak{B} = [t^3, t^2, t, 1]$. Find the $\mathfrak{B}\mathfrak{B}$-representation for

a) L.   b) $L \circ L$.   c) $L \circ L \circ L$.   d) $L \circ L \circ L \circ L$.

12 The bridge between matrix multiplication and the composition of linear transformations provided by Theorem 3.11 gives us insight as to why matrix multiplication is associative but not commutative. Explain.

13 We have defined earlier the notions of *row equivalence* and *column equivalence* for matrices; we now define *matrix equivalence*: A is *equivalent* to B if there exist nonsingular matrices P and Q such that $A = PBQ$. We write $A \stackrel{E}{=} B$.

a) Prove that matrix equivalence is an equivalence relation.
b) Prove: Let L be a linear transformation from V into W, and let A and B both be matrix representations for L. Then $A \stackrel{E}{=} B$. And conversely.

14 Let M be the subspace of R^{22} consisting of all matrices of the form

$$\begin{bmatrix} a & b \\ -b & a \end{bmatrix},$$

160 LINEAR TRANSFORMATIONS 3.6

and let f be the function defined on the complex number system, say C, by

$$f(z) = \begin{bmatrix} a & b \\ -b & a \end{bmatrix}, \quad \text{where} \quad z = a + bi.$$

a) Find $f(2 + 3i)$, $f(1)$, $f(i)$.
b) Prove that f is a one-to-one function having all of M as its range.
c) Prove: $f(cz) = cf(z)$, where c is real.
d) Prove: $f(z_1 + z_2) = f(z_1) + f(z_2)$.
e) Prove: $f(z_1 z_2) = f(z_1) \cdot f(z_2)$.
f) Assuming that M is a linear algebra (over the real number system) it follows from parts (b) through (e) that C is also. Explain.
g) Find the inverse of

$$\begin{bmatrix} a & b \\ -b & a \end{bmatrix}$$

by using complex number arithmetic.
h) Why is it now immediate that any two matrices in M commute?
i) Why is it now immediate that Θ_{22} is the only singular matrix in M?
j) Find a basis for C.

15 Let L be a linear transformation from V into V. If $L \circ L = L$, then L is said to be a *projection* in V.

a) In the usual geometric sense we project an arrow having the label $[x_1, x_2, x_3]'$ onto the $x_1 x_2$-plane by sending that arrow into the one having the label $[x_1, x_2, 0]'$. (Draw a picture.) Show that this "geometric projection" is a projection in the sense given above.
b) Let L be defined on R^{31} by

$$L(X) = \begin{bmatrix} 1 & 0 & 0 \\ 0 & 0 & 0 \\ 0 & 0 & 0 \end{bmatrix} X.$$

Prove that L is a projection and make a drawing showing how L "projects" arrows.
c) Same as (b) where L is defined on R^{21} by

$$L(X) = \begin{bmatrix} \frac{1}{2} & \frac{1}{2} \\ \frac{1}{2} & \frac{1}{2} \end{bmatrix} X.$$

d) Let $L[p(t)] = p(0)$, where $p(t)$ is in P^∞. Prove that L is a projection.
e) Note that none of the projections considered thus far is one-to-one. Prove that the identity transformation, $I(X) = X$, is the only one-to-one projection. [*Hint:* Taking L as any one-to-one projection, we may write $L = L \circ (L \circ L^{-1}) = (L \circ L) \circ L^{-1} = \cdots$.]
f) Prove: If A is the matrix of a projection, then so is $I - A$.
g) Prove: If A is the matrix of a projection, then its range is the kernel of the transformation having the matrix $I - A$.
h) Consider (by making suitable sketches) the theorems under (f) and (g) with respect to the transformation under (b).
i) Prove: If L_1 and L_2 are both projections in V and $L_1 \circ L_2 = L_2 \circ L_1$, then $L_1 \circ L_2$ is also a projection in V.

CHAPTER 4

THE DETERMINANT FUNCTION

4.1 MATHEMATICAL INDUCTION

We begin this chapter with a seeming digression, considering a topic of more general mathematical importance. In the next section, we shall make specific application of this material.

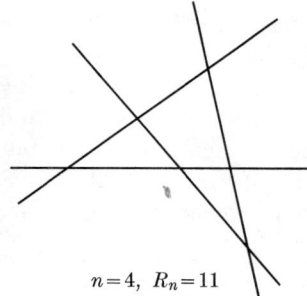 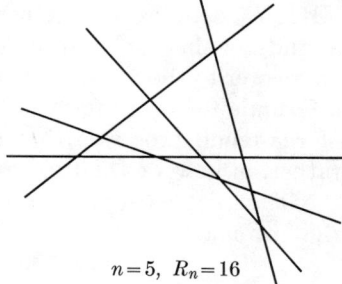

$n=4, \ R_n=11$ $\qquad$ $n=5, \ R_n=16$

Fig. 4.1

Certain mathematical problems are most conveniently studied by breaking them down into infinitely many cases: a first case, a second case, and so on. Having first solved the problem for a number of these special cases, we can then often deduce from such solutions a solution to the general problem. An illustration will make this clear.

Suppose that n lines are positioned upon the plane in a *nonexceptional manner*. We mean by this that there are no parallel pairs and that no three lines pass through the same point (see Fig. 4.1). We then pose the following problem: Into how many regions is the plane divided by these

n lines? This question resolves itself quite naturally into an infinite sequence of questions:

Into how many regions is the plane divided by one line?
Into how many regions is the plane divided by two lines?
And so on.

Such questions—at least the early ones—are easily answered empirically: we have only to make a suitable drawing and then count the regions. Suppose, then, that we have accumulated the following data:

n	R_n
1	2
2	4
3	7
4	11
5	16

(Here R_n denotes the number of regions obtained using n lines.) Through a study of this table, we hope to obtain by *inductive reasoning* (guessing) an answer to the general problem, a suitable answer to that problem being a formula for R_n in terms of n. Now, our interest here is not in the kind of reasoning processes that are useful in deducing such a formula, but rather, in how we shall recognize a correct formula if we do find it.

Thus, let us suppose that we have succeeded in constructing for R_n the formula

$$R_n = \frac{1}{2}(n^2 + n + 2) + (n-1)(n-2)(n-3)(n-4)(n-5)\frac{297-22}{5!}.$$

To give support to this conjecture, we evaluate this formula for $n = 1, 2, 3, 4, 5$:

$\frac{1}{2}(1 + 1 + 2) + 0 \cdot$ (other factors) $= 2,$
$\frac{1}{2}(\ 4 + 2 + 2) + 0 = 4,$
$\frac{1}{2}(\ 9 + 3 + 2) + 0 = 7,$
$\frac{1}{2}(16 + 4 + 2) + 0 = 11,$
$\frac{1}{2}(25 + 5 + 2) + 0 = 16.$

The very values listed in our table! Being naturally disinclined toward coincidences, we are tempted to conclude that our formula is valid for all n. But let us try one more case. For $n = 6$ we have

$$\frac{1}{2}(36 + 6 + 2) + (5)(4)(3)(2)(1)\frac{297 - 22}{5!} = 22 + (297 - 22) = 297.$$

We have made no sketch for the case $n = 6$; however, the answer 297 is obviously too large. Thus, our conjectured formula is incorrect.

The point of the foregoing discussion is this: a proposed formula for R_n can never be validated by checking it against a finite set of data; in other words, even if we were to extend our table to $n = 1000$ and then were able to deduce a formula that worked for $n = 1$ through 1000, there would still be the nagging doubt that the formula might not work for $n = 1001$. How then are such matters settled? The present section is devoted to answering this question; some preparation is required.

Definition 4.1. A nonempty set of numbers I is said to be *inductive* if it has the following property:

If x is in I, then so is $(x + 1)$.

Example 1. Let A be the set of all rational numbers, and suppose that x is any member of A. Then $(x + 1)$, being the sum of two rationals, is a rational number. Thus, $(x + 1)$ is in A. It follows (Definition 4.1) that A is an inductive set.

Example 2. Let B be the set of all integers less than 37, and suppose that x is any member of B except 36. Then $(x + 1)$ is an integer less than 37 and, consequently, $(x + 1)$ is in B. However, we cannot say that $(x + 1)$ is in B whenever x is, for 36 is in B, while $(36 + 1)$ is not. Thus, B is *not* an inductive set. □

Suppose now that we are confronted with a set I about which two facts are known: I is inductive and I contains the number 1. It then follows, recalling the definition of an inductive set, that every natural number belongs to I:

Since 1 is in I, then $1 + 1 = 2$ must be in I.
Since 2 is in I, then $2 + 1 = 3$ must be in I.
Since 3 is in I, then $3 + 1 = 4$ must be in I.
And so on.

If, in addition, we are told that I contains only natural numbers, then we may conclude that I is exactly the set of all natural numbers.

The above reasoning, because of the phrase "And so on," is not mathematically rigorous. However, it does serve to motivate the following proposition: If an *inductive set of natural numbers* contains the number

1, then that set is identical with the set of natural numbers. (Such fundamental propositions as this are considered in courses dealing with the foundations of mathematics.)

It is this characterization of the set of natural numbers that permits us to answer the question raised earlier. We have the following situation: A proposition in n is claimed to be true for every natural number; we wish to prove it so. To this end we let I be the set of all natural numbers for which the proposition is true. (That there are such numbers is implied by the fact that we would not have advanced the proposition in the first place if we did not know of specific instances for which it was true.) We then try to prove two facts about this set I: that 1 is in I and that I is inductive. If we are successful, we may then conclude that I is identical with the set of natural numbers. In other words, we will have proved that the proposition is true for all natural numbers. (This mode of proof is said to be a proof by *mathematical induction*.)

Example 3. Prove that

$$1 + 2 + 3 + \cdots + n = \tfrac{1}{2}n(n+1)$$

is a valid formula for all n. (Unless it is noted otherwise, the letter "n" is used to denote a natural number.)

Solution. Let I be the set of all n for which the given formula is valid. We have only to prove that I is an inductive set containing 1.

1. Replacing n by 1, we obtain the true statement

$$1 = \tfrac{1}{2}(1)(2).$$

Thus, 1 is in I.

2. To show that I is inductive we proceed as follows: Let k be any number in I. (This is usually called the *induction hypothesis*.) This means that

$$1 + 2 + 3 + \cdots + k = \tfrac{1}{2}k(k+1).$$

But then we may write

$$\begin{aligned} 1 + 2 + 3 + \cdots + (k+1) &= (1 + 2 + 3 + \cdots + k) + (k+1) \\ &= \tfrac{1}{2}k(k+1) + (k+1) \\ &= \tfrac{1}{2}(k+1)(k+2). \end{aligned}$$

In other words, if k is in I, then it follows that

$$1 + 2 + 3 + \cdots + (k+1) = \tfrac{1}{2}(k+1)(k+2).$$

But this is our formula with n replaced with $(k+1)$. Thus, we have proved that $(k+1)$ is in I. It follows that I is inductive.

Example 4. We have proved much earlier the matrix identity $(AB)' = B'A'$. We subsequently took it for granted that this result could be extended to any number of factors. We can now prove such an extension quite formally. The proposition is this: If $A_1, A_2, \ldots, A_n$ are any matrices for which the product $A_1 A_2 \cdots A_n$ is defined, then we have

$$(A_1 A_2 \cdots A_n)' = A_n' \cdots A_2' A_1'.$$

We prove this proposition by *mathematical induction:*

Let I be the set of all n for which the proposition is valid. We have only to show that I is an inductive set containing 1.

1. Evidently 1 is in I, for in this case our identity reduces to

$$(A_1)' = A_1',$$

the parentheses now being superfluous.

2. Suppose next that k is in I. This means that the identity is true for k factors. We then consider the case of $(k+1)$ factors, writing

$$\begin{aligned}
\{A_1 A_2 \cdots A_{k+1}\}' &= \{(A_1 A_2 \cdots A_k)(A_{k+1})\}' \\
&= (A_{k+1})'(A_1 A_2 \cdots A_k)' \quad \text{(By Theorem 1.11)} \\
&= (A_{k+1}')(A_k' \cdots A_2' A_1') \quad \text{(Since } k \text{ is in } I.) \\
&= A_{k+1}' \cdots A_2' A_1'.
\end{aligned}$$

Thus, our identity is true for $(k+1)$ factors. It follows that $(k+1)$ is in I and, consequently, that I is inductive. ☐

In practice, a proof by mathematical induction rarely makes explicit reference to the set I. Rather, the proof is regarded as complete if the following two statements are proved:

1. The given proposition is true for $n = 1$.
2. If the proposition is true for the kth case, then it is true for the $(k+1)$st case.

Closely associated with the notion of a proof by mathematical induction is that of an *inductive definition.* We close this section with an example illustrating this form of definition.

Example 5. A sequence $a_1, a_2, a_3, \ldots$ is introduced by the following *inductive definition:*

$$a_1 = 3 \quad \text{and} \quad a_{n+1} = \tfrac{1}{2} a_n + n.$$

Starting with $a_1 = 3$, successive terms in the sequence are easily determined:

$$\begin{aligned}
a_2 &= a_{1+1} = \tfrac{1}{2} a_1 + 1 = \tfrac{1}{2}(3) + 1 = \tfrac{5}{2}, \\
a_3 &= a_{2+1} = \tfrac{1}{2} a_2 + 2 = \tfrac{1}{2}(\tfrac{5}{2}) + 2 = \tfrac{13}{4},
\end{aligned}$$

and so on. A critical reader might now challenge the "and so on." Can we be sure that every a_n can be reached by successive computations? The answer is "yes." For consider the following:

Let I be the set of all natural numbers for which a_n is defined. Then 1 is in I, for it is given explicitly that $a_1 = 3$. Suppose next that k is in I. Thus, a_k is defined. Now, since the computation $\frac{1}{2}a_k + k$ can always be performed, it follows that a_{k+1} is defined. Thus, I is an inductive set. It follows, recalling the definition of I, that a_n is defined for all n.

The foregoing discussion is made more meaningful by considering a less trivial illustration: A sequence $b_1, b_2, b_3, \ldots$, is purported to be defined inductively by

$$b_1 = 3 \quad \text{and} \quad b_{n+1} = \frac{5b_n + (-1)^n}{(-1)^n b_n - 2}.$$

The reader may verify that

$$b_2 = -\tfrac{14}{5} \quad \text{and} \quad b_3 = \tfrac{65}{24}.$$

It would seem that we could go on and find any b_n. However, this need not be the case. For suppose that we were to come to a term b_k such that

$$(-1)^k b_k - 2 = 0.$$

It would then be impossible to go on and compute b_{k+1}. The point is this: It is not immediate in this case that our seeming inductive definition is in fact a valid definition for a sequence $b_1, b_2, b_3, \ldots$.

EXERCISE SET 4.1

1 An IQ test contains the following question: A certain list of numbers begins with

$$1, 4, 9, 16, 25.$$

What is the next number that should appear? A "troublemaker" gives the answer $\sqrt{17}$. Write a formula for a sequence $a_1, a_2, a_3, \ldots$ that justifies this answer.

2 Let $a_n = n^2 - n + 41$. It is asserted that a_n is a prime number for every natural number n. Prove or *disprove* this assertion. [*Hint:* Try a number larger than 40.]

3 Prove that $1 + 3 + 5 + \cdots + (2n - 1) = n^2$ for all n.

4 Prove that

$$\begin{bmatrix} 1 & 1 \\ 0 & 1 \end{bmatrix}^n = \begin{bmatrix} 1 & n \\ 0 & 1 \end{bmatrix}$$

for all n.

5 Let A and B be square matrices of the same order.
  a) Prove: If $AB = BA$, then $A^n B = B A^n$.
  b) Prove: If $AB = BA$, then $(AB)^n = A^n B^n$.
  c) Disprove: If $(AB)^2 = A^2 B^2$, then $AB = BA$.

6 A numerical valued function f is defined inductively over the set of all square matrices by the following two-part formula: If A is of order 1, then
$$f(A) = a_{11}.$$
If A is of order $(n+1)$, then
$$f(A) = \text{(sum of the entries lying on the perimeter of } A) - f(A^*),$$
where A^* is the nth-order matrix derived from A by deleting its first row and its first column. Now, find $f(A)$ for each of the given A's.

  a) $A = [23]$

  b) $A = \begin{bmatrix} 5 & -2 \\ 7 & 0 \end{bmatrix}$

  c) $A = \begin{bmatrix} 1 & 3 & 0 & 2 \\ 2 & 2 & 4 & -5 \\ 4 & -3 & 2 & 0 \\ 0 & 1 & 4 & 3 \end{bmatrix}$

7 Recall the problem at the beginning of this section. Prove that $R_n = \frac{1}{2}(n^2 + n + 2)$ for all n. [*Hint:* An $(n+1)$st line will cross each of the first n lines. In doing so, it must pass through $(n+1)$ regions. Thus, $R_{n+1} = R_n + \cdots$.]

8 Prove: If $A_1, A_2, \ldots, A_n$ are all nonsingular matrices of the same order, then $(A_1 A_2 \cdots A_n)^{-1} = A_n^{-1} \cdots A_2^{-1} A_1^{-1}$.

9 A sequence of numbers is defined inductively by
$$a_1 = \sqrt{2} \quad \text{and} \quad a_{n+1} = \sqrt{2 + a_n}.$$
  a) Prove that $a_n < 2$ for all n.
  b) Prove that $a_{n+1} > a_n$ for all n. [*Hint:* Show that $(a_{n+2} - a_{n+1}) = $ (positive number)$(a_{n+1} - a_n)$.]

10 Definition (for this exercise only): A square matrix is said to be 1-rowed if the sum of the entries in any row is 1.
  a) Let
$$A = \begin{bmatrix} 1 & 0 & 0 \\ \frac{1}{2} & 0 & \frac{1}{2} \\ 1 & 2 & -2 \end{bmatrix}.$$
  Verify that A, A^2, A^3, and A^4 are all 1-rowed.
  b) Prove: If A is 1-rowed, then A^n is 1-rowed. [*Hint:* Let $U = [1, 1, \ldots, 1]'$ and first prove the following: A matrix B is 1-rowed if and only if $BU = U$. Then write $A^{k+1} U = A(A^k U) = \cdots$.]

11 *Tower of Hanoi problem.* A set of n washers is stacked on one of three vertical pegs. Reading up that peg, each successive washer is smaller than the one preceding it. The problem is this: By moving washers one at a time from peg to peg—but in such a way that no larger washer is ever on top of a smaller one—can the original stack be transferred to one of the other two pegs?

a) The answer is "yes." Prove this by using mathematical induction. [*Hint:* Consider a special case. Suppose that you know how to transfer a stack of eight such washers. Could you not use that knowledge to solve the problem for nine washers?]

b) Let M_n denote the smallest number of moves needed to solve the problem for n washers. Find M_1, M_2, M_3, and M_4. Guess a formula for M_n, and then prove it. [*Hint:* Your experience in part (a) should be useful.]

4.2 THE DETERMINANT FUNCTION

It is easy—but perhaps uninteresting—to construct functions that assign numerical values to matrices. For example, given a matrix A, we could sum its columns and then multiply those sums together. Such a sequence of computations might then be taken as a "formula" for a certain matrix-to-number function f. Consider the illustration

$$f\left(\begin{bmatrix} 1 & 2 & 3 \\ 4 & 5 & 6 \end{bmatrix}\right) = 5 \cdot 7 \cdot 9 = 315.$$

A more elaborate function of this type can be constructed by using an inductive definition:

1. To first-order matrices the function g assigns the value of the single element making up the matrix, that is, $g([c]) = c$.

2. For matrices of order $(n+1)$ we have the following rule:

$g(A) = $ (the matrix product of A's last row times its last column) $- g(A^*)$,

where A^* is the matrix of order n that remains on striking out the last row and column of A. Consider the illustration

$$g\left(\begin{bmatrix} 5 & 2 & 4 & 0 \\ 3 & 1 & 2 & 1 \\ 2 & 0 & -1 & 4 \\ 1 & 4 & 3 & 1 \end{bmatrix}\right)$$

$$= (1 \cdot 0) + (4 \cdot 1) + (3 \cdot 4) + (1 \cdot 1) - g\left(\begin{bmatrix} 5 & 2 & 4 \\ 3 & 1 & 2 \\ 2 & 0 & -1 \end{bmatrix}\right)$$

$$= 17 - \left\{(2 \cdot 4) + (0 \cdot 2) + (-1)(-1) - g\left(\begin{bmatrix} 5 & 2 \\ 3 & 1 \end{bmatrix}\right)\right\}$$

$$= 8 + \{(3 \cdot 2) + (1 \cdot 1) - g([5])\}$$

$$= 15 - 5 = 10.$$

It should be noted that g is defined only for square matrices. (Why?)

4.2 THE DETERMINANT FUNCTION

The two functions just described have no significance; they are just curiosities. Indeed, few matrix-to-number functions play an important role in mathematics. However, some do. This chapter will be devoted to the study of one such function. Like g above, its domain is restricted to the set of square matrices, and also like g, it is defined inductively.

Definition 4.2. The *determinant function*, denoted by Δ, is defined as follows:

1. $\Delta([a_{11}]) = a_{11}$.
2. If $A = [a_{ij}]_{n+1, n+1}$, then

$$\Delta(A) = \sum_{k=1}^{n+1} (-1)^{k+1} a_{k1} \Delta(A_{k1}),$$

where A_{k1} is the n by n matrix that remains on striking out the kth row and the first column of A.

Example 1. Let

$$A = \begin{bmatrix} 1 & 2 \\ 3 & 4 \end{bmatrix}.$$

Find $\Delta(A)$.

Solution. Taking $n = 1$ in our recursion formula, we have

$$\begin{aligned} \Delta(A) &= \sum_{k=1}^{2} (-1)^{k+1} a_{k1} \Delta(A_{k1}) \\ &= (-1)^2 a_{11} \Delta(A_{11}) + (-1)^3 a_{21} \Delta(A_{21}) \\ &= 1 \cdot \Delta([4]) - 3 \cdot \Delta([2]). \end{aligned}$$

We next appeal to the explicit definition of Δ given for 1 by 1 matrices:

$$\Delta([4]) = 4 \quad \text{and} \quad \Delta([2]) = 2.$$

Thus,

$$\Delta(A) = (1 \cdot 4) - (3 \cdot 2) = -2.$$

In practice, the second-order case is handled by the explicit formula

$$\Delta\left(\begin{bmatrix} a_{11} & a_{12} \\ a_{21} & a_{22} \end{bmatrix}\right) = a_{11}a_{22} - a_{21}a_{12},$$

which the reader may verify.

Example 2. Let

$$A = \begin{bmatrix} 2 & 5 & -3 \\ 1 & 2 & 7 \\ -3 & 2 & 4 \end{bmatrix}.$$

Find $\Delta(A)$.

Solution. Taking $n = 2$ in our recursion formula, we have

$$\Delta(A) = \sum_{k=1}^{3} (-1)^{k+1} a_{k1} \Delta(A_{k1})$$
$$= (-1)^2 a_{11} \Delta(A_{11}) + (-1)^3 a_{21} \Delta(A_{21}) + (-1)^4 a_{31} \Delta(A_{31})$$
$$= 2 \cdot \Delta(A_{11}) - 1 \cdot \Delta(A_{21}) - 3\Delta(A_{31}).$$

The evaluation of Δ for the resulting 2 by 2 matrices may be carried out by a second application of the recursion formula or by using the formula given at the end of Example 1. Either way, the reader should verify that

$$\Delta(A_{11}) = -6,$$
$$\Delta(A_{21}) = 26,$$
$$\Delta(A_{31}) = 41.$$

Finally,
$$\Delta(A) = -161.$$

For 3 by 3 matrices the evaluation of Δ can also be carried out by using the following "formula":

1. Repeat the first two columns of A to its right:

$$\begin{bmatrix} 2 & 5 & -3 \\ 1 & 2 & 7 \\ -3 & 2 & 4 \end{bmatrix} \begin{matrix} 2 & 5 \\ 1 & 2 \\ -3 & 2 \end{matrix}$$

2. Slash the resulting array with arrows as shown:

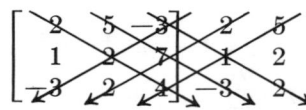

3. Then $\Delta(A)$ is equal to the sum of the "products under the arrows of negative slope" minus the sum of the "products under the arrows of positive slope." Thus,

$$\Delta(A) = (16 - 105 - 6) - (18 + 28 + 20) = -161.$$

The reader should verify that this method is valid in general. (See Exercise 12.) *For higher-order matrices there is no analogous scheme.* □

The evaluation of Δ for high-order matrices is, in general, a tedious business. However, there is an interesting special case in which the evaluation of $\Delta(A)$ becomes trivial regardless of the order of A.

4.2 THE DETERMINANT FUNCTION

Example 3. Let
$$A = \begin{bmatrix} 2 & 3 & 4 & -6 & 7 \\ 0 & 3 & 5 & 7 & -3 \\ 0 & 0 & -1 & 2 & 9 \\ 0 & 0 & 0 & 5 & 8 \\ 0 & 0 & 0 & 0 & 4 \end{bmatrix}.$$

Find $\Delta(A)$.

Solution. Taking $n = 4$ in our recursion formula, we have
$$\Delta(A) = 2 \cdot \Delta(A_{11}) - 0 + 0 - 0 + 0.$$
Similarly,
$$\Delta(A_{11}) = 3 \cdot \Delta[(A_{11})_{11}] - 0 + 0 - 0,$$
where
$$(A_{11})_{11} = \begin{bmatrix} -1 & 2 & 9 \\ 0 & 5 & 8 \\ 0 & 0 & 4 \end{bmatrix}.$$

Thus, so far,
$$\Delta(A) = 2 \cdot 3 \cdot \Delta\left(\begin{bmatrix} -1 & 2 & 9 \\ 0 & 5 & 8 \\ 0 & 0 & 4 \end{bmatrix}\right).$$

Continuing, we find that $\Delta(A)$ is simply the product of the elements appearing on the main diagonal of A, that is,
$$\Delta(A) = (2)(3)(-1)(5)(4) = -120.$$

(If A is lower triangular, then the same result holds, but this is not quite so evident.)

Theorem 4.1. If A is triangular, then $\Delta(A) = a_{11}a_{22} \cdots a_{nn}$.

Proof. (We consider only the more difficult lower triangular case.) Our proof is by mathematical induction:

1. By definition, $\Delta([a_{11}]) = a_{11}$.
2. Next, suppose that the given formula holds for all lower triangular matrices of order n. Then, taking A as a lower triangular matrix of order $(n+1)$, we may write
$$\Delta(A) = \sum_{k=1}^{n+1} (-1)^{k+1} a_{k1} \Delta(A_{k1})$$
$$= a_{11}\Delta(A_{11}) + \sum_{k=2}^{n+1} (-1)^{k+1} a_{k1} \Delta(A_{k1}).$$

The first term has been singled out here for the following reason: although each matrix A_{k1} is lower triangular and of order n—thus, our induction hypothesis is applicable—it is the case that for $k > 1$

the first row of A_{k1} contains only zeros. Thus, its main diagonal contains at least one zero. It follows, using our induction hypothesis, that for $k > 1$ we have $\Delta(A_{k1}) = 0 \cdot$ (other diagonal factors) $= 0$. But for $k = 1$, we have $\Delta(A_{11}) = a_{22}a_{33} \cdots a_{n+1,n+1}$. Finally, then,

$$\Delta(A) = a_{11}(a_{22}a_{33} \cdots a_{n+1,n+1}) + \sum_{k=2}^{n+1} 0$$
$$= a_{11}a_{22} \cdots a_{n+1,n+1}.$$

And this is what we had to prove.

EXERCISE SET 4.2

1 Find $\Delta(A)$, where A is the given matrix.

a) [755]   b) $\begin{bmatrix} 1 & 0 & 0 \\ 4 & 2 & 0 \\ 6 & 5 & 3 \end{bmatrix}$ c) $\begin{bmatrix} 3 & 0 & 0 \\ 0 & 4 & 0 \\ 0 & 0 & 5 \end{bmatrix}$ d) $\begin{bmatrix} 1 & 0 & 2 & 0 \\ 0 & 2 & 0 & 3 \\ 3 & 0 & 4 & 0 \\ 0 & 4 & 0 & 5 \end{bmatrix}$

In Exercises 2 through 8, A is the matrix

$$\begin{bmatrix} 3 & 2 & 0 \\ 3 & 1 & 1 \\ 7 & -1 & 2 \end{bmatrix}.$$

2 Find $\Delta(A)$.
3 Let $B = A'$. Find $\Delta(B)$, compare with $\Delta(A)$, and make a conjecture.
4 Same as 3, except that $B = A^{-1}$.
5 Same as 3, except that $B = EA$, where $E = [2 \to (7)2]_3$.
6 Same as 3, except that $B = EA$, where $E = [1 \leftrightarrow 3]_3$.
7 Same as 3, except that $B = EA$, where $E = [2 \to 2 + (14)3]_3$.
8 Let

$$B = \begin{bmatrix} 1 & 2 & 1 \\ 0 & 3 & 4 \\ 1 & 0 & -5 \end{bmatrix}.$$

Find $\Delta(B)$, find $\Delta(AB)$, compare with $\Delta(A) \cdot \Delta(B)$, and make a conjecture.

9 Let

$$A = \begin{bmatrix} 1 & 2 & 3 \\ 4 & 5 & 7 \\ 0 & 0 & 0 \end{bmatrix}.$$

Find $\Delta(A)$ and make a conjecture.

10 Let

$$A = \begin{bmatrix} 1 & 2 & 3 \\ 5 & 7 & -1 \\ 10 & 20 & 30 \end{bmatrix}.$$

Find $\Delta(A)$ and make a conjecture.

11 Let
$$A = \begin{bmatrix} 1 & 0 & 2 & 3 \\ 0 & 1 & 1 & 2 \\ 2 & 3 & 7 & 12 \\ 3 & -1 & 5 & 7 \end{bmatrix} \quad \text{and} \quad B = \begin{bmatrix} 1 & 0 & 0 & 2 \\ 1 & 2 & 5 & 5 \\ 0 & 1 & 2 & 2 \\ 1 & 1 & 2 & 4 \end{bmatrix}.$$

a) Find rank (A) and rank (B).
b) Find $\Delta(A)$ and $\Delta(B)$.
c) Make a conjecture.

12 At the end of Example 2 there was given a formula for the direct evaluation of the determinant of a 3 by 3 matrix. Prove that formula. [*Hint:* First apply that "formula" to $[a_{ij}]_{33}$, thus obtaining a "real formula" for $\Delta(A)$.]

4.3 BASIC PROPERTIES

For nontriangular matrices of high order the evaluation of $\Delta(A)$ is facilitated by exploiting certain properties of the determinant function. We first list these properties and show how they are applied. In the next section we shall consider the theory out of which they arise.

Property 1. Let A^* be the matrix derived from A by multiplying each element in a certain row by c. Then $\Delta(A^*) = c \cdot \Delta(A)$.

Property 2. Let A^* be the matrix derived from A by interchanging two rows. Then $\Delta(A^*) = -\Delta(A)$.

Property 3. Let A^* be the matrix derived from A by adding a multiple of one row to any other row. Then $\Delta(A^*) = \Delta(A)$.

Property 4. $\Delta(A') = \Delta(A)$.

For the moment we consider only the first three properties. Taken collectively, their message is this: If a matrix is premultiplied by an elementary matrix, then the determinant of the resulting matrix differs from that of the original matrix in a predictable way. This fact—plus the fact that we are good at evaluating the determinant of a triangular matrix—suggests the following technique for evaluating the determinant of a nontriangular matrix: Given the matrix A_1, find matrices A_2, A_3, $\ldots$, A_p such that A_p is upper triangular and such that each A_j differs from the preceding one by a single row operation. Now, evaluate $\Delta(A_p)$. Then $\Delta(A_{p-1})$ is either equal to a certain multiple of $\Delta(A_p)$, or to the negative of $\Delta(A_p)$, or to $\Delta(A_p)$ itself. Thus, $\Delta(A_{p-1})$ is easily found. Similarly, $\Delta(A_{p-2})$ is easily obtained once $\Delta(A_{p-1})$ is known. And so on, until finally $\Delta(A_1)$ is found. In practice, this procedure is abbreviated by simply performing row operations on A_1 while simultaneously compensating for any change occurring in the value of the determinant. An example will make this clear.

Example 1. Let
$$A = \begin{bmatrix} 1 & 2 & 1 & -1 \\ 1 & 2 & -2 & 3 \\ 2 & 1 & 1 & 3 \\ 0 & -1 & 3 & 2 \end{bmatrix}.$$

Find $\Delta(A)$.

Solution

$$\Delta(A) = \Delta\left(\begin{bmatrix} 1 & 2 & 1 & -1 \\ 0 & 0 & -3 & 4 \\ 2 & 1 & 1 & 3 \\ 0 & -1 & 3 & 2 \end{bmatrix}\right) \quad (R_2 \to R_2 - R_1)$$

$$= -\Delta\left(\begin{bmatrix} 1 & 2 & 1 & -1 \\ 2 & 1 & 1 & 3 \\ 0 & 0 & -3 & 4 \\ 0 & -1 & 3 & 2 \end{bmatrix}\right) \quad (R_2 \leftrightarrow R_3)$$

$$= -\Delta\left(\begin{bmatrix} 1 & 2 & 1 & -1 \\ 0 & -3 & -1 & 5 \\ 0 & 0 & -3 & 4 \\ 0 & -1 & 3 & 2 \end{bmatrix}\right) \quad (R_2 \to R_2 - 2R_1)$$

$$= \tfrac{1}{3}\Delta\left(\begin{bmatrix} 1 & 2 & 1 & -1 \\ 0 & -3 & -1 & 5 \\ 0 & 0 & -3 & 4 \\ 0 & 3 & -9 & -6 \end{bmatrix}\right) \quad (R_4 \to -3R_4)$$

$$= \tfrac{1}{3}\Delta\left(\begin{bmatrix} 1 & 2 & 1 & -1 \\ 0 & -3 & -1 & 5 \\ 0 & 0 & -3 & 4 \\ 0 & 0 & -10 & -1 \end{bmatrix}\right) \quad (R_4 \to R_4 + R_2)$$

$$= \tfrac{1}{30}\Delta\left(\begin{bmatrix} 1 & 2 & 1 & -1 \\ 0 & -3 & -1 & 5 \\ 0 & 0 & -30 & 40 \\ 0 & 0 & -10 & -1 \end{bmatrix}\right) \quad (R_3 \to 10R_3)$$

$$= -\tfrac{1}{90}\Delta\left(\begin{bmatrix} 1 & 2 & 1 & -1 \\ 0 & -3 & -1 & 5 \\ 0 & 0 & -30 & 40 \\ 0 & 0 & 30 & 3 \end{bmatrix}\right) \quad (R_4 \to -3R_4)$$

$$= -\tfrac{1}{90}\Delta\left(\begin{bmatrix} 1 & 2 & 1 & -1 \\ 0 & -3 & -1 & 5 \\ 0 & 0 & -30 & 40 \\ 0 & 0 & 0 & 43 \end{bmatrix}\right) \quad (R_4 \to R_4 + R_3)$$

$$= -\tfrac{1}{90}(1)(-3)(-30)(43) = -43.$$

The particular sequence of row operations used to reduce A to triangular form is one of personal preference. A wish to avoid fractions dictated certain of the row operations used in the foregoing calculation. On the other hand, if fractions are not "feared," then all applications of Property 1 can be avoided. (It should be appreciated by the reader that we must perform a certain number of determinant calculations to see how the method works; such experience would then aid us in programming a computer to do such calculations.)

Example 2. Let
$$A = \begin{bmatrix} 1 & 2 & 4 & 3 \\ 2 & 3 & 2 & 4 \\ 0 & 0 & 7 & 0 \\ 1 & 2 & 5 & 2 \end{bmatrix}.$$

Find $\Delta(A)$.

Solution. Properties 2 and 4 can be used to exploit the fact that A has an "almost zero" row:

$$\Delta(A) = -\Delta\left(\begin{bmatrix} 0 & 0 & 7 & 0 \\ 2 & 3 & 2 & 4 \\ 1 & 2 & 4 & 3 \\ 1 & 2 & 5 & 2 \end{bmatrix}\right) \quad (R_1 \leftrightarrow R_3)$$

$$= -\Delta\left(\begin{bmatrix} 0 & 2 & 1 & 1 \\ 0 & 3 & 2 & 2 \\ 7 & 2 & 4 & 5 \\ 0 & 4 & 3 & 2 \end{bmatrix}\right) \quad (A \to A')$$

$$= -7 \cdot \Delta\left(\begin{bmatrix} 2 & 1 & 1 \\ 3 & 2 & 2 \\ 4 & 3 & 2 \end{bmatrix}\right) \quad \text{(Definition 4.2)}$$

$$= -7 \cdot \Delta\left(\begin{bmatrix} 2 & 1 & 1 \\ -1 & 0 & 0 \\ 4 & 3 & 2 \end{bmatrix}\right) \quad (R_2 \to R_2 - 2R_1)$$

$$= (-7)(-1) = 7. \quad \square$$

So far we have learned some properties of the determinant function and we have acquired some skill in evaluating it. But now, two questions arise: What use has the determinant function? Where did it come from? We answer only the second question in this section; in subsequent sections, some applications will be given.

The determinant function has its origin in the study of linear systems having n equations and n unknowns. By solving such systems, in general, it was discovered that the solutions—for different choices of n—followed a certain pattern. It was this pattern that was "captured" in the determinant function. For example, consider the system

$$\begin{cases} a_{11}x_1 + a_{12}x_2 + a_{13}x_3 = b_1, \\ a_{21}x_1 + a_{22}x_2 + a_{23}x_3 = b_2, \\ a_{31}x_1 + a_{32}x_2 + a_{33}x_3 = b_3. \end{cases}$$

The augmented matrix of this system may be reduced in the usual way, and then a solution may be expressed in terms of the a_{ij} and the b_j. We write this result only for x_1:

$$x_1 = \frac{b_1 a_{22} a_{33} - b_1 a_{23} a_{32} + b_2 a_{13} a_{32} - b_2 a_{12} a_{33} + b_3 a_{12} a_{23} - b_3 a_{13} a_{22}}{a_{11} a_{22} a_{33} - a_{11} a_{23} a_{32} + a_{21} a_{13} a_{32} - a_{21} a_{12} a_{33} + a_{31} a_{12} a_{23} - a_{31} a_{13} a_{22}}.$$

But—and the determinant function was defined so that this would be the case—it turns out that this result has the following compact form:

$$x_1 = \frac{\Delta\left(\begin{bmatrix} b_1 & a_{12} & a_{13} \\ b_2 & a_{22} & a_{23} \\ b_3 & a_{32} & a_{33} \end{bmatrix}\right)}{\Delta\left(\begin{bmatrix} a_{11} & a_{12} & a_{13} \\ a_{21} & a_{22} & a_{23} \\ a_{31} & a_{32} & a_{33} \end{bmatrix}\right)}.$$

Analogous results hold for x_2 and x_3. Indeed—and this is the point—given n equations in n unknowns, each unknown can be expressed as a quotient, the numerator and denominator being determinants of certain nth-order matrices. The actual formulas for the general case, and how they are derived, will be taken up in Section 4.6.

It should also be noted, while speaking on the origins of the determinant function, that early mathematicians used the word "function" in a very restricted way: a *function* was a number-to-number correspondence and it had to have a "reasonable" formula. Thus, they dealt with the determinant function without really recognizing it for what it was: a matrix-to-number function. Consequently, they could not have invented the notation (and terminology) that we have been using. Now it happens that the traditional notation is the one in common use; we shall now introduce it: we write

$$|A| \quad \text{for} \quad \Delta(A).$$

Recall Example 1. The problem posed there could have been presented in the following form: Evaluate

$$\begin{vmatrix} 1 & 2 & 1 & -1 \\ 1 & 2 & -2 & 3 \\ 2 & 1 & 1 & 3 \\ 0 & -1 & 3 & 2 \end{vmatrix}.$$

The pair of vertical bars does double duty: it indicates the operation of the determinant function and it also serves as the "frame" for the matrix. Also, the symbol "$|A|$" is itself referred to as "a determinant." Thus, for example, it is correct to say: "If a determinant has two equal rows, then its value is zero." This is in keeping with the usual custom—this is again traceable to the early confusion about what a function was—that mathematicians have of calling a symbol indicating the operation of a certain function by that function's name. For example, the mark "log 2" is called "a logarithm"; the mark "$\sqrt[3]{17}$" is called a "cube root"; and so on.

EXERCISE SET 4.3

1 Evaluate the given determinant using the technique set forth in Example 1.

a) $\begin{vmatrix} 2 & 4 & -2 & 0 \\ 0 & 1 & 2 & 3 \\ 1 & 5 & 2 & -2 \\ 0 & 3 & 7 & 1 \end{vmatrix}$
b) $\begin{vmatrix} 2 & 4 & 6 & 8 \\ 3 & -3 & 3 & -3 \\ 50 & 60 & 70 & 80 \\ -1 & -1 & -1 & -1 \end{vmatrix}$

2 Evaluate the given determinant by first exploiting the "almost zero" row or column. (Recall Example 2.)

a) $\begin{vmatrix} 1 & 1 & 0 & 5 \\ 2 & 3 & 0 & 4 \\ 9 & 9 & 9 & 9 \\ 0 & -2 & 0 & 1 \end{vmatrix}$
b) $\begin{vmatrix} 2 & 4 & 7 & 6 & 8 \\ 3 & -3 & 6 & 3 & -3 \\ 0 & 0 & 5 & 0 & 0 \\ 50 & 60 & 4 & 70 & 80 \\ -1 & -1 & 3 & -1 & -1 \end{vmatrix}$

3 Evaluate the given determinants in order and then make a conjecture.

$$\begin{vmatrix} 1 & 1 \\ 1 & 2 \end{vmatrix}, \quad \begin{vmatrix} 1 & 1 & 1 \\ 1 & 2 & 3 \\ 1 & 3 & 6 \end{vmatrix}, \quad \begin{vmatrix} 1 & 1 & 1 & 1 \\ 1 & 2 & 3 & 4 \\ 1 & 3 & 6 & 10 \\ 1 & 4 & 10 & 20 \end{vmatrix}.$$

[*Hint:* Consider

$$a_{ij} = \binom{i+j-2}{i-1},$$

where

$$\binom{n}{k} = \frac{n!}{k!(n-k)!}.$$

In Exercises 4 through 7 the given equation is to be verified without using Properties 1 through 4. Thus, the reader is being asked to establish those properties for certain special cases.

4 $\begin{vmatrix} a_1 & a_2 & a_3 \\ b_1 & b_2 & b_3 \\ kc_1 & kc_2 & kc_3 \end{vmatrix} = k \begin{vmatrix} a_1 & a_2 & a_3 \\ b_1 & b_2 & b_3 \\ c_1 & c_2 & c_3 \end{vmatrix}$

5 $\begin{vmatrix} c_1 & c_2 & c_3 \\ b_1 & b_2 & b_3 \\ a_1 & a_2 & a_3 \end{vmatrix} = - \begin{vmatrix} a_1 & a_2 & a_3 \\ b_1 & b_2 & b_3 \\ c_1 & c_2 & c_3 \end{vmatrix}$

6 $\begin{vmatrix} a_1 + kb_1 & a_2 + kb_2 & a_3 + kb_3 \\ b_1 & b_2 & b_3 \\ c_1 & c_2 & c_3 \end{vmatrix} = \begin{vmatrix} a_1 & a_2 & a_3 \\ b_1 & b_2 & b_3 \\ c_1 & c_2 & c_3 \end{vmatrix}$

7 $\begin{vmatrix} a_1 & a_2 & a_3 \\ b_1 & b_2 & b_3 \\ c_1 & c_2 & c_3 \end{vmatrix} = \begin{vmatrix} a_1 & b_1 & c_1 \\ a_2 & b_2 & c_2 \\ a_3 & b_3 & c_3 \end{vmatrix}$

8 Use Properties 2 and 4 to show that

$$\begin{vmatrix} a_1 & b_1 & c_1 \\ a_2 & b_2 & c_2 \\ a_3 & b_3 & c_3 \end{vmatrix} = \begin{vmatrix} a_2 & c_2 & b_2 \\ a_1 & c_1 & b_1 \\ a_3 & c_3 & b_3 \end{vmatrix}.$$

9 Prove that $\Delta(cA) = c^n \Delta(A)$, where A is n by n.

10 a) Evaluate

$$\begin{vmatrix} 0 & 1 & -5 \\ -1 & 0 & 2 \\ 5 & -2 & 0 \end{vmatrix}.$$

b) Evaluate

$$\begin{vmatrix} 0 & -1 & 1 & -3 \\ 1 & 0 & -2 & -1 \\ -1 & 2 & 0 & -4 \\ 3 & 1 & 4 & 0 \end{vmatrix}.$$

c) Prove: If $A' = -A$, and the order of A is odd, then $\Delta(A) = 0$. (A matrix that has the property that $A' = -A$ is said to be *skew symmetric*.)

d) Prove: If A is a fourth-order, skew symmetric matrix, then

$$|A| = (a_{12}a_{34} + a_{13}a_{24} + a_{14}a_{23})^2.$$

11 Read Appendix C.

4.4 THEORY OF DETERMINANTS

In this section, and in the two sections that follow, we shall state and prove a number of theorems about the determinant function. The four properties considered in the preceding section will find a place in this theoretical development; in addition, certain other results will arise. This entire treatment must be regarded as only an introduction to the determinant function; however, a mastery of these results will provide the reader with an adequate background for following mathematical discussions in which the determinant function plays a role.

Theorem 4.2. *Property 1.* If A^* is derived from A by multiplying its pth row by c, then $|A^*| = c|A|$. (In terms of elementary matrices, we

have
$$|[p \to (c)p] \cdot A| = c|A|.$$

However, Theorem 4.2 says just a little bit more: the multiplier c is permitted to be zero, whereas c may not be zero in the row-multiplying elementary matrix.)

Proof. *By mathematical induction.* If $A = [a_{11}]$, then we have directly that
$$|A^*| = |ca_{11}| = ca_{11} = c|a_{11}| = c|A|.$$

Suppose next that the formula of the theorem holds for nth-order matrices (our induction hypothesis), and that A and A^* are of order $(n+1)$. Then we may begin by writing
$$|A^*| = \sum_{k=1}^{n+1} (-1)^{k+1} a_{k1}^* |A_{k1}^*|.$$

Next, we separate out the pth term of this sum:
$$|A^*| = (-1)^{p+1} a_{p1}^* |A_{p1}^*| + \sum_{k \ne p} (-1)^{k+1} a_{k1}^* |A_{k1}^*|$$
$$= \alpha + \beta.$$

The case for α: We observe first that $a_{p1}^* = ca_{p1}$. Next, we note that $A_{p1}^* = A_{p1}$, for the deleted pth row contains the only elements that distinguish A^* from A. It follows that
$$\alpha = (-1)^{p+1} ca_{p1} |A_{p1}|.$$

The case for β: We observe first that for $k \ne p$ we have $a_{k1}^* = a_{k1}$. But $A_{k1}^* \ne A_{k1}$. For in the construction of A_{k1}^* we have not deleted the pth row of A^*. However, they differ only in that one row of A_{k1}^* is c times the corresponding row of A_{k1}. Now these matrices are of order n; hence, our induction hypothesis is applicable. It follows that
$$\beta = \sum_{k \ne p} (-1)^{k+1} a_{k1} \{c|A_{k1}|\}.$$

Finally, combining these results, we have
$$|A^*| = \alpha + \beta = c\{(-1)^{p+1} a_{p1} |A_{p1}| + \sum_{k \ne p} (-1)^{k+1} a_{k1} |A_{k1}|\}$$
$$= c \sum_{k=1}^{n+1} (-1)^k a_{k1} |A_{k1}|$$
$$= c|A|.$$

Theorem 4.3. *Property 2.* If A^* is derived from A by interchanging its pth and qth row, then $|A^*| = -|A|$. (In terms of elementary matrices we have $|[p \leftrightarrow q] \cdot A| = -|A|$.)

Proof

1. Suppose that $q > p$. Then there is a positive integer k such that $q = p + k$. Thus, letting R_i denote the ith row of A, we may write

$$A = \begin{bmatrix} R_1 \\ \vdots \\ R_p \\ \vdots \\ R_{p+k} \\ \vdots \\ R_n \end{bmatrix} \quad \text{and} \quad A^* = \begin{bmatrix} R_1 \\ \vdots \\ R_{p+k} \\ \vdots \\ R_p \\ \vdots \\ R_n \end{bmatrix}.$$

   Now start with A. Move R_p down through the matrix by successively interchanging rows. In k moves, R_p will be in the place originally occupied by R_{p+k}, and R_{p+k} will be one row above its original position. Thus, R_{p+k} can make its way to the position originally occupied by R_p in only $(k-1)$ moves. Altogether, $(2k-1)$ interchanges of adjacent rows will transform A into A^*. Now suppose that the formula of the theorem was known to hold in the special case where the rows interchanged were adjacent. It would then follow that $|A^*| = -|A|$. (Why?)

2. We prove the special case: Let $q = p + 1$. Again we use mathematical induction. Here, the first case for which the theorem has meaning is where $n = 2$. There is one possible interchange:

$$A = \begin{bmatrix} a_{11} & a_{12} \\ a_{21} & a_{22} \end{bmatrix} \quad \text{and} \quad A^* = \begin{bmatrix} a_{21} & a_{22} \\ a_{11} & a_{12} \end{bmatrix}.$$

   By direct computation, we have $|A^*| = -|A|$.

   Next, suppose that the formula holds for matrices of order n (our induction hypothesis), and that A and A^* are of order $n+1$. Then we begin by writing

$$|A^*| = \sum_{k=1}^{n+1} (-1)^{k+1} a_{k1}^* |A_{k1}^*|$$
$$= (p\text{th term}) + ((p+1)\text{st term}) + (\text{remaining terms})$$
$$= \alpha + \beta + \gamma.$$

   Reasoning analogous to that used in the preceding proof permits the following replacements:

$$\alpha = (-1)^{p+1} a_{p1}^* |A_{p1}^*| = (-1)^{p+1} a_{p+1,1} |A_{p+1,1}|,$$
$$\beta = (-1)^{p+2} a_{p+1,1}^* |A_{p+1,1}^*| = (-1)^{p+2} a_{p1} |A_{p1}|,$$
$$\gamma = \sum_{k \neq p, p+1} (-1)^{k+1} a_{k1}^* |A_{k1}^*| = \sum_{k \neq p, p+1} (-1)^{k+1} a_{k1} \{-|A_{k1}|\}.$$

Finally,
$$\begin{aligned}|A^*| &= \beta + \alpha + \gamma \\ &= -\{(-1)^{p+1}a_{p1}|A_{p1}|\} - \{(-1)^{p+2}a_{p+1,1}|A_{p+1,1}|\} \\ &\quad - \left\{\sum_{k\neq p,p+1}(-1)^{k+1}a_{k1}|A_{k1}|\right\} \\ &= -\sum_{k=1}^{n+1}(-1)^{k+1}a_{k1}|A_{k1}| = -|A|.\end{aligned}$$

Corollary. If two rows of A are identical, then $|A| = 0$.

Proof. Let A^* be derived from A by interchanging its identical rows. Then, by our theorem, $|A^*| = -|A|$. But A^* is still the same as A! Thus,
$$|A^*| = |A|.$$
It follows that
$$|A| = -|A|.$$
And, consequently, that
$$|A| = 0.$$

Theorem 4.4. Let A, B, and C be nth-order matrices that are related as follows:
1. The three matrices are identical except for their pth rows.
2. The pth row of C is the vector sum of the pth rows of A and B.

Then
$$|C| = |A| + |B|.$$

Proof. A proof by mathematical induction may be constructed following the pattern set in the proof of Theorem 4.2. □

This particular property of the determinant function has not been considered before; a numerical illustration will make it more vivid:

$$\begin{vmatrix} 1 & 2 & 3 & 4 \\ 5 & 6 & 7 & 8 \\ 6 & 3 & 2 & 4 \\ 3 & 4 & 5 & 6 \end{vmatrix} = \begin{vmatrix} 1 & 2 & 3 & 4 \\ 5 & 6 & 7 & 8 \\ 2 & 0 & -3 & 5 \\ 3 & 4 & 5 & 6 \end{vmatrix} + \begin{vmatrix} 1 & 2 & 3 & 4 \\ 5 & 6 & 7 & 8 \\ 4 & 3 & 5 & -1 \\ 3 & 4 & 5 & 6 \end{vmatrix}.$$

We shall have only one use for Theorem 4.4: to help us prove Property 3.

Theorem 4.5. *Property 3.* Suppose that A^* is derived from A by adding c times its pth row to its qth row. Then $|A^*| = |A|$. (In terms of elementary matrices, we have $|[q \to q + (c)p] \cdot A| = |A|$.)

Proof. Again using a row notation for our matrices, we write

$$A = \begin{bmatrix} R_1 \\ \vdots \\ R_p \\ \vdots \\ R_q \\ \vdots \\ R_n \end{bmatrix}, \quad B = \begin{bmatrix} R_1 \\ \vdots \\ R_p \\ \vdots \\ cR_p \\ \vdots \\ R_n \end{bmatrix}, \quad \text{and} \quad A^* = \begin{bmatrix} R_1 \\ \vdots \\ R_p \\ \vdots \\ R_q + cR_p \\ \vdots \\ R_n \end{bmatrix}.$$

Then Theorem 4.4 is applicable, where A^* plays the role of C. Thus, $|A^*| = |A| + |B|$. But $|B| = 0$! (Why?) □

We close this section by considering an application of the equal-row property (the corollary to Theorem 4.3) to analytic geometry. This application shows again—recall the discussion at the end of the last section—the use of the determinant function as a notational device.

Example 1. Consider the equation

$$\begin{vmatrix} y & 3 & 8 & 5 \\ x^2 & 4 & 9 & 1 \\ x & 2 & 3 & -1 \\ 1 & 1 & 1 & 1 \end{vmatrix} = 0.$$

By "evaluating" the determinant, this equation can be brought into the form

$$ay + bx^2 + cx + d = 0,$$

or,

$$y = \alpha x^2 + \beta x + \gamma,$$

provided that $a \neq 0$. This last equation (and hence also the original one) is that of a parabola having a vertical axis of symmetry. Now return to the determinant form of the equation. If we substitute any one of the points (2, 3), (3, 8), or (−1, 5) into the determinant, then we obtain a determinant having two identical columns. It follows that the resulting determinant has the value zero. (We are using here the corollary to Theorem 4.3 plus the fact that $|A'| = |A|$. We shall prove this latter fact in the next section.) But this means that each of these points is on the parabola.

This argument permits the following general remark: The vertical parabola passing through the (noncollinear) points (x_1, y_1), (x_2, y_2), and (x_3, y_3) is given by the equation

$$\begin{vmatrix} y & y_1 & y_2 & y_3 \\ x^2 & x_1^2 & x_2^2 & x_3^2 \\ x & x_1 & x_2 & x_3 \\ 1 & 1 & 1 & 1 \end{vmatrix} = 0.$$

EXERCISE SET 4.4

1 Verify (without using Theorem 4.4) that
$$\begin{vmatrix} 1 & 0 & 2 \\ 2 & 3 & 1 \\ 7 & 3 & 4 \end{vmatrix} + \begin{vmatrix} 1 & 0 & 2 \\ 2 & 3 & 1 \\ -5 & 8 & 5 \end{vmatrix} = \begin{vmatrix} 1 & 0 & 2 \\ 2 & 3 & 1 \\ 2 & 11 & 9 \end{vmatrix}.$$

2 Return to the proof of Theorem 4.2. Write out the various steps of that proof for the special case where

$$A^* = \begin{bmatrix} a_{11} & a_{12} & a_{13} & a_{14} \\ a_{21} & a_{22} & a_{23} & a_{24} \\ 7a_{31} & 7a_{32} & 7a_{33} & 7a_{34} \\ a_{41} & a_{42} & a_{43} & a_{44} \end{bmatrix}.$$

3 Prove directly from the definition of Δ that $\Delta(A) = 0$ whenever A has a row of zeros.

4 Prove, using results obtained in this section, the following: If the ith row of A is a multiple of its jth row, then $|A| = 0$.

5 Find in determinant form:
  a) An equation for the line through $(3, 2)$ and $(4, 7)$.
  b) An equation for the circle through $(2, 1)$, $(3, 7)$, and $(-1, 5)$.

6 a) Evaluate

$$\begin{vmatrix} 0 & 1 & 0 \\ 1 & 0 & 0 \\ 0 & 0 & 1 \end{vmatrix}.$$

b) Evaluate

$$\begin{vmatrix} 0 & 0 & 1 & 0 \\ 0 & 0 & 0 & 1 \\ 1 & 0 & 0 & 0 \\ 0 & 1 & 0 & 0 \end{vmatrix}.$$

  c) Definition: A *permutation matrix* is a square matrix that can be transformed into the identity matrix by row interchanges. Prove: If A is a permutation matrix, then $|A| = \pm 1$. [*Hint:* $A = E_n \cdots E_2 E_1 I$, where each E_j is a row interchange.]

7 Prove Theorem 4.4.

8 a) Verify the following formula for differentiating a second-order *determinant of functions*:

$$\frac{d}{dt}\begin{vmatrix} a_{11}(t) & a_{12}(t) \\ a_{21}(t) & a_{22}(t) \end{vmatrix} = \begin{vmatrix} a'_{11}(t) & a_{12}(t) \\ a'_{21}(t) & a_{22}(t) \end{vmatrix} + \begin{vmatrix} a_{11}(t) & a'_{12}(t) \\ a_{21}(t) & a'_{22}(t) \end{vmatrix}.$$

  b) Write a formula analogous to that of part (a) for third-order determinants.
  c) Let $A = [a_{ij}(t)]_{nn}$ and let $A^{(q)}$ be the matrix derived from A by differentiating the entries in its qth column. Prove (by mathematical induction) that

$$\frac{d}{dt}|A| = \sum_{k=1}^{n} |A^{(k)}|.$$

[*Hint:* $[A_{k1}]^{(q)} = [A^{(q+1)}]_{k1}.$]

9 *An exercise in analytic geometry.*

a) Place a triangle T on the xy-plane in such a way that one of its vertices is at the origin. Taking the vertices in counterclockwise order, label them $(0, 0)$, (a, b), and (c, d). Now, show that the area of T is

$$A(T) = \frac{1}{2} \begin{vmatrix} a & b \\ c & d \end{vmatrix}.$$

[*Hint:* Enclose T in a rectangle.]

b) Let T be the triangle whose vertices, taken in a counterclockwise order, are (x_1, y_1), (x_2, y_2), and (x_3, y_3). Show that

$$A(T) = \frac{1}{2} \begin{vmatrix} 1 & x_1 & y_1 \\ 1 & x_2 & y_2 \\ 1 & x_3 & y_3 \end{vmatrix}.$$

[*Hint:* Join vertices to the origin, thus constructing three triangles of the type considered in part (a).]

c) Prove: Three points in a plane are collinear if and only if the determinant of part (b) has the value 0.

10 Read Appendix D.

4.5 MORE THEORY

If we are dealing with a function f whose domain and range consist of elements that can be "multiplied," then it is natural to ask whether or not

$$f(x \cdot y) = f(x) \cdot f(y).$$

The reader has met this question before, and he should recall some of the answers:

$$\sqrt{xy} = \sqrt{x} \cdot \sqrt{y}, \quad \text{but} \quad \sin xy \neq \sin x \cdot \sin y,$$

$$\lim_{t \to a} f(t)g(t) = \lim_{t \to a} f(t) \cdot \lim_{t \to a} g(t), \quad \text{but} \quad \frac{d}{dt}f(t)g(t) \neq \frac{d}{dt}f(t) \cdot \frac{d}{dt}g(t),$$

and so on. We now ask: Is the determinant of a product equal to the product of the determinants? That is, is $|AB| = |A| \cdot |B|$?

Is it reasonable to suppose that this is so? First, the complicated row-by-column matrix multiplication is followed by the equally complicated determinant evaluation. Then the other way round: the determinant of each matrix is evaluated separately and then the results are multiplied, the multiplication now being that of ordinary numbers. It goes against our intuition; the results can hardly, except by accident, be the same. But they are!

Before proving this assertion, we pause to recall certain facts about elementary matrices. Let A be any n by n matrix. Then, by appropriate row operations, A may be reduced to the row echelon matrix R. Now either the last row of R is a zero row or it is not. If it is not, then—because the first nonzero entry of a given row must occur to the right of the first nonzero entry in the preceding row—the ones that "begin" each row must lie on the main diagonal. That is, R must be the identity matrix. In other words, using the language of elementary matrices, we are assured that there exist elementary matrices $E_1, E_2, \ldots, E_k$ such that

$$E_k \cdots E_2 E_1 A = R,$$

where either $R = I$ or the last row of R is a zero row. It then follows that

$$A = E_1^{-1} E_2^{-1} \cdots E_k^{-1} R.$$

But the inverse of an elementary matrix is itself an elementary matrix. Thus, we have the following result: a square matrix either may be factored into elementary matrices or it has a factorization in which the last matrix has zeros in its bottom row and the preceding factors are all elementary matrices. (This result might properly be viewed as a corollary to Theorem 1.9. It was not mentioned in Chapter 1 because we had then no use for such a result. Now we do.)

Theorem 4.6. For any two nth-order matrices A and B, we have

$$|AB| = |A| \cdot |B|.$$

Proof

1. The formula is certainly true if A is an elementary matrix. For example, suppose that A is a row multiplier, that is,

$$A = \begin{bmatrix} 1 & & & & & & \\ & 1 & & & & & \\ & & \ddots & & & & \\ & & & 1 & & & \\ & & & & c & & \\ & & & & & 1 & \\ & & & & & & \ddots \\ & & & & & & & 1 \end{bmatrix}.$$

Then $|A| = c$. But also, by Theorem 4.2, $|AB| = c|B|$. Thus,

$$|AB| = c|B| = |A| \cdot |B|.$$

Similarly, using Theorems 4.3 and 4.5, the reader may verify that the formula holds when A is either of the other two kinds of elementary matrices.

2. Suppose next that A has a factorization into elementary matrices. Then our problem is to prove that

$$|E_k \cdots E_2 E_1 B| = |E_k \cdots E_2 E_1| \cdot |B|.$$

We do this by mathematical induction. The case for $k = 1$ has already been established (part 1). Next, suppose that the formula is true when there are k elementary matrices (our induction hypothesis). Then we may write

$$\begin{aligned}
|E_{k+1} \cdots E_2 E_1 B| &= |E_{k+1}(E_k \cdots E_2 E_1 B)| & \\
&= |E_{k+1}| \cdot |E_k \cdots E_2 E_1 B| & \text{(Part 1)} \\
&= |E_{k+1}|(|E_k \cdots E_2 E_1| \cdot |B|) & \text{(Induction hypothesis)} \\
&= (|E_{k+1}| \cdot |E_k \cdots E_2 E_1|)|B| & \text{(Why?)} \\
&= |E_{k+1} \cdots E_2 E_1| \cdot |B|. & \text{(Part 1)}
\end{aligned}$$

3. Finally, suppose that $A = E_k \cdots E_2 E_1 R$, where the last row of R is a zero row. Then, using part 2, we may write

$$|A| = |E_k \cdots E_2 E_1| \cdot |R| = |E_k \cdots E_2 E_1| \cdot 0 = 0.$$

Thus,

$$|A| \cdot |B| = 0 \cdot |B| = 0.$$

On the other hand,

$$\begin{aligned}
|AB| &= |(E_k \cdots E_2 E_1)(RB)| & \\
&= |E_k \cdots E_2 E_1| \cdot |RB| & \text{(Part 2)} \\
&= |E_k \cdots E_2 E_1| \cdot 0 & \text{(Why?)} \\
&= 0. \quad \square &
\end{aligned}$$

Aside from its intrinsic interest, Theorem 4.6 can be put to good use: we use it to prove Property 4, and also to establish a connection between the determinant of a matrix and the notion of linear dependence.

Theorem 4.7. *Property 4.* For any nth-order matrix, we have

$$|A'| = |A|.$$

Proof. Let R be the row reduction of A. Thus, there exist matrices $E_1, E_2, \ldots, E_k$ such that

$$A = E_1 E_2 \cdots E_k R.$$

We first make two observations:

1. The matrix R (whether equal to I or not) is an upper triangular matrix. Thus, R' is lower triangular. But also, R and R' have the same main diagonal. It follows, using Theorem 4.1, that
$$|R'| = |R|.$$

2. If E is a row multiplier, then $E' = E$. If E is a row interchanger, then $E' = E$. But if
$$E = [i \to i + (c)j], \quad \text{then} \quad E' = [j \to j + (c)i].$$
However, it is still true (use Property 3) that
$$|E'| = |I| = |E|.$$
Thus, an elementary matrix and its transpose both have the same determinant.

3. We may now write
$$\begin{aligned}
|A'| &= |(E_1 E_2 \cdots E_k R)'| \\
&= |R' E_k' \cdots E_2' E_1'| \\
&= |R'| \cdot |E_k'| \cdots |E_2'| \cdot |E_1'| & \text{(Theorem 4.6)} \\
&= |R| \cdot |E_k| \cdots |E_2| \cdot |E_1| & \text{(By 1 and 2)} \\
&= |E_1| \cdot |E_2| \cdots |E_k| \cdot |R| & \text{(Why?)} \\
&= |E_1 E_2 \cdots E_k R| & \text{(Theorem 4.6)} \\
&= |A|.
\end{aligned}$$

(Actually, we have been using here an extension of Theorem 4.6 to any number of factors. See Exercise 10.)

Theorem 4.8. For square matrices the following statements are all equivalent:

1. $|A| = 0$.
2. A is singular.
3. The equation $AX = \Theta$ has nontrivial (other than Θ) solutions.
4. The rows of A are linearly dependent.
5. A does not have a factorization into elementary matrices.

Proof. To proclaim that certain statements are equivalent is an abbreviated way of listing a collection of theorems of the form: *If P, then Q.* Any of the statements may be taken for P, and any other for Q. Such equivalent statement theorems are proved in a circular fashion: we show that 1 implies 2, that 2 implies 3, and so on, ending finally with a proof that the last statement implies the first one.

1. (If $|A| = 0$, then A has no inverse.) By contradiction: Suppose that there exists a B such that $AB = I$. Then, using Theorem 4.6, we may write
$$1 = |I| = |AB| = |A| \cdot |B| = 0 \cdot |B| = 0.$$

2. (If A has no inverse, then the equation $AX = \Theta$ has nontrivial solutions.) Let L_A be the linear transformation from R^{n1} into R^{n1} defined by
$$L_A(X) = AX.$$

If A (as a matrix) has no inverse, then L_A (as a transformation) has no inverse. It follows (Theorem 3.4) that the kernel of L_A has dimension exceeding zero. Thus, the kernel of L_A—equivalently, the solution space of $AX = \Theta$—is an infinite set.

3. (If $AX = \Theta$ has nontrivial solutions, then the rows of A are linearly dependent.) If the kernel of L_A has dimension exceeding zero, then—calling on the Dimension Theorem—the range of L_A has dimension less than n. It follows that the column rank of A (hence the row rank) is less than n. Thus, the rows of A are linearly dependent.

4. (If the rows of A are linearly dependent, then A does not have a factorization into elementary matrices.) By contradiction: Suppose that
$$A = E_1 E_2 \cdots E_k.$$
Then
$$A = (E_1 E_2 \cdots E_k) I.$$

Now the rows of I are independent. It follows, by Theorem 2.18, that the rows of A are independent.

5. (If A has no factorization into elementary matrices, then $|A| = 0$.) We may write—recall the discussion preceding Theorem 4.6—
$$A = E_1 E_2 \cdots E_k R,$$
where the last row of R is a zero row. Then
$$|A| = |E_1 E_2 \cdots E_k| \cdot |R| = 0.$$

Example 1. For what values of λ will the set
$$S = \{[\lambda, -2, 2], [1, \lambda - 3, -1], [3, -3, \lambda - 1]\}$$
be linearly dependent?

Solution. Consider the matrix

$$A = \begin{bmatrix} \lambda & -2 & 2 \\ 1 & (\lambda - 3) & -1 \\ 3 & -3 & (\lambda - 1) \end{bmatrix}.$$

By Theorem 4.8, the rows of A will be linearly dependent if, and only if, $|A| = 0$. Now

$$|A| = \lambda \begin{vmatrix} (\lambda - 3) & -1 \\ -3 & (\lambda - 1) \end{vmatrix} - 1 \begin{vmatrix} -2 & 2 \\ -3 & (\lambda - 1) \end{vmatrix} + 3 \begin{vmatrix} -2 & 2 \\ (\lambda - 3) & -1 \end{vmatrix}$$
$$= \lambda^3 - 4\lambda^2 - 4\lambda + 16$$
$$= (\lambda + 2)(\lambda - 2)(\lambda - 4).$$

It follows that $\lambda = -2, 2,$ or 4. As a check, consider S when $\lambda = -2$:

$$S = \{[-2, -2, 2], [1, -5, -1], [3, -3, -3]\}.$$

Note that $[3, -3, -3] = [1, -5, -1] - [-2, -2, 2]$.

EXERCISE SET 4.5

1 Let

$$A = \begin{bmatrix} 1 & 2 & -2 \\ 0 & 5 & 1 \\ 1 & 2 & 3 \end{bmatrix} \quad \text{and} \quad B = \begin{bmatrix} 1 & 2 & 0 \\ -2 & 1 & 0 \\ 0 & 3 & 1 \end{bmatrix}.$$

Verify that $|AB| = |A| \cdot |B|$.

2 a) Prove that $|A^{-1}| = 1/|A|$.
b) Prove that $|BA| = |AB|$.
c) Let A and B be as in Exercise 1. Find $|A^{-1}|$ and $|BA|$.

3 For what values of λ will the set

$$S = \{[\lambda, 1, 2], [1, \lambda, -1], [2, -1, \lambda]\}$$

be linearly dependent?

4 In Section 1.3 we considered an algorithm for finding the inverse of a matrix. When the algorithm failed to work, we assumed that the given matrix had no inverse. That assumption is now justified by Theorem 4.8. Explain.

5 As a result of Theorem 4.7, all of our "row theorems" are now "column theorems." For example, combine Theorems 4.3 and 4.7 to prove the following: If B is derived from A by interchanging two columns, then $|B| = -|A|$.

6 Theorem 4.9. If A and B are nth-order matrices such that $AB = I$, then also $BA = I$.

Prove Theorem 4.9 in the following manner:
a) First show that $AB = I$ implies that B^{-1} exists.
b) Then show that $B^{-1} = A$.

7 Let L be the linear transformation from R^{31} into R^{31} defined by

$$L(X) = \begin{bmatrix} 1 & 0 & 0 \\ 1 & -4 & 3 \\ 2 & -14 & 9 \end{bmatrix} X.$$

For what values of λ will the equation $L(X) = \lambda X$ have solutions other than $X = \Theta$? [*Hint:* First write the equation in the form $AX = \Theta$.]

8 Prove: Let A and B be nth-order matrices and suppose that B is nonsingular. Then

$$\Delta(B^{-1}AB) = \Delta(A).$$

9 For what values of λ will the set

$$S = \{\lambda t^2 + t + 1, \, t^2 + \lambda t + 1, \, t^2 + t + \lambda\}$$

be linearly independent?

10 Prove, by mathematical induction, that

$$|A_1 A_2 \cdots A_k| = |A_1| \cdot |A_2| \cdots |A_k|.$$

11 Let $A = [1]_{nn}$. Prove that the equation $AX = nX$ has infinitely many solutions, where X is in R^{n1}.

12 a) Show that

$$\begin{vmatrix} 1 & 2 & 0 & 0 \\ 3 & 4 & 0 & 0 \\ 0 & 0 & 5 & 6 \\ 0 & 0 & 7 & 8 \end{vmatrix} = \begin{vmatrix} 1 & 2 \\ 3 & 4 \end{vmatrix} \cdot \begin{vmatrix} 5 & 6 \\ 7 & 8 \end{vmatrix}.$$

b) Show that

$$\begin{vmatrix} 1 & 2 & 1 & 0 & 0 \\ 3 & -1 & 2 & 0 & 0 \\ 2 & 1 & 4 & 0 & 0 \\ 0 & 0 & 0 & 3 & 2 \\ 0 & 0 & 0 & 4 & 2 \end{vmatrix} = \begin{vmatrix} 1 & 2 & 1 \\ 3 & -1 & 2 \\ 2 & 1 & 4 \end{vmatrix} \cdot \begin{vmatrix} 3 & 2 \\ 4 & 2 \end{vmatrix}.$$

c) Make a conjecture.

13 a) Prove: If the rank of A is r, then (1) any square submatrix of A having more than r rows has a zero determinant, and (2) there exists at least one rth-order submatrix whose determinant is different from zero. [*Hint:* What can be said about the rows of a submatrix having more than r rows? What can be said about the columns of a submatrix having r independent rows?]

b) Prove the converse of the proposition under part (a).

4.6 A FORMULA FOR A^{-1}

The determinant of A has been defined as a certain sum, that sum involving the elements in the first column of A. By interchanging the first column with the jth, we produce a matrix B such that $|A| = -|B|$. Thus, the evaluation of $|A|$ can be obtained easily from that of $|B|$. The evaluation of $|B|$—appealing directly to our definition—makes use of the first column of B; but the first column of B is the jth column of A. All of this suggests that there should be a formula which expresses $|A|$ directly in terms of the elements in its jth column. There is such a formula; however, its expression requires that we first be familiar with the notion of a *cofactor*.

Definition 4.3. Given the nth-order matrix A, the function *cofactor* is defined over the entries in A by

$$\operatorname{cof}(a_{ij}) = (-1)^{i+j}|A_{ij}|,$$

where A_{ij} is the matrix derived from A by deleting its ith row and its jth column.

Example 1. Let

$$A = \begin{bmatrix} 1 & 7 & -3 \\ 2 & 6 & 4 \\ 5 & -2 & 1 \end{bmatrix}.$$

Then

$$\operatorname{cof}(a_{23}) = (-1)^{2+3}|A_{23}|$$

$$= (-1)\begin{vmatrix} 1 & 7 \\ 5 & -2 \end{vmatrix}$$

$$= 37.$$

Theorem 4.10. The formula

$$|A| = \sum_{k=1}^{n} a_{kp} \operatorname{cof}(a_{kp})$$

is valid for $p = 1, 2, \ldots, n$. (The term "cofactor" or "complementary factor" is thus given meaning.)

Proof. By successive interchanges with columns to its left, move the pth column of A into the first-column position. Call the resulting matrix A^*. Since $(p - 1)$ column interchanges are required, we have

$$|A| = (-1)^{p-1}|A^*|$$

$$= (-1)^{p-1} \sum_{k=1}^{n} (-1)^{k+1} a_{k1}^* |A_{k1}^*|.$$

We make two observations:

1. $a_{k1}^* = a_{kp}$, $\quad k = 1, 2, \ldots, n$, $\quad$ (Why?)
2. $A_{k1}^* = A_{kp}$, $\quad k = 1, 2, \ldots, n$. $\quad$ (Why?)

It follows that

$$|A| = (-1)^{p-1} \sum_{k=1}^{n} (-1)^{k+1} a_{kp} |A_{kp}|$$

$$= \sum_{k=1}^{n} a_{kp} \{(-1)^{p+k} |A_{kp}|\}$$

$$= \sum_{k=1}^{n} a_{kp} \operatorname{cof}(A_{kp}). \quad \square$$

In using the formula of Theorem 4.10 (having chosen a particular p), we say that the determinant is being *expanded along its pth column*. (In a completely analogous way, we may expand a determinant along its pth row. See Exercise 6.) Now suppose that we make the following mistake in performing a pth-column expansion: we multiply the elements in the pth-column of A by the cofactors of the corresponding elements in the qth column of A. What is the result? Well, we do not get the right answer (in general); however, we do get something more interesting than just the wrong answer.

Theorem 4.11. *The mixed columns theorem.* If $p \neq q$, then

$$\sum_{k=1}^{n} a_{kp} \operatorname{cof}(a_{kq}) = 0.$$

Proof. Let A^* be derived from A by replacing its qth column by a copy of its pth column. We make three observations:

1. $|A^*| = 0.$ $\quad$ (Why?)
2. $a_{kq}^* = a_{kp}$, $\quad k = 1, 2, \ldots, n$. $\quad$ (Why?)
3. $A_{kq}^* = A_{kq}$, $\quad k = 1, 2, \ldots, n$. $\quad$ (Why?)

It follows from (3) that

$$\operatorname{cof}(a_{kq}^*) = \operatorname{cof}(a_{kq}).$$

We may now write

$$0 = |A^*| = \sum_{k=1}^{n} a_{kq}^* \operatorname{cof}(a_{kq}^*)$$

$$= \sum_{k=1}^{n} a_{kp} \operatorname{cof}(a_{kq}).$$

Theorem 4.12. The inverse of A is given by

$$A^{-1} = \frac{1}{|A|} [\text{cof } (a_{ij})]'_{nn}.$$

(Note that the formula makes no sense if $|A| = 0$. But then, as we have seen in Theorem 4.8, the matrix A has no inverse.)

Proof. Let

$$B = \frac{1}{|A|} [\text{cof } (a_{ij})]'_{nn}.$$

Then

$$b_{ij} = \frac{\text{cof } (a_{ji})}{|A|}.$$

Now consider the product $C = BA$:

$$c_{ij} = \sum_{k=1}^{n} b_{ik} a_{kj}$$

$$= \sum_{k=1}^{n} \frac{\text{cof } (a_{ki})}{|A|} \cdot a_{kj}$$

$$= \frac{1}{|A|} \sum_{k=1}^{n} a_{kj} \text{ cof } (a_{ki})$$

$$= \begin{cases} 1, & \text{if } i = j \\ 0, & \text{if } i \neq j \end{cases} \qquad \begin{array}{l} \text{(Theorem 4.10)} \\ \text{(Theorem 4.11)} \end{array}$$

$$= \delta_{ij}.$$

Thus, $C = I$. Since $BA = I$, then also (Theorem 4.9) $AB = I$. Therefore $B = A^{-1}$.

Example 2. Let

$$A = \begin{bmatrix} 1 & 2 & 3 \\ 2 & 1 & 2 \\ 4 & 3 & 2 \end{bmatrix}.$$

The cofactors are computed as follows:

$$\text{cof } (a_{11}) = (-1)^{1+1} \begin{vmatrix} 1 & 2 \\ 3 & 2 \end{vmatrix} = -4,$$

$$\text{cof } (a_{12}) = (-1)^{1+2} \begin{vmatrix} 2 & 2 \\ 4 & 2 \end{vmatrix} = 4,$$

$$\text{cof } (a_{13}) = (-1)^{1+3} \begin{vmatrix} 2 & 1 \\ 4 & 3 \end{vmatrix} = 2,$$

and so on. Finally, as the reader may verify,

$$[\text{cof } (a_{ij})]_{33} = \begin{bmatrix} -4 & 4 & 2 \\ 5 & -10 & 5 \\ 1 & 4 & -3 \end{bmatrix}.$$

Thus,

$$[\text{cof } (a_{ij})]'_{33} = \begin{bmatrix} -4 & 5 & 1 \\ 4 & -10 & 4 \\ 2 & 5 & -3 \end{bmatrix}.$$

Next we compute $|A|$ by using—we already have the necessary cofactors—the first-row expansion:

$$|A| = (1)(-4) + (2)(4) + (3)(2) = 10.$$

It follows, appealing to the formula of Theorem 4.12, that

$$A^{-1} = \frac{1}{|A|} [\text{cof } (a_{ij})]'_{33} = \begin{bmatrix} -\frac{2}{5} & \frac{1}{2} & \frac{1}{10} \\ \frac{2}{5} & -1 & \frac{2}{5} \\ \frac{1}{5} & \frac{1}{2} & -\frac{3}{10} \end{bmatrix}. \quad \square$$

The formula of Theorem 4.12 is not as efficient a way for finding A^{-1} as the method given in Section 1.4, except, perhaps, for matrices of order 2 or 3. Consider the computations implied: If A is of order n, then one nth-order determinant must be evaluated, followed by the evaluation of n^2 determinants of order $(n-1)$. However, this formula does provide us with an easy way to confirm the assertion made earlier, namely, that the solution of a linear system having n equations in n unknowns can be expressed in terms of quotients of certain determinants.

Theorem 4.13. *Cramer's rule.* Consider the system

$$\begin{cases} a_{11}x_1 + a_{12}x_2 + \cdots + a_{1n}x_n = b_1, \\ a_{21}x_1 + a_{22}x_2 + \cdots + a_{2n}x_n = b_2, \\ \vdots \\ a_{n1}x_1 + a_{n2}x_2 + \cdots + a_{nn}x_n = b_n. \end{cases}$$

If the coefficient matrix A is nonsingular, then the system has exactly one solution. The individual unknowns are given by

$$x_j = \frac{|A^{(j)}|}{|A|}, \quad j = 1, 2, \ldots, n,$$

where $A^{(j)}$ is the matrix derived from A by replacing its jth column by $B = [b_1, b_2, \ldots, b_n]'$.

Proof. If $AX = B$ and A is nonsingular, then

$$X = A^{-1}B$$
$$= \left\{\frac{1}{|A|} [\text{cof } (a_{ij})]'_{nn}\right\} B$$
$$= \frac{1}{|A|} (A^*B), \quad \text{where} \quad a^*_{ij} = \text{cof } (a_{ji}).$$

It then follows that

$$x_j = \frac{1}{|A|} (j\text{th row of } A^*)(\text{the column } B)$$
$$= \frac{1}{|A|} \sum_{k=1}^{n} a^*_{jk} b_k$$
$$= \frac{1}{|A|} \sum_{k=1}^{n} \text{cof } (a_{kj}) \cdot b_k.$$

Now recall the definition of $A^{(j)}$. We make two observations:

1. $a^{(j)}_{kj} = b_k, \quad k = 1, 2, \ldots, n$.
2. $A^{(j)}_{kj} = A_{kj}, \quad k = 1, 2, \ldots, n$. (Why?)

Thus,
$$\text{cof } (a^{(j)}_{kj}) = \text{cof } (a_{kj}).$$

Finally, returning to x_j, we have

$$x_j = \frac{1}{|A|} \sum_{k=1}^{n} \text{cof } (a^{(j)}_{kj}) \cdot a^{(j)}_{kj}$$
$$= \frac{|A^{(j)}|}{|A|}.$$

Example 3. Cramer's rule gives us a convenient way to find just one of the unknowns in a given system. For example, suppose that we have cause to know x_3, where

$$\begin{cases} x_1 + 2x_2 + x_3 + 2x_4 = 3, \\ 2x_1 + 3x_2 + 2x_3 + x_4 = 4, \\ x_2 + 2x_3 = 0, \\ 3x_1 + 5x_3 + 3x_4 = 1. \end{cases}$$

Then

$$A = \begin{bmatrix} 1 & 2 & 1 & 2 \\ 2 & 3 & 2 & 1 \\ 0 & 1 & 2 & 0 \\ 3 & 0 & 5 & 3 \end{bmatrix} \quad \text{and} \quad A^{(3)} = \begin{bmatrix} 1 & 2 & 3 & 2 \\ 2 & 3 & 4 & 1 \\ 0 & 1 & 0 & 0 \\ 3 & 0 & 1 & 3 \end{bmatrix}.$$

We use a third-row expansion for each determinant:

$$|A| = a_{32} \text{ cof } (a_{32}) + a_{33} \text{ cof } (a_{33})$$

$$= 1 \cdot (-1)^5 \begin{vmatrix} 1 & 1 & 2 \\ 2 & 2 & 1 \\ 3 & 5 & 3 \end{vmatrix} + 2 \cdot (-1)^6 \begin{vmatrix} 1 & 2 & 2 \\ 2 & 3 & 1 \\ 3 & 0 & 3 \end{vmatrix}$$

$$= - \begin{vmatrix} 1 & 1 & 2 \\ 0 & 0 & -3 \\ 3 & 5 & 3 \end{vmatrix} + 2 \begin{vmatrix} 1 & 2 & 1 \\ 2 & 3 & -1 \\ 3 & 0 & 0 \end{vmatrix}$$

$$= -(-3)(-2) + 2(3)(-5)$$

$$= -36.$$

Similarly, $|A^{(3)}| = 18$. Therefore,

$$x_3 = -\tfrac{18}{36} = -\tfrac{1}{2}. \quad \square$$

From a computational point of view, the use of Cramer's rule cannot be recommended, except, perhaps, in the case of a system having just two equations. Consider what is required: To solve a system having n equations, we must evaluate $(n+1)$ determinants, each of order n. The same problem, using the row-reduction method of Section 1.5, requires only that we reduce one n by $(n+1)$ matrix to row echelon form. However, from a theoretical viewpoint, Cramer's rule is most useful: it permits us to "solve" the "general systems" met in proofs and derivations by simply invoking the determinant function.

EXERCISE SET 4.6

1 Let

$$A = \begin{bmatrix} 5 & 2 & 0 \\ 1 & 2 & 3 \\ 3 & -1 & 4 \end{bmatrix}.$$

a) Evaluate $\sum_{k=1}^{3} a_{2k} \text{ cof } (a_{2k})$ and $\sum_{k=1}^{3} a_{3k} \text{ cof } (a_{2k})$.

b) Evaluate $\sum_{k=1}^{3} a_{k3} \text{ cof } (a_{k3})$ and $\sum_{k=1}^{3} a_{k1} \text{ cof } (a_{k3})$.

2 Use the formula of Theorem 4.12 to find A^{-1}.

a) $A = \begin{bmatrix} 1 & 2 \\ 3 & 4 \end{bmatrix}$

b) The A of Exercise 1.

3 Given the system
$$\begin{cases} 5x_1 + 2x_2 = 3, \\ x_1 + 2x_2 + 3x_3 = 4, \\ 3x_1 - x_2 + 4x_3 = 5, \end{cases}$$
solve for x_2.

4 Cramer's rule is applicable to other than systems of numerical equations. Find the polynomials $p(t)$ and $q(t)$ such that
$$\begin{cases} 2p(t) + 3q(t) = t^2 + 2t + 2, \\ 3p(t) + 4q(t) = 2t^2 + 7. \end{cases}$$

5 a) Show that one solution of the system
$$\begin{cases} a_1 x_1 + a_2 x_2 + a_3 x_3 = 0 \\ b_1 x_1 + b_2 x_2 + b_3 x_3 = 0 \end{cases}$$
is given by
$$x_1 = \begin{vmatrix} a_2 & a_3 \\ b_2 & b_3 \end{vmatrix}, \quad x_2 = -\begin{vmatrix} a_1 & a_3 \\ b_1 & b_3 \end{vmatrix}, \quad x_3 = \begin{vmatrix} a_1 & a_2 \\ b_1 & b_2 \end{vmatrix}.$$

b) Write an analogous formula for the case of a homogeneous system having three equations and four unknowns.

6 State and prove the row analogs of Theorems 4.10 and 4.11.

7 Let $A = [a_{ij}]_{nn}$ and let $B = [\text{cof }(a_{ij})]'_{nn}$.
a) If A is symmetric, then so is B.
b) Prove: If A is singular, then $AB = \Theta$.
c) Prove: $|B| = |A|^{n-1}$.

8 Let
$$A = \begin{bmatrix} -\frac{1}{3} & -\frac{2}{3} & -\frac{2}{3} \\ \frac{2}{3} & \frac{1}{3} & -\frac{2}{3} \\ \frac{2}{3} & -\frac{2}{3} & \frac{1}{3} \end{bmatrix} \quad \text{and} \quad B = [\text{cof }(a_{ij})]_{33}.$$

a) Write out the matrix B.
b) Find A^{-1}.
c) Prove or disprove: There is no nonsingular matrix A whose inverse is its transpose.

4.7 THE SECOND DIAGONALIZATION PROBLEM

In Section 1.6 we considered the following problem: Given a symmetric matrix A, find a second matrix B such that $B'AB$ is a diagonal matrix. Motivation for considering such a problem arose in connection with quadratic forms. We now take up a second diagonalization problem, and in the next section we shall give a reason for its existence.

Problem. Given a square matrix A, find a second matrix B such that $B^{-1}AB$ is diagonal.

How do we begin? One approach—an approach that is frequently useful in mathematics—is as follows: We assume that the problem has a solution. We then show that any such diagonalizing matrix must have some other property, say property P. Next, contemplating property P, we realize that the problem of finding a matrix having property P is one that we know how to solve. This realization sends us scurrying back to the argument that led us to property P: if that argument is reversible, then a matrix having property P must be a diagonalizing matrix. If this is the case, then we have been successful in translating a problem that we do not know how to solve into one that we do know how to solve. Furthermore, if it turns out that the new problem has no solution, then we can conclude that the original problem has no solution. Such a program—as we shall now see—is successful in the case of the problem just posed.

Given an nth-order matrix A, let B be any matrix such that

$$B^{-1}AB = D,$$

where D is a diagonal matrix. It then follows that $AB = BD$. Or, looking at the columns of these matrices, we may write

(the jth column of AB) = (the jth column of BD), $\quad j = 1, 2, \ldots, n$.

To express this fact nonverbally we must introduce some notation. Let

$$D = \begin{bmatrix} \lambda_1 & & & \\ & \lambda_2 & & \\ & & \ddots & \\ & & & \lambda_n \end{bmatrix},$$

and let $B = [B^1, B^2, \ldots, B^n]$, where B^j is the jth column of B. It then follows—using only the definition of matrix multiplication—that

(the jth column of AB) = AB^j

and

(the jth column of BD) = $\lambda_j B^j$.

Thus, our column equations assume the following form:

$$AB^j = \lambda_j B^j, \quad j = 1, 2, \ldots, n.$$

Let us review our argument: we passed from

1. $B^{-1}AB = D$

to the equation
$$2.\ AB = BD,$$
and then to
$$3.\ AB^j = \lambda_j B^j, \qquad j = 1, 2, \ldots, n.$$

We shall now try to go the other way: suppose that we have a set of n by 1 column matrices, say $\{B^1, B^2, \ldots, B^n\}$, and a set of n scalars, say $\{\lambda_1, \lambda_2, \ldots, \lambda_n\}$, and that condition 3 holds. We may then construct the matrices

$$B = [B^1, B^2, \ldots, B^n] \quad \text{and} \quad D = \begin{bmatrix} \lambda_1 & & & \\ & \lambda_2 & & \\ & & \ddots & \\ & & & \lambda_n \end{bmatrix}.$$

Condition 3 then implies that the jth column of AB is identical with the jth column of BD. Thus, condition 2 follows. The passage from condition 2 to condition 1 seems immediate. However, it is so only if we are armed with the knowledge that B has an inverse. This is not implicit in condition 3. However, one way of ensuring that B has an inverse is to insist that its columns are linearly independent. We are now able to enunciate our "property P."

Given a square matrix A, suppose that B is a second matrix such that each column B^j has associated with it a scalar λ_j such that $AB^j = \lambda_j B^j$. Further, suppose that the columns of B are linearly independent. Then we say that B has *property P with respect to A*.

Theorem 4.14. Given a square matrix A, suppose that there exists a matrix B such that $B^{-1}AB$ is a diagonal matrix. Then B has property P with respect to A. Conversely, if B has property P with respect to A, then $B^{-1}AB$ is a diagonal matrix.

Proof. The proof of this theorem is contained in the foregoing discussion. □

The problem of finding a matrix B such that $B^{-1}AB$ is diagonal may thus be solved (if it is solvable at all) in the following manner: Find n linearly independent column vectors that satisfy an equation of the form

$$AX = \lambda X.$$

Then take these vectors (in any order) as the columns of B.

Example 1. Let
$$A = \begin{bmatrix} 2 & 3 \\ 6 & -1 \end{bmatrix}.$$
Find B such that $B^{-1}AB$ is a diagonal matrix.

Solution. We seek independent column vectors B^1 and B^2 that satisfy an equation of the form
$$AX = \lambda X.$$
Equivalently, we may write
$$(A - \lambda I)X = \Theta.$$

Now if the matrix $(A - \lambda I)$ has an inverse, then the only possible solution for X is the zero vector. A zero vector cannot play the role of either B^1 or B^2. (Why?) Thus, in order that such an equation have *nontrivial* solutions, we must require that λ be so chosen that the matrix $(A - \lambda I)$ has no inverse. This will be the case if (Theorem 4.8) the determinant of $(A - \lambda I)$ is zero. We are thus led to the equation
$$|A - \lambda I| = 0.$$
Now
$$A - \lambda I = \begin{bmatrix} 2 & 3 \\ 6 & -1 \end{bmatrix} - \begin{bmatrix} \lambda & 0 \\ 0 & \lambda \end{bmatrix} = \begin{bmatrix} (2-\lambda) & 3 \\ 6 & (-1-\lambda) \end{bmatrix},$$
and
$$|A - \lambda I| = (2-\lambda)(-1-\lambda) - 18$$
$$= \lambda^2 - \lambda - 20$$
$$= (\lambda + 4)(\lambda - 5).$$

Thus, the equation $|A - \lambda I| = 0$ has two solutions: $\lambda_1 = -4$ and $\lambda_2 = 5$. It follows that there are exactly two equations of the form $(A - \lambda I)X = \Theta$ that have nontrivial solutions. We solve each of them:

1. When $\lambda = -4$, we have
$$\begin{bmatrix} 6 & 3 \\ 6 & 3 \end{bmatrix} X = \Theta.$$
The corresponding system reduces to $x_1 + \frac{1}{2}x_2 = 0$. It follows that all solutions are given by
$$X = \begin{bmatrix} -\frac{1}{2}t \\ t \end{bmatrix}.$$

2. When $\lambda = 5$, we have
$$\begin{bmatrix} -3 & 3 \\ 6 & -6 \end{bmatrix} X = \Theta.$$
It follows that
$$X = \begin{bmatrix} t \\ t \end{bmatrix}.$$

It is evident that any representative pair of vectors (one chosen from each solution space) will be linearly independent. Thus, taking convenient

values for t, we set
$$B^1 = \begin{bmatrix} 1 \\ -2 \end{bmatrix} \quad \text{and} \quad B^2 = \begin{bmatrix} 1 \\ 1 \end{bmatrix},$$
and take
$$B = \begin{bmatrix} 1 & 1 \\ -2 & 1 \end{bmatrix}.$$

The reader may now verify that
$$\begin{bmatrix} 1 & 1 \\ -2 & 1 \end{bmatrix}^{-1} \begin{bmatrix} 2 & 3 \\ 6 & -1 \end{bmatrix} \begin{bmatrix} 1 & 1 \\ -2 & 1 \end{bmatrix} = \begin{bmatrix} -4 & 0 \\ 0 & 5 \end{bmatrix}.$$

Example 2. Same as Example 1, where
$$A = \begin{bmatrix} 1 & -1 & 1 \\ 4 & 0 & -1 \\ 4 & -2 & 1 \end{bmatrix}.$$

Solution. We give the computations here without further comment:
$$|A - \lambda I| = \begin{vmatrix} (1-\lambda) & -1 & 1 \\ 4 & -\lambda & -1 \\ 4 & -2 & (1-\lambda) \end{vmatrix}$$
$$= -\lambda^3 + 2\lambda^2 + \lambda - 2$$
$$= -(\lambda + 1)(\lambda - 1)(\lambda - 2).$$

1. ($\lambda = -1$)
$$\begin{bmatrix} 2 & -1 & 1 \\ 4 & 1 & -1 \\ 4 & -2 & 2 \end{bmatrix} X = \Theta, \quad \text{or} \quad \begin{bmatrix} 1 & 0 & 0 \\ 0 & 1 & -1 \\ 0 & 0 & 0 \end{bmatrix} X = \Theta.$$

Then we have
$$X = \begin{bmatrix} 0 \\ t \\ t \end{bmatrix}.$$

Take
$$B^1 = \begin{bmatrix} 0 \\ 1 \\ 1 \end{bmatrix}.$$

2. ($\lambda = 1$)
$$\begin{bmatrix} 0 & -1 & 1 \\ 4 & -1 & -1 \\ 4 & -2 & 0 \end{bmatrix} X = \Theta, \quad \text{or} \quad \begin{bmatrix} 1 & 0 & -\frac{1}{2} \\ 0 & 1 & -1 \\ 0 & 0 & 0 \end{bmatrix} X = \Theta.$$

Then we have
$$X = \begin{bmatrix} \frac{1}{2}t \\ t \\ t \end{bmatrix}.$$

Take
$$B^2 = \begin{bmatrix} 1 \\ 2 \\ 2 \end{bmatrix}.$$

3. ($\lambda = 2$)
$$\begin{bmatrix} -1 & -1 & 1 \\ 4 & -2 & -1 \\ 4 & -2 & -1 \end{bmatrix} X = \Theta, \quad \text{or} \quad \begin{bmatrix} 1 & 0 & -\frac{1}{2} \\ 0 & 1 & -\frac{1}{2} \\ 0 & 0 & 0 \end{bmatrix} X = \Theta.$$

Then we have
$$X = \begin{bmatrix} \frac{1}{2}t \\ \frac{1}{2}t \\ t \end{bmatrix}.$$

Take
$$B^3 = \begin{bmatrix} 1 \\ 1 \\ 2 \end{bmatrix}.$$

Finally, a solution to our problem is given by
$$B = \begin{bmatrix} 0 & 1 & 1 \\ 1 & 2 & 1 \\ 1 & 2 & 2 \end{bmatrix}.$$

(It is tacitly assumed here that we have checked on the independence of B's columns. In Section 4.9 we shall prove that this is always the case whenever there are n distinct values for λ and each B^j corresponds to a different λ.)

Example 3. Same as Example 1, where
$$A = \begin{bmatrix} 1 & 1 & -2 \\ 0 & 1 & 0 \\ 0 & -1 & 3 \end{bmatrix}.$$

Solution
$$|A - \lambda I| = \begin{vmatrix} (1-\lambda) & 1 & -2 \\ 0 & (1-\lambda) & 0 \\ 0 & -1 & (3-\lambda) \end{vmatrix} = (1-\lambda)^2(3-\lambda).$$

Here there are only two choices for λ. How shall we obtain *three* columns for B?

1. ($\lambda = 1$)
$$\begin{bmatrix} 0 & 1 & -2 \\ 0 & 0 & 0 \\ 0 & -1 & 2 \end{bmatrix} X = \Theta.$$

The system degenerates to $x_2 = 2x_3$. It follows that

$$X = \begin{bmatrix} s \\ 2t \\ t \end{bmatrix} = s \begin{bmatrix} 1 \\ 0 \\ 0 \end{bmatrix} + t \begin{bmatrix} 0 \\ 2 \\ 1 \end{bmatrix}.$$

Thus, the solution space is two-dimensional. It follows that we can find two linearly independent columns corresponding to this single choice of λ. Indeed, we may take

$$B^1 = \begin{bmatrix} 1 \\ 0 \\ 0 \end{bmatrix} \quad \text{and} \quad B^2 = \begin{bmatrix} 0 \\ 2 \\ 1 \end{bmatrix}.$$

2. ($\lambda = 3$)
$$\begin{bmatrix} -2 & 1 & -2 \\ 0 & -2 & 0 \\ 0 & -1 & 0 \end{bmatrix} X = \Theta.$$

Then we have
$$X = \begin{bmatrix} -t \\ 0 \\ t \end{bmatrix}.$$

Take
$$B^3 = \begin{bmatrix} 1 \\ 0 \\ -1 \end{bmatrix}.$$

Finally, a solution to our problem is given by

$$B = \begin{bmatrix} 1 & 0 & 1 \\ 0 & 2 & 0 \\ 0 & 1 & -1 \end{bmatrix}.$$

Example 4. Same as Example 1, where

$$A = \begin{bmatrix} 3 & 1 & 0 \\ 0 & 3 & 0 \\ 0 & 0 & 2 \end{bmatrix}.$$

(Surely it will not take much effort to diagonalize this "almost diagonal" matrix.)

Solution

$$|A - \lambda I| = \begin{vmatrix} (3-\lambda) & 1 & 0 \\ 0 & (3-\lambda) & 0 \\ 0 & 0 & (2-\lambda) \end{vmatrix} = (3-\lambda)^2(2-\lambda).$$

1. ($\lambda = 3$)

$$\begin{bmatrix} 0 & 1 & 0 \\ 0 & 0 & 0 \\ 0 & 0 & -1 \end{bmatrix} X = \Theta. \quad \text{Then} \quad X = \begin{bmatrix} t \\ 0 \\ 0 \end{bmatrix}.$$

2. ($\lambda = 2$)

$$\begin{bmatrix} 1 & 1 & 0 \\ 0 & 1 & 0 \\ 0 & 0 & 0 \end{bmatrix} X = \Theta. \quad \text{Then} \quad X = \begin{bmatrix} 0 \\ 0 \\ t \end{bmatrix}.$$

Both solution spaces are one dimensional—it is thus impossible to find three independent vectors. This means that there is no matrix B that has property P with respect to A. It follows (Theorem 4.14) that there is no diagonalizing matrix for A.

Example 5. Same as Example 1, where

$$A = \begin{bmatrix} 2 & 3 \\ -3 & 2 \end{bmatrix}.$$

Solution

$$|A - \lambda I| = \begin{vmatrix} 2-\lambda & 3 \\ -3 & 2-\lambda \end{vmatrix} = \lambda^2 - 4\lambda + 13.$$

By the quadratic formula,

$$\lambda = \frac{4 \pm \sqrt{-36}}{2} = 2 \pm 3i.$$

The fact that the roots of $|A - \lambda I| = 0$ are complex numbers does not in itself preclude the possibility of A having a diagonalizing matrix, provided that we permit the use of matrices having complex number entries. We proceed as before.

1. ($\lambda = 2 + 3i$)

$$\begin{bmatrix} -3i & 3 \\ -3 & -3i \end{bmatrix} X = \Theta,$$

$$\begin{bmatrix} -3i & 3 & 0 \\ -3 & -3i & 0 \end{bmatrix} \stackrel{\text{(row)}}{=} \begin{bmatrix} 1 & i & 0 \\ -3 & -3i & 0 \end{bmatrix} \stackrel{\text{(row)}}{=} \begin{bmatrix} 1 & i & 0 \\ 0 & 0 & 0 \end{bmatrix}.$$

It follows that

$$X = \begin{bmatrix} -it \\ t \end{bmatrix}.$$

Take
$$B^1 = \begin{bmatrix} 1 \\ i \end{bmatrix}.$$

2. ($\lambda = 2 - 3i$)
$$\begin{bmatrix} 3i & 3 \\ -3 & 3i \end{bmatrix} X = \Theta.$$

Proceeding as in part 1, we obtain
$$X = \begin{bmatrix} it \\ t \end{bmatrix}.$$

Take
$$B^2 = \begin{bmatrix} 1 \\ -i \end{bmatrix}.$$

Finally, a solution to our problem is given by
$$B = \begin{bmatrix} 1 & 1 \\ i & -i \end{bmatrix}.$$

EXERCISE SET 4.7

1 Return to Examples 3 and 5, and in each case verify that $B^{-1}AB$ is indeed a diagonal matrix.

2 Let $A = \begin{bmatrix} 4 & -5 & -2 \\ 2 & 3 & 4 \\ -2 & 2 & 1 \end{bmatrix}$.

a) **Definition 4.4.** The equation $|A - \lambda I| = 0$ is called the *characteristic equation* of A.

Find the characteristic equation of the given A.

b) **Definition 4.5.** A scalar that satisfies the characteristic equation of A is called a *characteristic value* of A. (The term *eigenvalue* is also used.)

Find all characteristic values for the given A.

c) **Definition 4.6.** If λ is a characteristic value of A and X is a nonzero column vector such that $AX = \lambda X$, then X is said to be a *characteristic vector of A corresponding to λ*. (The term *eigenvector* is also used.)

Find a characteristic vector corresponding to each of the characteristic values of the given A.

d) Find a matrix B such that $B^{-1}AB$ is diagonal.

3 Find a matrix B (provided that one exists) such that $B^{-1}AB$ is diagonal, where A is the given matrix.

a) $\begin{bmatrix} 2 & 1 \\ 5 & 2 \end{bmatrix}$ b) $\begin{bmatrix} 0 & 0 & 0 \\ 1 & 1 & 1 \\ 2 & 2 & 2 \end{bmatrix}$ c) $\begin{bmatrix} 0 & 0 & 0 \\ 1 & 0 & 0 \\ 1 & 1 & 0 \end{bmatrix}$

4. The characteristic vectors corresponding to a given characteristic value λ all satisfy the homogeneous equation $(A - \lambda I)X = \Theta$. The remaining solution of this equation is the zero vector. Thus, even though the zero vector is excluded as a characteristic vector (see Definition 4.6), we nevertheless tacitly include it when we use the phrase *the space of characteristic vectors corresponding to* λ.

   a) Verify that $\lambda = 0$ is a characteristic value of $A = [1]_{44}$.
   b) Find a basis for the space of characteristic vectors corresponding to $\lambda = 0$.

5. **Definition 4.7.** Let A and B be square matrices of the same order. If there exists a nonsingular matrix P such that $B = P^{-1}AP$, then B is said to be *similar* to A. We write $B \stackrel{s}{=} A$.

   a) Let
   $$A = \begin{bmatrix} 3 & 4 \\ 2 & 5 \end{bmatrix} \quad \text{and} \quad B = \begin{bmatrix} 1 & 0 \\ 0 & 7 \end{bmatrix}.$$

   Verify that $B \stackrel{s}{=} A$.
   b) Prove that *matrix similarity* is an equivalence relation.
   c) Let
   $$C = \begin{bmatrix} 2 & -1 \\ -5 & 6 \end{bmatrix}.$$

   Verify that $C \stackrel{s}{=} A$. [*Hint:* First verify that $B \stackrel{s}{=} C$.]
   d) Prove: If $B \stackrel{s}{=} C$, then B and C have the same characteristic values. [*Hint:* Show that $|B - \lambda I| = |C - \lambda I|$.]
   e) Prove: If L is a linear transformation from V into V, and $\mathcal{B}$ and $\mathcal{C}$ are bases for V, then $L_{\mathcal{B}\mathcal{B}} \stackrel{s}{=} L_{\mathcal{C}\mathcal{C}}$.

6. a) Let L be defined on R^{31} by $L(X) = AX$, where A is the matrix of Example 2. Let $\mathcal{B} = [B^1, B^2, B^3]$, where the B^i are those found under Example 2. Now, find the $\mathcal{B}\mathcal{B}$-representation of L.
   b) Let L be defined on R^{n1} by $L(X) = AX$, where A is n by n. Now prove the following: If the vectors of a basis $\mathcal{B}$ are all characteristic vectors for A, then the $\mathcal{B}\mathcal{B}$-representation for L is a diagonal matrix. [*Hint:* Show that the jth column of $L_{\mathcal{B}\mathcal{B}}$ has all zero entries except, perhaps, for the jth entry.]

7. **Definition 4.8.** Let L be a linear transformation from V into V, where V is any vector space that is not zero dimensional. Then a scalar λ for which the equation $L(X) = \lambda X$ has solutions other than $X = \Theta$ is said to be a *characteristic value* of L. A nonzero vector that satisfies this equation is said to be a *characteristic vector of L corresponding to* λ.

   a) Let L be the linear transformation from P^2 into P^2 defined by
   $$L[p(t)] = p'(0)t + p(t - 1).$$

   Using matrix methods, find all characteristic values of L and bases for the corresponding spaces of characteristic vectors.

b) Same as part (a) without using matrix methods [*Hint:* Set $L(at^2 + bt + c) = \lambda(at^2 + bt + c)$.]

8 Let L be defined on R^{21} by

$$L\left(\begin{bmatrix} x_1 \\ x_2 \end{bmatrix}\right) = \begin{bmatrix} x_1 + x_2 \\ x_2 \end{bmatrix}.$$

a) Make an "arrow drawing" to depict this linear transformation, that is, sketch a coordinate system and then place on it several arrows representing different choices of X along with arrows representing the corresponding values of $L(X)$.

b) As a result of your experience in part (a), guess a vector X such that the arrows for X and $L(X)$ point in the same direction. Are there many such X?

c) Returning to the formula for L, find the characteristic values of L and bases for the corresponding spaces of characteristic vectors.

9 Find B such that $B^{-1}AB$ is diagonal, where A is the given matrix. [*Hint:* See Example 5.]

a) $\begin{bmatrix} 12 & 5 \\ -5 & 4 \end{bmatrix}$
b) $\begin{bmatrix} 1 & 0 & 0 \\ 0 & 12 & 5 \\ 0 & -5 & 4 \end{bmatrix}$

10 Prove or disprove:

a) If E is an elementary matrix, then the characteristic values of EA are the same as those of A.

b) If A is singular, then there is no B such that $B^{-1}AB$ is diagonal.

c) If A and B have the same characteristic values, then they are similar.

11 Prove or disprove:

a) A square matrix and its transpose have the same characteristic values.

b) If A is nonsingular, then its characteristic values are simply the reciprocals of those of A^{-1}.

c) AB and BA have the same set of characteristic values, where A and B are any nth-order matrices. [*Hint:* If $ABX = \lambda X$, then $BA(BX) = \cdots$.]

12 Let a, b, c, and d be integers. Prove: If $a + b = c + d$, then the characteristic values of

$$\begin{bmatrix} a & b \\ c & d \end{bmatrix}$$

are integers. Indeed, we may take $\lambda_1 = a + b$ and $\lambda_2 = a - c$.

13 Let A be as in Exercise 2. Find A^{10}. [*Hint:* A diagonal matrix is easy to raise to the tenth power.]

14 Let A be the matrix of Exercise 2.

a) Write out the characteristic equation for A. Then, replace λ by X, the constant term c by cI_3, and the zero on the right side by Θ_{33}. The resulting equation is a *matrix equation*. Verify that A satisfies this equation. (We say that A *satisfies its own characteristic equation*.)

b) Take the matrix equation of part (a), substitute A for X, multiply both sides by A^{-1}, and then solve for A^{-1}.

c) Use the formula derived in part (b) to find A^{-1}.

15 A continuation of Exercise 14, part (a). The following theorem is called the Cayley-Hamilton theorem: Every nth-order matrix satisfies its own characteristic equation. This remarkable theorem is most easy to prove for the special case where A is similar to a diagonal matrix. Prove this special case using the following pattern.

a) Prove (by mathematical induction): If

$$B^{-1}AB = \begin{bmatrix} \lambda_1 & & & \\ & \lambda_2 & & \\ & & \ddots & \\ & & & \lambda_n \end{bmatrix}, \quad \text{then} \quad B^{-1}A^k B = \begin{bmatrix} \lambda_1^k & & & \\ & \lambda_2^k & & \\ & & \ddots & \\ & & & \lambda_n^k \end{bmatrix}.$$

b) Prove: If $p(A) = a_n A^n + \cdots + a_1 A + a_0 I$, then

$$B^{-1} p(A) B = \begin{bmatrix} p(\lambda_1) & & & \\ & p(\lambda_2) & & \\ & & \ddots & \\ & & & p(\lambda_n) \end{bmatrix}.$$

c) Prove: If A is similar to a diagonal matrix, then A satisfies its own characteristic equation. [*Hint:* Take p, in part (b), as the *characteristic polynomial* for A.

4.8 AN APPLICATION TO LINEAR DIFFERENTIAL EQUATIONS

Given any number a, suppose that f is a differentiable function having the property that $f'(t) = af(t)$. Then f is necessarily from that family of functions defined by the formula $f(t) = ce^{at}$. We reason as follows: Let $g(t) = f(t)e^{-at}$. Then

$$\begin{aligned} g'(t) &= f'(t)e^{-at} + f(t)(-a)e^{-at} \\ &= e^{-at}[f'(t) - af(t)] \\ &= 0. \end{aligned}$$

It follows (by a theorem of calculus) that g is a constant function. Thus, there is a number c such that $g(t) = c$ for all t. But then, recalling the definition of g, we have $f(t) = ce^{at}$. Conversely, if $f(t) = ce^{at}$, then a direct computation shows that $f'(t) = af(t)$. In summary, then, we have proved the following theorem: All solutions of the differential equation $\dot{x}(t) = ax(t)$ are given by the formula $f(t) = ce^{at}$. [We write $\dot{x}(t)$ instead of $x'(t)$ because we shall shortly be dealing with matrices of functions:

4.8 AN APPLICATION TO LINEAR DIFFERENTIAL EQUATIONS

there the prime notation would conflict with our notation for matrix transposition.]

In this section, we shall be interested in the following more general problem: Given the matrix of numbers $A = [a_{ij}]_{nn}$, we seek functions $x_1(t), x_2(t), \ldots, x_n(t)$ such that

$$\begin{cases} \dot{x}_1(t) = a_{11}x_1(t) + a_{12}x_2(t) + \cdots + a_{1n}x_n(t), \\ \dot{x}_2(t) = a_{21}x_1(t) + a_{22}x_2(t) + \cdots + a_{2n}x_n(t), \\ \;\;\vdots \\ \dot{x}_n(t) = a_{n1}x_1(t) + a_{n2}x_2(t) + \cdots + a_{nn}x_n(t). \end{cases}$$

[When $n = 1$, this system of differential equations reduces to the special case first considered: $\dot{x}_1(t) = a_{11}x_1(t)$.] Exploiting the hybrid arithmetic of matrices of numbers and matrices of functions, we may write this system as a single matrix equation:

$$\begin{bmatrix} \dot{x}_1(t) \\ \dot{x}_2(t) \\ \vdots \\ \dot{x}_n(t) \end{bmatrix} = \begin{bmatrix} a_{11} & a_{12} & \cdots & a_{1n} \\ a_{21} & a_{22} & \cdots & a_{2n} \\ \vdots & \vdots & & \vdots \\ a_{n1} & a_{n2} & \cdots & a_{nn} \end{bmatrix} \begin{bmatrix} x_1(t) \\ x_2(t) \\ \vdots \\ x_n(t) \end{bmatrix},$$

or,

$$\dot{X}(t) = AX(t),$$

or,

$$\dot{X} = AX.$$

Example 1. Find all triples of functions that satisfy the system

$$\begin{cases} \dot{x}_1(t) = -x_1(t), \\ \dot{x}_2(t) = x_2(t), \\ \dot{x}_3(t) = 2x_3(t). \end{cases}$$

Solution. This is the easy case: each equation involves only one unknown function. This means that each equation may be solved individually (using our earlier result):

$$\begin{cases} x_1(t) = c_1 e^{-t}, \\ x_2(t) = c_2 e^{t}, \\ x_3(t) = c_3 e^{2t}. \end{cases}$$

This is the *complete solution;* particular solutions arise on choosing particular values for c_1, c_2, and c_3.

It is instructive to express this system and its solution in matrix notation: all solutions of the matrix differential equation

$$\dot{X} = \begin{bmatrix} -1 & 0 & 0 \\ 0 & 1 & 0 \\ 0 & 0 & 2 \end{bmatrix} X$$

are given by
$$X = \begin{bmatrix} c_1 e^{-t} \\ c_2 e^{t} \\ c_3 e^{2t} \end{bmatrix}.$$

The point is this: The feature that made this system so easy to solve was that the matrix describing the system is diagonal.

Example 2. Solve the system
$$\begin{cases} \dot{x}_1 = x_1 - x_2 + x_3, \\ \dot{x}_2 = 4x_1 - x_3, \\ \dot{x}_3 = 4x_1 - 2x_2 + x_3. \end{cases}$$

(That our functions are functions of t will be tacitly understood.)

Solution. A fundamental technique used in solving differential equations is that of substituting new unknown functions for the given unknown functions (a change-of-variable technique). For example, supposing that this system has a solution, we may introduce new functions u_1, u_2, and u_3 by the formulas
$$\begin{cases} x_1 = u_2 + u_3, \\ x_2 = u_1 + 2u_2 + u_3, \\ x_3 = u_1 + 2u_2 + 2u_3. \end{cases}$$

(The reason for these particular formulas will ultimately be made clear.) It then follows, recalling the linearity properties of differentiation, that
$$\begin{cases} \dot{x}_1 = \phantom{\dot{u}_1 + 2}\dot{u}_2 + \dot{u}_3, \\ \dot{x}_2 = \dot{u}_1 + 2\dot{u}_2 + \dot{u}_3, \\ \dot{x}_3 = \dot{u}_1 + 2\dot{u}_2 + 2\dot{u}_3. \end{cases}$$

Substitution into the original system leads to a new system involving the functions u_1, u_2, and u_3:
$$\begin{cases} (\dot{u}_2 + \dot{u}_3) = (u_2 + u_3) - (u_1 + 2u_2 + u_3) + (u_1 + 2u_2 + 2u_3), \\ (\dot{u}_1 + 2\dot{u}_2 + \dot{u}_3) = 4(u_2 + u_3) - (u_1 + 2u_2 + 2u_3), \\ (\dot{u}_1 + 2\dot{u}_2 + 2\dot{u}_3) = 4(u_2 + u_3) - 2(u_1 + 2u_2 + u_3) + (u_1 + 2u_2 + 2u_3). \end{cases}$$

Thus, we may solve the original system in the following manner: First, solve this new system for the functions u_1, u_2, and u_3. Then, returning to the equations giving the x_j's in terms of the u_j's, we can write the solution for the original system. But this new system is more complicated than the one we started with! What is gained by this roundabout approach? Well, it happens that this new system may be drastically simplified.

First, subtract twice the first equation from the third:
$$\dot{u}_1 = 2(u_2 + u_3) - (u_1 + 2u_2 + 2u_3),$$
$$\dot{u}_1 = -u_1.$$

4.8 AN APPLICATION TO LINEAR DIFFERENTIAL EQUATIONS

Then subtract the second equation from the third:

$$\dot{u}_3 = -2(u_1 + 2u_2 + u_3) + 2(u_1 + 2u_2 + 2u_3),$$
$$\dot{u}_3 = 2u_3.$$

Finally, substitute this expression for $\dot{u}_3$ into the first equation:

$$\dot{u}_2 + 2u_3 = (u_2 + u_3) - (u_1 + 2u_2 + u_3) + (u_1 + 2u_2 + 2u_3),$$
$$\dot{u}_2 = u_2.$$

Thus, the original u-system is equivalent to the system

$$\begin{cases} \dot{u}_1 = -u_1, \\ \dot{u}_2 = u_2, \\ \dot{u}_3 = 2u_3. \end{cases}$$

This is the *easy case* of Example 1; its solution is given by

$$\begin{cases} u_1 = c_1 e^{-t}, \\ u_2 = c_2 e^{t}, \\ u_3 = c_3 e^{2t}. \end{cases}$$

It then follows—recall the change-of-variable formulas—that the original x-system has the solution

$$\begin{cases} x_1 = \phantom{c_1 e^{-t} + {}} c_2 e^{t} + c_3 e^{2t}, \\ x_2 = c_1 e^{-t} + 2c_2 e^{t} + c_3 e^{2t}, \\ x_3 = c_1 e^{-t} + 2c_2 e^{t} + 2c_3 e^{2t}. \end{cases}$$

Now what happened back there? Why did this change-of-variable technique work so nicely? Where did this particular change of variable come from? Switching to matrix notation makes all things clear: We wish to solve

$$\dot{X} = \begin{bmatrix} 1 & -1 & 1 \\ 4 & 0 & -1 \\ 4 & -2 & 1 \end{bmatrix} X.$$

We introduce new unknowns by the formula

$$X = \begin{bmatrix} 0 & 1 & 1 \\ 1 & 2 & 1 \\ 1 & 2 & 2 \end{bmatrix} U.$$

But then

$$\dot{X} = \begin{bmatrix} 0 & 1 & 1 \\ 1 & 2 & 1 \\ 1 & 2 & 2 \end{bmatrix} \dot{U}.$$

Substituting for X and $\dot{X}$ in the original equation leads to a new equation in U:

$$\begin{bmatrix} 0 & 1 & 1 \\ 1 & 2 & 1 \\ 1 & 2 & 2 \end{bmatrix} \dot{U} = \begin{bmatrix} 1 & -1 & 1 \\ 4 & 0 & -1 \\ 4 & -2 & 1 \end{bmatrix} \begin{bmatrix} 0 & 1 & 1 \\ 1 & 2 & 1 \\ 1 & 2 & 2 \end{bmatrix} U,$$

or,

$$\dot{U} = \begin{bmatrix} 0 & 1 & 1 \\ 1 & 2 & 1 \\ 1 & 2 & 2 \end{bmatrix}^{-1} \begin{bmatrix} 1 & -1 & 1 \\ 4 & 0 & -1 \\ 4 & -2 & 1 \end{bmatrix} \begin{bmatrix} 0 & 1 & 1 \\ 1 & 2 & 1 \\ 1 & 2 & 2 \end{bmatrix} U.$$

But, recalling Example 2 of Section 4.7, this equation reduces to

$$\dot{U} = \begin{bmatrix} -1 & 0 & 0 \\ 0 & 1 & 0 \\ 0 & 0 & 2 \end{bmatrix} U,$$

and its solution is given by

$$U = \begin{bmatrix} c_1 e^{-t} \\ c_2 e^{t} \\ c_3 e^{2t} \end{bmatrix}.$$

Finally,

$$X = \begin{bmatrix} 0 & 1 & 1 \\ 1 & 2 & 1 \\ 1 & 2 & 2 \end{bmatrix} U = \begin{bmatrix} c_2 e^{t} + c_3 e^{2t} \\ c_1 e^{-t} + 2c_2 e^{t} + c_3 e^{2t} \\ c_1 e^{-t} + 2c_2 e^{t} + 2c_3 e^{2t} \end{bmatrix}. \quad \square$$

The method used in Example 2 works provided that the coefficient matrix of the system is similar to a diagonal matrix. We reason as follows: Given the system

$$\dot{X} = AX,$$

suppose that we have found a matrix B such that $B^{-1}AB = D$, where D is diagonal. Then we may *diagonalize our equation* with the substitutions $X = BU$ and $\dot{X} = B\dot{U}$:

$$\dot{X} = AX,$$
$$(B\dot{U}) = A(BU),$$
$$\dot{U} = (B^{-1}AB)U,$$
$$\dot{U} = DU,$$
$$\dot{U} = \begin{bmatrix} \lambda_1 & & & \\ & \lambda_2 & & \\ & & \ddots & \\ & & & \lambda_n \end{bmatrix} U.$$

4.8 AN APPLICATION TO LINEAR DIFFERENTIAL EQUATIONS

This last equation is easily solved:

$$U = \begin{bmatrix} c_1 e^{\lambda_1 t} \\ c_2 e^{\lambda_2 t} \\ \vdots \\ c_n e^{\lambda_n t} \end{bmatrix}.$$

But then so is the original system: $X = BU$. Or, upon expanding this matrix formula, we have

$$\begin{cases} x_1 = b_{11} c_1 e^{\lambda_1 t} + b_{12} c_2 e^{\lambda_2 t} + \cdots + b_{1n} c_n e^{\lambda_n t}, \\ x_2 = b_{21} c_1 e^{\lambda_1 t} + b_{22} c_2 e^{\lambda_2 t} + \cdots + b_{2n} c_n e^{\lambda_n t}, \\ \vdots \\ x_n = b_{n1} c_1 e^{\lambda_1 t} + b_{n2} c_2 e^{\lambda_2 t} + \cdots + b_{nn} c_n e^{\lambda_n t}. \end{cases}$$

Note that some or all of the λ_j's may be complex numbers. Thus, our final result is meaningful (in general) only to the reader who has dealt with complex exponential functions.

So far we have concentrated on finding a formula that gives all solutions of the equation $\dot{X} = AX$: the so-called *complete solution*. In practice, however, we often wish to determine a *particular solution*, such particular solutions being defined by the placement of additional restrictions on the unknown functions. We close this section with a typical problem of this type.

Example 3. Find the functions that satisfy the system of Example 2 and at the same time satisfy the following *initial conditions*:

$$x_1(0) = 5, \quad x_2(0) = 6, \quad \text{and} \quad x_3(0) = 7.$$

Solution. The complete solution of this system is given by

$$X(t) = \begin{bmatrix} 0 & 1 & 1 \\ 1 & 2 & 1 \\ 1 & 2 & 2 \end{bmatrix} \begin{bmatrix} c_1 e^{-t} \\ c_2 e^{t} \\ c_3 e^{2t} \end{bmatrix}.$$

Now the particular solution that we want has the property that

$$X(0) = \begin{bmatrix} x_1(0) \\ x_2(0) \\ x_3(0) \end{bmatrix} = \begin{bmatrix} 5 \\ 6 \\ 7 \end{bmatrix}.$$

Thus, replacing t by 0 in the complete solution leads to the equation

$$\begin{bmatrix} 5 \\ 6 \\ 7 \end{bmatrix} = \begin{bmatrix} 0 & 1 & 1 \\ 1 & 2 & 1 \\ 1 & 2 & 2 \end{bmatrix} \begin{bmatrix} c_1 \\ c_2 \\ c_3 \end{bmatrix}.$$

It follows that

$$\begin{bmatrix} c_1 \\ c_2 \\ c_3 \end{bmatrix} = \begin{bmatrix} 0 & 1 & 1 \\ 1 & 2 & 1 \\ 1 & 2 & 2 \end{bmatrix}^{-1} \begin{bmatrix} 5 \\ 6 \\ 7 \end{bmatrix} = \begin{bmatrix} -2 & 0 & 1 \\ 1 & 1 & -1 \\ 0 & -1 & 1 \end{bmatrix} \begin{bmatrix} 5 \\ 6 \\ 7 \end{bmatrix} = \begin{bmatrix} -3 \\ 4 \\ 1 \end{bmatrix}.$$

Finally, then, the desired particular solution is given by

$$X(t) = \begin{bmatrix} 0 & 1 & 1 \\ 1 & 2 & 1 \\ 1 & 2 & 2 \end{bmatrix} \begin{bmatrix} -3e^{-t} \\ 4e^{t} \\ e^{2t} \end{bmatrix}.$$

Or, in expanded form,

$$\begin{cases} x_1(t) = \phantom{-3e^{-t} +{}} 4e^{t} + e^{2t}, \\ x_2(t) = -3e^{-t} + 8e^{t} + e^{2t}, \\ x_3(t) = -3e^{-t} + 8e^{t} + 2e^{2t}. \end{cases}$$

EXERCISE SET 4.8

1 Solve the system

$$\begin{cases} \dot{x}_1 = 3x_1 + 2x_2, \\ \dot{x}_2 = 4x_1 + x_2. \end{cases}$$

2 Solve $\dot{X} = AX$, where A is the matrix of Exercise 2, Exercise set 4.7.

3 Consider the equation $\dot{X} = AX$, where

$$A = \begin{bmatrix} 4 & -2 & 1 \\ 4 & -2 & 2 \\ 6 & -6 & 5 \end{bmatrix}.$$

a) Find the complete solution.
b) Find the particular solution for which $X(0) = [2, -2, 4]'$.
c) Find the particular solution for which $\dot{X}(0) = [4, 5, 6]'$.
d) Find the particular solution for which $x_1(0) = 2$, $x_2(0) = 3$, and $\dot{x}_3(0) = 4$.

4 Let V be the vector space whose vectors are 3 by 1 matrices of differentiable functions. (For example, $[2t, e^t, \sin t]'$ is in V.) And let L be the linear transformation defined on V by

$$L(X) = \dot{X} - \begin{bmatrix} 1 & 1 & 0 \\ 0 & 2 & 1 \\ 0 & 0 & 3 \end{bmatrix} X.$$

a) Find the kernel of L. What is its dimension?
b) Show that

$$B = \begin{bmatrix} t - 4 \\ -7t + 1 \\ -15t + 11 \end{bmatrix}$$

is in the range of L. [*Hint:* Let $A = [a_1 t + a_2, b_1 t + b_2, c_1 t + c_2]'$ and set $L(A) = B$.]

c) Solve the equation $L(X) = B$. [*Hint:* Recall Theorem 3.7.]

5 Theorem 4.15. Let V be the space whose vectors are n by 1 columns of differentiable functions. Let X^* denote the complete solution of the equation $\dot{X} = AX$, and let B^* be any particular solution of $\dot{X} = AX + B$. Then the complete solution of this latter equation is given by $X = X^* + B^*$.

a) Use Theorem 4.15 to solve

$$\dot{X} = \begin{bmatrix} 3 & 2 \\ 4 & 1 \end{bmatrix} X + \begin{bmatrix} 4 \\ -3 \end{bmatrix}.$$

b) Solve $\dot{x} = 3x + \sin t$. [*Hint:* For a particular solution try a function of the form $a \sin t + b \cos t$.]

c) Prove Theorem 4.15. [*Hint:* Introduce L by the formula $L(X) = \dot{X} - AX$ and apply Theorem 3.7.]

6 a) Solve the second-order differential equation $\ddot{x} - \dot{x} - 2x = 0$. [*Hint:* Set $x_1 = x$ and $x_2 = \dot{x}$; then solve the system

$$\begin{cases} \dot{x}_1 = x_2, \\ \dot{x}_2 = 2x_1 + x_2, \end{cases}$$

for x_1.]

b) Solve $\ddot{x} - \dot{x} - 2x = 2e^t + 5$. [*Hint:* For a particular solution try a function of the form $ae^t + b$.]

7 Solve the *third-order* differential equation $\dddot{x} - 6\ddot{x} + 11\dot{x} - 6x = 0$. Hint: Set $x_1 = x$, $x_2 = \dot{x}$, and $x_3 = \ddot{x}$; then solve the system

$$\begin{cases} \dot{x}_1 = x_2, \\ \dot{x}_2 = x_3, \\ \dot{x}_3 = 6x_3 - 11x_2 + 6x_1. \end{cases}$$

4.9 JORDAN MATRICES

We have seen in the last section that the system of differential equations characterized by the matrix equation $\dot{X} = A\dot{X}$ is easily solved if there is a matrix B such that $B^{-1}AB$ is diagonal. But such a B does not exist for every square matrix A. (See Example 4 of Section 4.7.) In this section we shall devote our attention to such troublesome A's.

Our discussion here will be both informal and incomplete: justification for all of our remarks would require a lengthy digression into the theory of differential equations and also a knowledge of certain more advanced aspects of linear algebra that are beyond the scope of this introductory text. We begin with an example.

Example 1. Solve the system
$$\begin{cases} \dot{x}_1 = 3x_1 + x_2, \\ \dot{x}_2 = 3x_2, \\ \dot{x}_3 = 2x_3. \end{cases}$$

Solution. We have immediately that
$$x_2 = c_2 e^{3t} \quad \text{and} \quad x_3 = c_3 e^{2t}.$$

To find x_1 we must work harder. Substituting the expression for x_2 into the first equation gives
$$\dot{x}_1 = 3x_1 + c_2 e^{3t}.$$

A particular solution for this equation—we now regard c_2 as being fixed—is found by "forcing" $\alpha t e^{3t}$ to satisfy the foregoing equation:
$$\frac{d}{dt}(\alpha t e^{3t}) = 3(\alpha t e^{3t}) + c_2 e^{3t},$$
$$3\alpha t e^{3t} + \alpha e^{3t} = 3\alpha t e^{3t} + c_2 e^{3t},$$
$$\alpha e^{3t} = c_2 e^{3t},$$
$$\alpha = c_2.$$

Thus, we see that the function $c_2 t e^{3t}$ is a particular solution of our equation. (That a particular solution should have been found among the family of functions $\alpha t e^{3t}$ may be attributed to a good guess on our part. A better reason is given in courses on differential equations. In any event, we have found a particular solution.) By Theorem 4.15, the complete solution is given by
$$x_1 = c_1 e^{3t} + c_2 t e^{3t},$$
for $c_1 e^{3t}$ gives all solutions of $\dot{x}_1 = 3x_1$. Finally, then, the given system has as its complete solution
$$\begin{cases} x_1 = c_1 e^{3t} + c_2 t e^{3t}, \\ x_2 = c_2 e^{3t}, \\ x_3 = c_3 e^{2t}. \end{cases} \square$$

It is instructive to look at the system of Example 1 in matrix form:
$$\dot{X} = \begin{bmatrix} 3 & 1 & 0 \\ 0 & 3 & 0 \\ 0 & 0 & 2 \end{bmatrix} X.$$

The matrix is that of Example 4, Section 4.7. It cannot be diagonalized. However, it is itself almost diagonal: its only nonzero entries other than those on the main diagonal occur on the diagonal above the main diagonal.

4.9 JORDAN MATRICES

Now it happens (given any square matrix A) that we can always find a matrix B such that $B^{-1}AB$ is in this "double diagonal" form (Theorem 4.16). As in Example 1, the system of differential equations corresponding to such a matrix is only slightly more difficult to solve than the diagonal case.

Thus, we have reason to be interested in the following matrix problem. Given a square matrix A, find a nonsingular matrix B such that $B^{-1}AB$ is diagonal *except*, perhaps, for some nonzero entries directly above the main diagonal. The remainder of the section is devoted to this problem.

Definition 4.9. First, it will be convenient to denote the nth-order matrix

$$\begin{bmatrix} \lambda & 1 & & & & \\ & \lambda & 1 & & & \\ & & \lambda & 1 & & \\ & & & \ddots & \ddots & \\ & & & & \lambda & 1 \\ & & & & & \lambda \end{bmatrix}.$$

by $G_n(\lambda)$. (Entries not shown are zeros.) Now suppose that we have k such matrices, and that we consider a matrix A having these k matrices "strung along" its main diagonal, all other entries in A being zero. That is,

$$A = \begin{bmatrix} G_{n_1}(\lambda_1) & & & \\ & G_{n_2}(\lambda_2) & & \\ & & \ddots & \\ & & & G_{n_k}(\lambda_k) \end{bmatrix}.$$

Then A is said to be a *Jordan matrix*.

Example 2. Pictures help. We consider some 5 by 5 Jordan matrices:

a) $\begin{bmatrix} 2 & 1 & 0 & 0 & 0 \\ 0 & 2 & 1 & 0 & 0 \\ 0 & 0 & 2 & 0 & 0 \\ 0 & 0 & 0 & 7 & 1 \\ 0 & 0 & 0 & 0 & 7 \end{bmatrix}.$ Here, $A = \begin{bmatrix} G_3(2) & \\ & G_2(7) \end{bmatrix}.$

b) $\begin{bmatrix} 2 & 1 & 0 & 0 & 0 \\ 0 & 2 & 0 & 0 & 0 \\ 0 & 0 & 2 & 1 & 0 \\ 0 & 0 & 0 & 2 & 1 \\ 0 & 0 & 0 & 0 & 2 \end{bmatrix}.$ Here, $A = \begin{bmatrix} G_2(2) & \\ & G_3(2) \end{bmatrix}.$

(In other words, the λ_j need not be distinct.)

c) $\begin{bmatrix} 1 & 0 & 0 & 0 & 0 \\ 0 & 2 & 0 & 0 & 0 \\ 0 & 0 & 3 & 0 & 0 \\ 0 & 0 & 0 & 4 & 0 \\ 0 & 0 & 0 & 0 & 5 \end{bmatrix}$. Here, $A = \begin{bmatrix} G_1(1) \\ G_1(2) \\ G_1(3) \\ G_1(4) \\ G_1(5) \end{bmatrix}$.

(In other words, a diagonal matrix is a Jordan matrix.)

Theorem 4.16. Given any square matrix A, there exists a matrix B such that $B^{-1}AB$ is a Jordan matrix J, each block of J corresponding to one of the characteristic values of A.

We will not prove this theorem. But we will consider a method for finding such a B. Suppose then, looking back from the final result, that we have a B such that $B^{-1}AB = J$, where J is a Jordan matrix. To simplify our discussion, suppose also that the first block along J's main diagonal is

$$G_3(\lambda_1) = \begin{bmatrix} \lambda_1 & 1 & 0 \\ 0 & \lambda_1 & 1 \\ 0 & 0 & \lambda_1 \end{bmatrix}.$$

Thus, we have

$$B^{-1}AB = \begin{bmatrix} \begin{matrix} \lambda_1 & 1 & 0 \\ 0 & \lambda_1 & 1 \\ 0 & 0 & \lambda_1 \end{matrix} & \\ & \ddots \end{bmatrix}, \quad \text{or,} \quad AB = B \begin{bmatrix} \begin{matrix} \lambda_1 & 1 & 0 \\ 0 & \lambda_1 & 1 \\ 0 & 0 & \lambda_1 \end{matrix} & \\ & \ddots \end{bmatrix}.$$

It follows that the jth column of AB is equal to the jth column of BJ. Consider just the first three columns; we are led to the equations

$$AB^1 = B \begin{bmatrix} \lambda_1 \\ 0 \\ \vdots \\ 0 \end{bmatrix} = \lambda_1 B^1,$$

$$AB^2 = B \begin{bmatrix} 1 \\ \lambda_1 \\ 0 \\ \vdots \\ 0 \end{bmatrix} = B^1 + \lambda_1 B^2, \quad AB^3 = B \begin{bmatrix} 0 \\ 1 \\ \lambda_1 \\ 0 \\ \vdots \\ 0 \end{bmatrix} = B^2 + \lambda_1 B^3.$$

Or, equivalently,
$$(A - \lambda_1 I)B^1 = \Theta,$$
$$(A - \lambda_1 I)B^2 = B^1,$$
$$(A - \lambda_1 I)B^3 = B^2.$$

If we had been dealing with $G_{n_1}(\lambda_1)$ instead of $G_3(\lambda_1)$, where $n_1 > 3$, then this list of equations would continue in the same pattern, ending with $(A - \lambda_1 I)B^{n_1} = B^{n_1-1}$.

Similarly, corresponding to the block $G_{n_2}(\lambda_2)$, we find that the next n_2 columns of B must satisfy the equations

$$(A - \lambda_2 I)B^{n_1+1} = \Theta,$$
$$(A - \lambda_2 I)B^{n_1+2} = B^{n_1+1},$$
$$\vdots$$
$$(A - \lambda_2 I)B^{n_1+n_2} = B^{n_1+n_2-1},$$

and so on, with each of the blocks $G_{n_j}(\lambda_j)$.

Our problem now is to somehow reverse this argument. Starting with A we would like to construct lists of column vector equations whose solutions would be the columns of B. The following elaborate scheme will do the job:

1. Find the characteristic values of A. Label them $\lambda_1, \lambda_2, \ldots, \lambda_n$. (The order is immaterial.)

2. Solve the characteristic equation corresponding to λ_1. The solution will involve variables $r, s, \ldots$. Generate an independent set of characteristic vectors for λ_1 by setting each variable in turn equal to 1 while holding all other variables equal to 0. Let the resulting vectors be denoted by

$$X^{11}, X^{12}, X^{13}, \ldots, X^{1p_1}.$$

Repeat this process for $\lambda_2, \lambda_3, \ldots, \lambda_h$, letting the characteristic vectors selected for λ_j be denoted by

$$X^{j1}, X^{j2}, \ldots, X^{jp_j}.$$

3. Pair each of the vectors obtained under 2 with the corresponding characteristic value. This gives rise to the following list of pairs:

$$(\lambda_1, X^{11}), (\lambda_1, X^{12}), \ldots, (\lambda_1, X^{1p_1}),$$
$$(\lambda_2, X^{21}), (\lambda_2, X^{22}), \ldots, (\lambda_2, X^{2p_2}),$$
$$\vdots$$
$$(\lambda_h, X^{h1}), (\lambda_h, X^{h2}), \ldots, (\lambda_h, X^{hp_h}).$$

4. Finally, each of these pairs leads to a set of consecutive columns for B. We proceed as follows: Start with (λ_1, X^{11}). Relabel X^{11} as B^1 and then solve the equation

$$(A - \lambda_1 I)X = B^1.$$

Take B^2 as that solution which arises on setting all nonbeginner variables equal to 0, and then pass on to the equation

$$(A - \lambda_1 I)X = B^2.$$

Solve this equation. Take B^3 as that solution which arises on setting all nonbeginner variables equal to 0, and then pass on to the equation

$$(A - \lambda_1 I)X = B^3.$$

We continue in this manner until we come to an equation which has no solution. (It may be the very first equation.) Thus, the pair (λ_1, X^{11}) will give rise to the vectors

$$B^1, B^2, \ldots, B^{n_1}.$$

Take these vectors as the first n_1 columns of B.

Next, take the pair (λ_1, X^{12}). Relabel X^{12} as B^{n_1+1} and, as before, determine the vectors $B^{n_1+2}, \ldots, B^{n_1+n_2}$ by solving (in succession) the equations

$$(A - \lambda_1 I)X = B^{n_1+1},$$
$$(A - \lambda_1 I)X = B^{n_1+2},$$
$$\vdots$$
$$(A - \lambda_1 I)X = B^{n_1+n_2-1}.$$

Take these vectors as the next n_2 columns of B.

Continuing in this way, considering each pair (λ_j, X^{jk}) in turn, we obtain all of the columns of a matrix B having the property that $B^{-1}AB$ is a Jordan matrix. An extraordinary program? Yes, but it is a program that is more difficult to describe than to put into practice.

Example 3. Let

$$A = \begin{bmatrix} 3 & 1 & 0 & 0 \\ -1 & 2 & 1 & 0 \\ 1 & 0 & 1 & 0 \\ 0 & 0 & 0 & 3 \end{bmatrix}.$$

Find B such that $B^{-1}AB$ is a Jordan matrix.

4.9 JORDAN MATRICES

Solution. We follow the four-step program just described, but leaving most of the details to the reader:

1. $|A - \lambda I| = (\lambda - 2)^3(\lambda - 3)$. Let $\lambda_1 = 2$ and $\lambda_2 = 3$.
2. a) The solutions of $(A - \lambda_1 I)X = \Theta$ are given by

$$X = \begin{bmatrix} t \\ -t \\ t \\ 0 \end{bmatrix}.$$

This solution space is evidently of dimension 1. Thus, a maximal independent set of characteristic vectors corresponding to λ_1 will consist of just one vector. We let

$$X^{11} = \begin{bmatrix} 1 \\ -1 \\ 1 \\ 0 \end{bmatrix}.$$

b) The solutions of $(A - \lambda_2 I)X = \Theta$ are given by

$$X = \begin{bmatrix} 0 \\ 0 \\ 0 \\ t \end{bmatrix}.$$

Again the solution space has dimension 1. We let

$$X^{21} = \begin{bmatrix} 0 \\ 0 \\ 0 \\ 1 \end{bmatrix}.$$

3. There are just two pairs to consider:

$$(\lambda_1, X^{11}) \text{ and } (\lambda_2, X^{21}).$$

4. a) Let $B^1 = X^{11}$ and then consider the equation

$$(A - \lambda_1 I)X = B^1.$$

Its solutions are given by

$$X = \begin{bmatrix} t+1 \\ -t \\ t \\ 0 \end{bmatrix}.$$

Now let

$$B^2 = \begin{bmatrix} 1 \\ 0 \\ 0 \\ 0 \end{bmatrix}$$

and consider next the equation

$$(A - \lambda_1 I)X = B^2.$$

Its solutions are given by

$$X = \begin{bmatrix} t \\ 1 - t \\ t \\ 0 \end{bmatrix}.$$

Let

$$B^3 = \begin{bmatrix} 0 \\ 1 \\ 0 \\ 0 \end{bmatrix}$$

and then consider the equation

$$(A - \lambda_1 I)X = B^3.$$

This equation has no solutions.

b) Let $B^4 = X^{21}$ and then consider the equation

$$(A - \lambda_2 I)X = B^4.$$

This equation has no solutions. (If there were solutions, it would mean that we had made a computational error; for we already have as many columns as B can hold.)

Finally, then, we take

$$B = [B^1, B^2, B^3, B^4] = \begin{bmatrix} 1 & 1 & 0 & 0 \\ -1 & 0 & 1 & 0 \\ 1 & 0 & 0 & 0 \\ 0 & 0 & 0 & 1 \end{bmatrix}.$$

Let us check this result:

$$B^{-1}A = \begin{bmatrix} 0 & 0 & 1 & 0 \\ 1 & 0 & -1 & 0 \\ 0 & 1 & 1 & 0 \\ 0 & 0 & 0 & 1 \end{bmatrix} \begin{bmatrix} 3 & 1 & 0 & 0 \\ -1 & 2 & 1 & 0 \\ 1 & 0 & 1 & 0 \\ 0 & 0 & 0 & 3 \end{bmatrix} = \begin{bmatrix} 1 & 0 & 1 & 0 \\ 2 & 1 & -1 & 0 \\ 0 & 2 & 2 & 0 \\ 0 & 0 & 0 & 3 \end{bmatrix}.$$

and
$$B^{-1}AB = \begin{bmatrix} 1 & 0 & 1 & 0 \\ 2 & 1 & -1 & 0 \\ 0 & 2 & 2 & 0 \\ 0 & 0 & 0 & 3 \end{bmatrix} \begin{bmatrix} 1 & 1 & 0 & 0 \\ -1 & 0 & 1 & 0 \\ 1 & 0 & 0 & 0 \\ 0 & 0 & 0 & 1 \end{bmatrix} = \begin{bmatrix} 2 & 1 & 0 & 0 \\ 0 & 2 & 1 & 0 \\ 0 & 0 & 2 & 0 \\ 0 & 0 & 0 & 3 \end{bmatrix}.$$

Note: Reversing the order of the characteristic values, taking $\lambda_1 = 3$ and $\lambda_2 = 2$, would have led to a matrix B such that

$$B^{-1}AB = \begin{bmatrix} 3 & 0 & 0 & 0 \\ 0 & 2 & 1 & 0 \\ 0 & 0 & 2 & 1 \\ 0 & 0 & 0 & 2 \end{bmatrix}.$$

EXERCISE SET 4.9

In Exercises 1 through 5, find a matrix B such that $B^{-1}AB = J$.

1 Let $A = \begin{bmatrix} 5 & 1 & -3 \\ 0 & 5 & 2 \\ 0 & 0 & 3 \end{bmatrix}$, $J = \begin{bmatrix} 3 & 0 & 0 \\ 0 & 5 & 1 \\ 0 & 0 & 5 \end{bmatrix}$.

2 Let $A = \begin{bmatrix} 2 & 1 & 1 & 1 \\ 0 & 2 & 1 & 1 \\ 0 & 0 & 2 & 1 \\ 0 & 0 & 0 & 2 \end{bmatrix}$, $J = \begin{bmatrix} 2 & 1 & 0 & 0 \\ 0 & 2 & 1 & 0 \\ 0 & 0 & 2 & 1 \\ 0 & 0 & 0 & 2 \end{bmatrix}$.

3 Let $A = \begin{bmatrix} 2 & 1 & 1 & 1 \\ 0 & 2 & 1 & 1 \\ 0 & 0 & 2 & 0 \\ 0 & 0 & 0 & 2 \end{bmatrix}$, $J = \begin{bmatrix} 2 & 0 & 0 & 0 \\ 0 & 2 & 1 & 0 \\ 0 & 0 & 2 & 1 \\ 0 & 0 & 0 & 2 \end{bmatrix}$.

4 Let $A = \begin{bmatrix} 3 & 0 & 0 & 0 \\ 1 & 3 & 0 & 0 \\ 0 & 0 & 4 & 0 \\ 0 & 0 & 1 & 4 \end{bmatrix}$, $J = \begin{bmatrix} 4 & 1 & 0 & 0 \\ 0 & 4 & 0 & 0 \\ 0 & 0 & 3 & 1 \\ 0 & 0 & 0 & 3 \end{bmatrix}$.

5 Let $A = \begin{bmatrix} 0 & 0 & 0 & 0 & 0 \\ 1 & 0 & 0 & 0 & 0 \\ 1 & 1 & 0 & 0 & 0 \\ 1 & 1 & 1 & 0 & 0 \\ 1 & 1 & 1 & 1 & 0 \end{bmatrix}$, $J = \begin{bmatrix} 0 & 1 & 0 & 0 & 0 \\ 0 & 0 & 1 & 0 & 0 \\ 0 & 0 & 0 & 1 & 0 \\ 0 & 0 & 0 & 0 & 1 \\ 0 & 0 & 0 & 0 & 0 \end{bmatrix}$.

6 Solve the system $\dot{X} = AX$, where A is the matrix of Exercise 1.

7 Same as Exercise 6, where A is the matrix of Exercise 5.

8 Let L be the linear transformation from P^2 into P^2 defined by the formula $L[p(t)] = p'(t)$.
   a) Does there exist an ordered basis $\mathcal{B}$ such that the $\mathcal{B}\mathcal{B}$-representation of L would be diagonal? (Think of the implications of having L represented by a diagonal matrix.)
   b) Find an ordered basis $\mathcal{B}$ such that the $\mathcal{B}\mathcal{B}$-representation of L is a Jordan matrix.

9 Solve $\ddot{x} - 2\dot{x} + x = 0$ using the method of Exercise 6, Section 4.8.

10 Solve $\dot{x} = 0$, first using a direct, natural approach and then by using the method of Exercise 7, Section 4.8.

11 This is a continuation of Exercise 15, Section 3.6. Let L be a projection in V and let A be a matrix representation for L.
   a) Prove: The only characteristic values possible for A are 0 and 1.
   b) Prove: A is similar to a diagonal matrix. [*Hint:* By Theorem 4.16 we are assured that A is similar to a Jordan matrix, say B. Show that $B^2 = B$ is possible only if B is diagonal.]
   c) Let L^* be defined on R^{21} by

$$L^*(X) = \begin{bmatrix} \frac{1}{2} & \frac{1}{2} \\ \frac{1}{2} & \frac{1}{2} \end{bmatrix} X.$$

Show that L^* is a projection and then determine a basis $\mathcal{B}$ such that the $\mathcal{B}\mathcal{B}$-formula for L^* is

$$Y_{\mathcal{B}} = \begin{bmatrix} 1 & 0 \\ 0 & 0 \end{bmatrix} X_{\mathcal{B}}.$$

4.10 THE VANDERMONDE DETERMINANT

Let A be any matrix generated by the formula

$$A = \begin{bmatrix} 1 & 1 & 1 \\ a & b & c \\ a^2 & b^2 & c^2 \end{bmatrix}.$$

Then the determinant of A will be some polynomial expression in a, b, and c. It is interesting—and also of theoretical importance—that this expression can be put into a very special factored form:

$$\begin{aligned}
|A| = \begin{vmatrix} 1 & 1 & 1 \\ a & b & c \\ a^2 & b^2 & c^2 \end{vmatrix} &= \begin{vmatrix} 1 & 0 & 0 \\ a & (b-a) & (c-a) \\ a^2 & (b^2-a^2) & (c^2-a^2) \end{vmatrix} \\
&= (b-a)(c^2-a^2) - (c-a)(b^2-a^2) \\
&= (b-a)(c-a)[(c+a) - (b+a)] \\
&= (b-a)(c-a)(c-b).
\end{aligned}$$

4.10 THE VANDERMONDE DETERMINANT

As the reader might guess, an analogous result holds for nth-order versions of A.

Definition 4.10. Let A_n be any matrix given by

$$A_n = \begin{bmatrix} 1 & 1 & \cdots & 1 \\ a_0 & a_1 & \cdots & a_n \\ a_0^2 & a_1^2 & \cdots & a_n^2 \\ \vdots & \vdots & & \vdots \\ a_0^n & a_1^n & \cdots & a_n^n \end{bmatrix}.$$

Then A_n is called a *Vandermonde matrix*, and $|A_n|$ is called a *Vandermonde determinant*.

Theorem 4.17. A Vandermonde determinant may be evaluated by the following formula:

$$|A_n| = [(a_1 - a_0)] \cdot [(a_2 - a_0)(a_2 - a_1)] \cdot [(a_3 - a_0)(a_3 - a_1)(a_3 - a_2)] \cdots [(a_n - a_0)(a_n - a_1) \cdots (a_n - a_{n-1})].$$

Proof. This proof is facilitated by using a product notation: we write $\prod_{k=m}^{n} b_k$ to denote the product $b_m \cdot b_{m+1} \cdot b_{m+2} \cdots b_n$. The formula of our theorem, which involves a *product of products*, may then be expressed compactly by writing

$$|A_n| = \prod_{j=1}^{n} \left[\prod_{i=0}^{j-1} (a_j - a_i) \right].$$

(The reader would profit by expanding this *double product* for various choices of n.) We now proceed to the proof; our method will be that of mathematical induction.

For $n = 1$, we have

$$\prod_{j=1}^{1} \left[\prod_{i=0}^{j-1} (a_j - a_i) \right] = \prod_{i=0}^{0} (a_1 - a_i) = a_1 - a_0 = \begin{vmatrix} 1 & 1 \\ a_0 & a_1 \end{vmatrix} = |A_1|.$$

Next, we assume that the formula is valid for n. To show that it continues to hold for $(n + 1)$, we make an inspired play: Let p be the polynomial function defined by

$$p(t) = \begin{vmatrix} 1 & 1 & \cdots & 1 \\ t & a_1 & \cdots & a_{n+1} \\ \vdots & \vdots & & \vdots \\ t^{n+1} & a_1^{n+1} & \cdots & a_{n+1}^{n+1} \end{vmatrix}.$$

We make two observations about p:

1. $p(t) = c(t - a_1)(t - a_2) \cdots (t - a_{n+1})$. (Why?)
2. The leading coefficient of $p(t)$ is given by

$$c = (-1)^{n+1} \begin{vmatrix} 1 & 1 & \cdots & 1 \\ a_1 & a_2 & \cdots & a_{n+1} \\ \vdots & \vdots & & \vdots \\ a_1^n & a_2^n & \cdots & a_{n+1}^n \end{vmatrix}.$$

Now c, except perhaps for sign, is an nth-order Vandermonde determinant. It is here that our induction assumption becomes applicable; we may write

$$c = (-1)^{n+1} \prod_{j=2}^{n+1} \left[\prod_{i=1}^{j-1} (a_j - a_i) \right].$$

(The product formula first given was based on the a's being distinguished by the subscripts $0, 1, 2, \ldots, n$. In the present application of that formula the subscripts are $1, 2, 3, \ldots, (n+1)$; hence, the shift in the indices.) Combining our two observations, we may write

$$p(t) = (-1)^{n+1} \prod_{j=2}^{n+1} \left[\prod_{i=1}^{j-1} (a_j - a_i) \right] \cdot (t - a_1)(t - a_2) \cdots (t - a_{n+1}).$$

It follows that

$$p(a_0) = (-1)^{n+1} \prod_{j=2}^{n+1} \left[\prod_{i=1}^{j-1} (a_j - a_i) \right] \cdot \prod_{j=1}^{n+1} (a_0 - a_j)$$

$$= \prod_{j=2}^{n+1} \left[\prod_{i=1}^{j-1} (a_j - a_i) \right] \cdot \prod_{j=1}^{n+1} (a_j - a_0)$$

$$= (a_1 - a_0) \cdot \prod_{j=2}^{n+1} \left[\prod_{i=1}^{j-1} (a_j - a_i) \right] \cdot \prod_{j=2}^{n+1} (a_j - a_0)$$

$$= (a_1 - a_0) \cdot \prod_{j=2}^{n+1} \left[(a_j - a_0) \prod_{i=1}^{j-1} (a_j - a_i) \right]$$

$$= (a_1 - a_0) \cdot \prod_{j=2}^{n+1} \left[\prod_{i=0}^{j-1} (a_j - a_i) \right]$$

$$= \prod_{j=1}^{n+1} \left[\prod_{i=0}^{j-1} (a_j - a_i) \right].$$

On the other hand, by the definition of p,

$$p(a_0) = A_{n+1}. \quad \square$$

As a first application of Theorem 4.17, we shall fulfill the promise made at the close of Example 2, in Section 4.7.

4.10 THE VANDERMONDE DETERMINANT

Theorem 4.18. Let A be an nth-order matrix having n distinct characteristic values, say $\lambda_1, \lambda_2, \ldots, \lambda_n$. Let $S = \{X^1, X^2, \ldots, X^n\}$ be a set of characteristic vectors for A, where X^j corresponds to λ_j. Then S is linearly independent.

Proof. Suppose that we have numbers $\alpha_1, \alpha_2, \ldots, \alpha_n$ such that

$$\alpha_1 X^1 + \alpha_2 X^2 + \cdots + \alpha_n X^n = \Theta.$$

Then we multiply both sides by A:

$$\alpha_1(AX^1) + \alpha_2(AX^2) + \cdots + \alpha_n(AX^n) = A\Theta.$$

But $AX^j = \lambda_j X^j$. Thus, we may write

$$\alpha_1 \lambda_1 X^1 + \alpha_2 \lambda_2 X^2 + \cdots + \alpha_n \lambda_n X^n = \Theta.$$

Further multiplications by A will lead to analogous results. Indeed, $(n-1)$ such multiplications give rise to the following set of equations:

$$\begin{aligned} \alpha_1 X^1 + & \alpha_2 X^2 + \cdots + \alpha_n X^n = \Theta, \\ \lambda_1(\alpha_1 X^1) + & \lambda_2(\alpha_2 X^2) + \cdots + \lambda_n(\alpha_n X^n) = \Theta, \\ \lambda_1^2(\alpha_1 X^1) + & \lambda_2^2(\alpha_2 X^2) + \cdots + \lambda_n^2(\alpha_n X^n) = \Theta, \\ & \vdots \\ \lambda_1^{n-1}(\alpha_1 X^1) + & \lambda_2^{n-1}(\alpha_2 X^2) + \cdots + \lambda_n^{n-1}(\alpha_n X^n) = \Theta. \end{aligned}$$

Equivalently,

$$\begin{bmatrix} 1 & 1 & \cdots & 1 \\ \lambda_1 & \lambda_2 & \cdots & \lambda_n \\ \lambda_1^2 & \lambda_2^2 & \cdots & \lambda_n^2 \\ \vdots & \vdots & & \vdots \\ \lambda_1^{n-1} & \lambda_2^{n-1} & \cdots & \lambda_n^{n-1} \end{bmatrix} \begin{bmatrix} \alpha_1 X^1 \\ \alpha_2 X^2 \\ \alpha_3 X^3 \\ \vdots \\ \alpha_n X^n \end{bmatrix} = \begin{bmatrix} \Theta \\ \Theta \\ \Theta \\ \vdots \\ \Theta \end{bmatrix}.$$

We are thus led to a Vandermonde matrix, and one that obviously has an inverse. (Why?) Multiplying through by this inverse gives

$$\begin{bmatrix} \alpha_1 X^1 \\ \alpha_2 X^2 \\ \vdots \\ \alpha_n X^n \end{bmatrix} = \begin{bmatrix} \Theta \\ \Theta \\ \vdots \\ \Theta \end{bmatrix}.$$

Thus, $\alpha_j X^j = \Theta$ for $j = 1, 2, \ldots, n$. But no $X^j = \Theta$. It follows that each $\alpha_j = 0$.

Corollary. As in the theorem, suppose that A has n distinct characteristic values. Then, for each λ_j the corresponding space of characteristic vectors has dimension 1.

Proof. *By contradiction.* Suppose that the space of characteristic vectors corresponding to one of the characteristic values, say λ_1, has dimension exceeding 1. (The characteristic values may be relabeled if necessary so that such a characteristic value is first.) Let S be as in the theorem; and let X be a second characteristic vector corresponding to λ_1, one that is *not* a multiple of X^1. Then, since S is a basis for R^{n1}, we may write

$$X = a_1 X^1 + a_2 X^2 + \cdots + a_n X^n,$$

where at least one a_i, for $i > 1$, is different from 0. Multiplying both sides by A, we obtain

$$\lambda_1 X = a_1 \lambda_1 X^1 + a_2 \lambda_2 X^2 + \cdots + a_n \lambda_n X^n.$$

It follows that $\lambda_1 \neq 0$. (Why?) Thus, we may divide by λ_1:

$$X = a_1 X^1 + a_2 \frac{\lambda_2}{\lambda_1} X^2 + \cdots + a_n \frac{\lambda_n}{\lambda_1} X^n.$$

Finally, since our representation for X is necessarily unique, we may conclude that

$$a_i \frac{\lambda_i}{\lambda_1} = a_i.$$

But then, contrary to our hypothesis, $\lambda_i = \lambda_1$. □

Our second application of Theorem 4.17 is with respect to a test for the linear independence of a set of functions. To provide motivation for this discussion, we turn again to the study of differential equations. In particular, we are interested in the following important theorem: *The solution space of the linear differential equation*

$$a_n x^{(n)}(t) + a_{n-1} x^{(n-1)}(t) + \cdots + a_1 x'(t) + a_0 x(t) = \Theta$$

is n-dimensional. This theorem may be exploited in the following way: To solve the given equation, find first a set of n linearly independent particular solutions, say $S = \{f_1, f_2, \ldots, f_n\}$. Then S is necessarily a basis for the solution space. It follows that the complete solution of the equation is given by

$$f = c_1 f_1 + c_2 f_2 + \cdots + c_n f_n.$$

We have, then, a reason for desiring a test that will establish the linear independence of a set of functions. Before presenting the test, we give a definition.

4.10 THE VANDERMONDE DETERMINANT

Definition 4.11. Let V be the vector space of all functions that are defined on $(-\infty, \infty)$, let $S = \{f_1, f_2, \ldots, f_n\}$ be a subset of V, and let W be the function defined by

$$W(t) = \begin{vmatrix} f_1(t) & f_2(t) & \cdots & f_n(t) \\ f_1'(t) & f_2'(t) & \cdots & f_n'(t) \\ \vdots & \vdots & & \vdots \\ f_1^{(n-1)}(t) & f_2^{(n-1)}(t) & \cdots & f_n^{(n-1)}(t) \end{vmatrix}.$$

Then W is said to be the *Wronskian of the set S*. (Note that we can speak of the Wronskian of S only if the members of S are differentiable at least $(n-1)$ times.)

Theorem 4.19. If the Wronskian of S is not the zero function, then the set S is a linearly independent subset of V.

Proof. Suppose that we have numbers $\alpha_1, \alpha_2, \ldots, \alpha_n$ such that

$$\alpha_1 f_1 + \alpha_2 f_2 + \cdots + \alpha_n f_n = \Theta.$$

Differentiating this equation $(n-1)$ times leads to the following set of equations:

$$\begin{cases} \alpha_1 f_1 & + \alpha_2 f_2 & + \cdots + \alpha_n f_n & = \Theta, \\ \alpha_1 f_1' & + \alpha_2 f_2' & + \cdots + \alpha_n f_n' & = \Theta, \\ \vdots & & & \\ \alpha_1 f_1^{(n-1)} & + \alpha_2 f_2^{(n-1)} & + \cdots + \alpha_n f_n^{(n-1)} & = \Theta. \end{cases}$$

Next, choose t_0 such that $W(t_0) \neq 0$, where W is the Wronskian of S. (We are using here our hypothesis). Now return to the first of the foregoing set of equations. Its message is that the function

$$g = \alpha_1 f_1 + \alpha_2 f_2 + \cdots + \alpha_n f_n$$

is the zero function. This means that $g(t) = 0$ for all t. In particular,

$$g(t_0) = \alpha_1 f_1(t_0) + \alpha_2 f_2(t_0) + \cdots + \alpha_n f_n(t_0) = 0.$$

Similarly, each of the foregoing functional equations gives rise to an analogous numerical equation. Thus, we may write

$$\begin{cases} \alpha_1 f_1(t_0) & + \alpha_2 f_2(t_0) & + \cdots + \alpha_n f_n(t_0) & = 0, \\ \alpha_1 f_1'(t_0) & + \alpha_2 f_2'(t_0) & + \cdots + \alpha_n f_n'(t_0) & = 0, \\ \vdots & & & \\ \alpha_1 f_1^{(n-1)}(t_0) & + \alpha_2 f_2^{(n-1)}(t_0) & + \cdots + \alpha_n f_n^{(n-1)}(t_0) & = 0. \end{cases}$$

Equivalently,

$$\begin{bmatrix} f_1(t_0) & f_2(t_0) & \cdots & f_n(t_0) \\ f_1'(t_0) & f_2'(t_0) & \cdots & f_n'(t_0) \\ \vdots & \vdots & & \vdots \\ f_1^{(n-1)}(t_0) & f_2^{(n-1)}(t_0) & \cdots & f_n^{(n-1)}(t_0) \end{bmatrix} \begin{bmatrix} \alpha_1 \\ \alpha_2 \\ \vdots \\ \alpha_n \end{bmatrix} = \begin{bmatrix} 0 \\ 0 \\ \vdots \\ 0 \end{bmatrix}.$$

Now the coefficient matrix has an inverse. (Why?) It follows that

$$\begin{bmatrix} \alpha_1 \\ \alpha_2 \\ \vdots \\ \alpha_n \end{bmatrix} = \begin{bmatrix} 0 \\ 0 \\ \vdots \\ 0 \end{bmatrix}.$$

Example 1. Prove that the set $S = \{e^{a_1 t}, e^{a_2 t}, \ldots, e^{a_n t}\}$ is linearly independent, provided that the a_k are all distinct. (This example provides the promised second application of Theorem 4.17.)

Solution. The Wronskian of S is given by

$$W(t) = \begin{vmatrix} e^{a_1 t} & e^{a_2 t} & \cdots & e^{a_n t} \\ a_1 e^{a_1 t} & a_2 e^{a_2 t} & \cdots & a_n e^{a_n t} \\ \vdots & \vdots & & \vdots \\ a_1^{n-1} e^{a_1 t} & a_2^{n-1} e^{a_2 t} & \cdots & a_n^{(n-1)} e^{a_n t} \end{vmatrix}.$$

Or,

$$W(t) = e^{a_1 t} \cdot e^{a_2 t} \cdots e^{a_n t} \begin{vmatrix} 1 & 1 & \cdots & 1 \\ a_1 & a_2 & \cdots & a_n \\ \vdots & \vdots & & \vdots \\ a_1^{n-1} & a_2^{n-1} & \cdots & a_n^{n-1} \end{vmatrix}.$$

Taking $t_0 = 0$, we have

$$W(0) = \begin{vmatrix} 1 & 1 & \cdots & 1 \\ a_1 & a_2 & \cdots & a_n \\ \vdots & \vdots & & \vdots \\ a_1^{n-1} & a_2^{n-1} & \cdots & a_n^{n-1} \end{vmatrix}.$$

It follows that $W(0) \neq 0$. (Why?) Thus, appealing to Theorem 4.19, we may conclude that S is linearly independent.

EXERCISE SET 4.10

1 Let A_4 be a fourth-order Vandermonde matrix.

   a) Use Theorem 4.17 to express $|A_4|$ in a factored form. (Show all factors.)

b) Derive the factored form of $|A_4|$ from the analogous factored form for a third-order Vandermonde determinant. [*Hint:* Mimic the appropriate part of the proof of Theorem 4.17.]

2 Let $B_n = [j^{i-1}]_{nn}$.
   a) Evaluate $|B_5|$.
   b) Show that $|B_n| = (1!)(2!) \cdots (n!)$.

3 Let A be any matrix generated by the formula

$$A = \begin{bmatrix} bcd & acd & abd & abc \\ 1 & 1 & 1 & 1 \\ a & b & c & d \\ a^2 & b^2 & c^2 & d^2 \end{bmatrix}.$$

Prove that A is nonsingular if and only if a, b, c, and d are all distinct. [*Hint:* First consider the case where none of the variables is zero.]

4 Let $S = \{1, t, t^2, \ldots, t^n\}$.
   a) Show that the Wronskian of S is the constant function $W(t) = (1!)(2!) \cdots (n!)$.
   b) Show that S is linearly independent.
   c) Show that $S^* = \{e^t, te^t, t^2 e^t, \ldots, t^n e^t\}$ is linearly independent. [*Hint:* Do not appeal to the Wronskian of S^*.]

5 a) Let $S = \{1, e^t, te^t, e^{2t}\}$. Prove that each member of S is a particular solution of the differential equation

$$f^{(4)}(t) - 4f'''(t) + 5f''(t) - 2f'(t) = 0.$$

   b) Show that S is linearly independent.
   c) Give the complete solution of the equation in part (a), assuming that this solution is a vector space of dimension 4.

6 Let $f(t) = t^2$ and let

$$g(t) = \begin{cases} 0, & t \leq 0, \\ t^2, & t > 0. \end{cases}$$

   a) Show that the set $S = \{f, g\}$ is linearly independent.
   b) Show that the Wronskian of S is the zero function.
   c) Prove or disprove the proposition converse to that of Theorem 4.19.

7 a) Let W be the Wronskian of $S = \{f_1, f_2, f_3\}$. Show that

$$W'(t) = \begin{vmatrix} f_1(t) & f_2(t) & f_3(t) \\ f_1'(t) & f_2'(t) & f_3'(t) \\ f_1'''(t) & f_2'''(t) & f_3'''(t) \end{vmatrix}.$$

[*Hint:* Recall Exercise 8 of Section 4.4.]

   b) Let W be the Wronskian of $S = \{f_1, f_2, \ldots, f_n\}$. Write a formula analogous to that of part (a) for $W'(t)$.

8 Prove: If a polynomial function of degree n takes on the value 0 for $n+1$ different choices of t, then that function must be the zero function. [*Hint:* Let $p(t) = a_n t^n + \cdots + a_1 t + a_0$ and suppose that
$$p(c_1) = p(c_2) = \cdots = p(c_{n+1}) = 0.$$
Then find a square matrix C such that
$$C[a_n, \ldots, a_1, a_0]' = \Theta.]$$

9 Let A and B be nth-order matrices.
  a) Prove: If A has n distinct characteristic values and $AB = BA$, then each characteristic vector of A is also a characteristic vector of B. [*Hint:* Multiply both sides of $AX = \lambda X$ by B.]
  b) Prove: If A and B both have the same n linearly independent characteristic vectors, then $AB = BA$.

CHAPTER 5

EUCLIDEAN SPACES

5.1 LENGTH AND ANGLE

An elementary physics text defines a *vector* as any quantity that has both magnitude and direction, and then proceeds to depict a vector by displaying a picture of an arrow. We have also from time to time likened a vector to an arrow, especially when a discussion could be motivated by using an arrow drawing. However, the vectors of many of our vector spaces—being matrices, polynomials, etc.—seem not all like arrows: they are not quantities that have both magnitude and direction. Not yet, that is. For it turns out—and this is the basis for this chapter—that the vectors of any vector space can be made to have magnitude and direction. Just how we "make" a vector have magnitude and direction will take some explaining; in this section we shall restrict our attention to the column spaces.

We begin our discussion by considering the vectors in R^{31}. These vectors, as we have seen before, are easily identified with a space of arrows

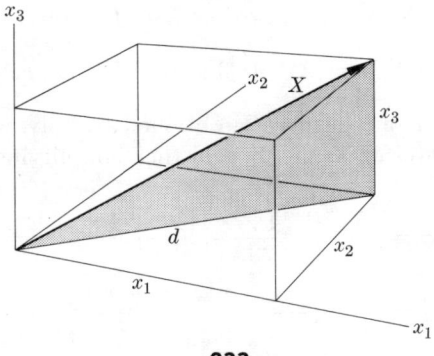

Figure 5.1

emanating from the origin of a coordinate system. Suppose, in particular, that the axes of our coordinate system are chosen so as to have a common scale and so as to be mutually perpendicular (see Fig. 5.1). Then the length of the arrow corresponding to a given X, denoted by $|X|$, is easily expressed in terms of the natural coordinates of S; we have only to apply (twice) the Pythagorean theorem:

$$|X|^2 = d^2 + x_3^2 = (x_1^2 + x_2^2) + x_3^2.$$

Therefore,

$$|X| = \sqrt{x_1^2 + x_2^2 + x_3^2}.$$

The question of direction is more difficult. What exactly do we mean by the direction of an arrow? A reasonable answer: The direction of an arrow is simply how that arrow points with respect to the axes of our coordinate system. How it points is reflected in the angles that the vector makes with the positive ends of those axes. Or, letting N^1, N^2, and N^3 have their usual meanings, the direction of X is known by the angles it makes with N^1, N^2, and N^3 (see Fig. 5.2).

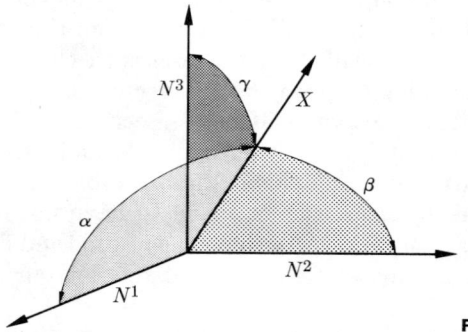

Figure 5.2

Thus, the question of the direction of an arrow depends ultimately on the idea of the angle between two arrows, and to find the angle between X and Y (see Fig. 5.3) we appeal to the *law of cosines*:

$$|X - Y|^2 = |X|^2 + |Y|^2 - 2|X||Y|\cos\phi.$$

A useful formula for our purpose is obtained by solving for $\cos\phi$, introducing the coordinates of X and Y, and then simplifying:

$$\cos\phi = \frac{|X|^2 + |Y|^2 - |Y - X|^2}{2|X||Y|}$$

$$= \frac{x_1 y_1 + x_2 y_2 + x_3 y_3}{\sqrt{x_1^2 + x_2^2 + x_3^2}\sqrt{y_1^2 + y_2^2 + y_3^2}}.$$

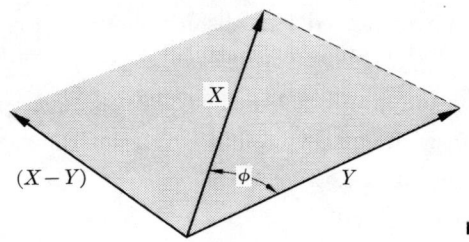

Figure 5.3

Finally, ϕ is taken as the angle between 0 and π (inclusive) having this cosine. [Note that two nonzero vectors X and Y are perpendicular if $x_1y_1 + x_2y_2 + x_3y_3 = 0$. Why?]

We now make an observation that is crucial: The formulas for $|X|$ and $\cos \phi$ have a "common denominator." To bring out this commonness, let "dot" denote the function defined by the formula

$$\text{dot}(X, Y) = x_1y_1 + x_2y_2 + x_3y_3.$$

(The domain of dot consists of all ordered pairs of vectors chosen from R^{31}; its range is the set of all real numbers.) We may now write

$$|X| = \sqrt{\text{dot}(X, X)} \quad \text{and} \quad \cos \phi = \frac{\text{dot}(X, Y)}{\sqrt{\text{dot}(X, X)} \sqrt{\text{dot}(Y, Y)}}.$$

The point is this: The notions of length and angle are both expressible in terms of this dot function. Since this function has an immediate generalization to column spaces of arbitrary dimension, the foregoing formulas can be used to extend the notions of length and angle to any column space.

Definition 5.1. Let S be the set of all ordered pairs of vectors taken from R^{n1}. The function *dot* is defined on S by

$$\text{dot}(X, Y) = x_1y_1 + x_2y_2 + \cdots + x_ny_n.$$

Equivalently, identifying a 1 by 1 matrix with its single entry, we may write $\text{dot}(X, Y) = X'Y$. The *length* of X is defined by

$$|X| = \sqrt{\text{dot}(X, X)}.$$

The *angle between X and Y* is the value of ϕ between 0 and π inclusive such that

$$\cos \phi = \frac{\text{dot}(X, Y)}{|X| \cdot |Y|}.$$

(We assume for the present—otherwise this definition of angle might not make sense for every pair of nonzero vectors—that the quotient defining $\cos \phi$ will never exceed 1 in absolute value.) Finally, we say that nonzero vectors X and Y are *orthogonal* if dot $(X, Y) = 0$.

Theorem 5.1. The function dot has the following basic properties:
1. dot $(X, X) \geq 0$, with equality holding only for the case $X = \Theta$.
2. dot $(Y, X) = $ dot (X, Y).
3. dot $(cX, Y) = c$ dot (X, Y).
4. dot $(X, Y + Z) = $ dot $(X, Y) + $ dot (X, Z).

Proof. We consider only Property 4; the other properties are proved just as easily and are thus left to the reader:

dot $(X, Y + Z) = X'(Y + Z) = X'Y + X'Z = $ dot $(X, Y) + $ dot (X, Z). □

Properties 2 and 4 remind us of ordinary multiplication: in Property 2 we have a commutative property and in Property 4, a distributive property. For this reason mathematicians have called this function the *dot product* and have used a multiplication symbol to denote it, writing

$$X \cdot Y \quad \text{for} \quad \text{dot } (X, Y).$$

We shall—because of its brevity—adopt this standard notation; however, it should be pointed out that the dot product fails to be like the product of ordinary arithmetic in a very significant way: the product of two numbers is a number; the dot product of two vectors is *not* a vector. (Some would argue that the name "dot product" is misleading, for in other usages of the word "product" we generally have an operation in which the product of two things is again a thing of the same species.)

Suppose now that we are presented with certain vectors from R^4[1], say X and Y, and are asked to find their lengths and the angle between them. Then—by applying Definition 5.1—we can easily comply with this request. However, it is reasonable that the entire operation should leave us quite cold: What does it all mean, really? The use of terms like "length" and "angle" imply certain spatial relationships; since we cannot visualize a four-dimensional space, it seems somehow inappropriate to use these terms. Well, this reaction is a natural one and not easy to counter. The following example will, perhaps, give the reader some insight into why this geometric language is indeed appropriate.

Example 1. A geometrical problem that nicely motivates the work of this section is that of finding the distance (in three-dimensional space) from a point to a line. To simplify the problem slightly, let us suppose that the line passes through the origin. Now, a vector formulation of this problem is readily obtained by appealing to arrow drawings. Suppose that the given line is parallel to the vector B; then it may be described by the

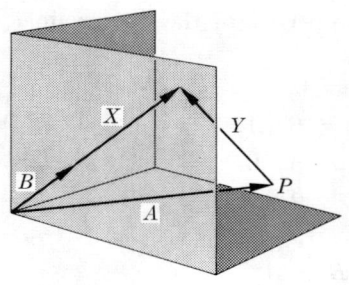

Figure 5.4   **Figure 5.5**

vector formula $X = tB$. Next, let A be the vector from the origin to the given point. Then $Y = X - A$ (see Fig. 5.4) is a vector that reaches from the point to the line. Since X is a function of t, so is Y. Thus—giving our intuition free rein—we may take t to represent time and imagine our configuration of vectors to be in motion: as t increases the vector X grows along the line taking Y with it (see Fig. 5.5.)

The problem may now be solved by answering either of the following questions:

1. For what value of t will Y be orthogonal to B?
2. For what value of t will Y have the smallest length?

Having found such a t, say t^*, then the corresponding Y is given by

$$Y^* = X^* - A = t^*B - A.$$

The desired distance from the point to the line is simply the length of Y^*.

How shall we find this t^*? We give two solutions, corresponding to the two questions raised above.

Solution 1. We have $Y = X - A$, where $X = tB$. It follows that

$$Y = tB - A.$$

Since the desired Y is to be orthogonal to B we adjoin the equation

$$B \cdot Y = 0.$$

Thus, the scalar t^* and the vector Y^* may be realized by solving the system

$$\begin{cases} Y = tB - A, \\ B \cdot Y = 0. \end{cases}$$

Eliminating Y by substitution into the second equation gives

$$B \cdot (tB - A) = 0.$$

238 EUCLIDEAN SPACES   5.1

Availing ourselves of the arithmetical properties of the dot product, we easily solve for t:
$$B \cdot (tB) - B \cdot A = 0,$$
$$t(B \cdot B) = B \cdot A,$$
$$t^* = \frac{A \cdot B}{B \cdot B}.$$

Finally,
$$Y^* = \left(\frac{A \cdot B}{B \cdot B}\right) B - A.$$

and the desired distance from point to line is given by $|Y^*|$.

Note that the equation-solving techniques used to determine t^* and Y^* are the same as those we would use if A, B, and Y denoted numbers rather than vectors. Thus, we see more clearly, perhaps, why mathematicians view the dot function as a "kind of multiplication." The reader should find it instructive to mimic the above solution using the notation dot (X, Y) for $X \cdot Y$, and making only those moves that are permitted by Theorem 5.1.

Solution 2. As before, we have $Y = tB - A$. Thus, $|Y| = |tB - A|$, or,
$$|Y| = \sqrt{(tB - A) \cdot (tB - A)}$$
$$= \sqrt{(B \cdot B)t^2 - 2(A \cdot B)t + (A \cdot A)}.$$

Now, since $\sqrt{f(t)}$ has its smallest value for the same t that minimizes $f(t)$, we are led to the problem of minimizing the function
$$f(t) = (B \cdot B)t^2 - 2(A \cdot B)t + (A \cdot A).$$

Appealing to calculus, we first find the derivative of f:
$$f'(t) = 2(B \cdot B)t - 2(A \cdot B).$$

Then the value of t which makes this derivative 0—the value which minimizes $f(t)$—is given by
$$t^* = \frac{A \cdot B}{B \cdot B}.$$

This is the same value as that obtained under the first solution. ☐

Each of the two solutions given under Example 1 made use of *coordinate-free* methods. That is, we found t^* by manipulating directly the given vectors A and B, making no appeal to their coordinates. Because of this, our solutions are equally valid whether the problem be posed in a two-dimensional space or in a three-dimensional one. In other words, if we are given a line and a point in a plane, then it still makes sense to ask how far the point is from the line. Again—assuming that the line passes through the origin—the given information will consist of two vectors A and B. And again the problem will be solved by first finding the scalar t^*—using either of the given solutions.

What if we had this problem to solve in a higher dimensional space? The reader may at first protest: How can we have this problem in a higher dimensional space? Lines and points in spaces of dimension greater than three cannot be visualized as having a "location" in that space; thus, it makes no sense to talk about the distance from a point to a line in such a space. Not so! For using our analytic definitions of length and orthogonality we have seen that the given geometrical problem can be translated into an analytic one involving certain vector equations. And we have solved that problem by methods which are independent of the dimension of the space. Thus, for example, even if A and B were from a four-dimensional space, we could still carry out (with Theorem 5.1 playing an important role) the manipulations required to find t^*.

The reader may protest further: All right, we can solve certain abstract vector problems in higher dimensional spaces, but that is all we are doing; there is really—despite the geometrical language—no geometrical problem from which the analytic one has arisen. A rebuttal: Suppose that we are working in a space of dimension $n > 3$. We are asked first, given vectors A and B, to find the scalar t and the vector Y that satisfy the system

$$\begin{cases} Y = tB - A, \\ B \cdot Y = 0. \end{cases}$$

We are then asked to find the value of t that minimizes the function

$$f(t) = \sqrt{(tB - A) \cdot (tB - A)},$$

and also for the corresponding minimum value. Now there is nothing about these two problems to suggest that the same value for t satisfies both problems, and that the length of the Y satisfying the system is the same as the minimum value obtained for $f(t)$. But they are: This is the implication of the coordinate-free methods used in Example 1. This means that the definitions we have used to define distance and angle in R^{n1} are successful in the sense that in any such column space the point on a line nearest to a given point off that line is so "located" that the vector joining the two points is orthogonal to that giving direction to the line. In other words, the generalized notions of orthogonality and distance seem to be related in higher dimensional column spaces in precisely the same way

EXERCISE SET 5.1

1. Let $A = [2, 1, 3\sqrt{2}, 2]'$, $B = [-2, 2, 0, 1]'$, and $C = A - B$.
   a) Find the angle between A and B.
   b) Find the angle between A and C.
   c) Find the angle between B and $-C$.
   d) Total these three angles.
   e) Make an arrow drawing that relates the result of part (d) to a basic fact of plane geometry. [*Note:* Although we are working in a four-dimensional space, the particular vectors A, B, and C all lie in a two-dimensional subspace of that space. Thus, arrow drawings are still meaningful.]

2. Let $A = [1, 0, -1, 2, 1]'$, $B = [-1, 1, 2, -1, 0]'$, and $C = A + B$.
   a) Show that the angle between A and C is the same as that between B and C.
   b) Make an arrow drawing relating this result to a basic fact of plane geometry.

3. a) Does the dot product satisfy the associative law $A \cdot (B \cdot C) = (A \cdot B) \cdot C$? Explain.
   b) Prove Property 2 of Theorem 5.1.

4. Derive the following properties as corollaries to Theorem 5.1:
   a) dot $(X + Y, Z) = $ dot $(X, Z) + $ dot (Y, Z).
   b) dot $(X, cY) = c$ dot (X, Y).

5. Rewrite the properties of Theorem 5.1 using the notation $X \cdot Y$ in place of dot (X, Y).

6. **Definition 5.2.** By the *direction of* X we shall mean the n-tuple $(\cos \alpha_1, \cos \alpha_2, \ldots, \cos \alpha_n)$, where α_j is the angle between X and N^j. The entries of this n-tuple are called the *direction cosines* of X.
   a) Use Definition 5.2 to find the direction of $X = [2, 1, -2, 4]'$.
   b) Prove: If $c > 0$, then the direction of cX is the same as that of X.
   c) Explain the italicized part of the following statement: The direction of X *is given by* the vector of unit length having the same direction as X. [*Hint:* Compare the direction cosines of X with the natural coordinates of $(1/|X|)X$.]
   d) Use part (c) to find the direction $B = [4, 2, -4, 8]'$.

7. **Definition 5.3.** By the *distance from* X *to* Y we shall mean the number $|X - Y|$.

Let L be the line in R^{41} described by $X = tA + B$, where $A = [1, 2, 0, -1]'$ and $B = [0, 1, 1, 2]'$. Let X^1, X^2, and X^3 be the three points on L corresponding to $t_1 = -1$, $t_2 = 0$, and $t_3 = 2$.

a) Show that the distance from X^1 to X^3 is equal to the distance from X^1 to X^2 plus that from X^2 to X^3.

b) Show that the direction of $X^3 - X^2$ is the same as that of $X^2 - X^1$.

8 a) How far is the "point" $A = [5, 2, 2, 4]'$ from the line $X = tB$, where $B = [1, -2, 1, 3]'$?

b) How far is A from the line $X = tB + C$, where $C = [13, 2, 9, 9]'$?

c) How far is A from the 2-plane $X = sD + tE$, where $D = [1, 0, 2, -1]'$ and $E = [0, 0, 1, 2]'$?

9 Let $A = [1, 0, 2, 1]'$ and $B = [1, 1, -1, 3]'$.

a) Solve the following system for A^1, A^2, and s.

$$\begin{cases} A = A^1 + A^2, \\ A^1 = sB, \\ A^2 \cdot B = 0. \end{cases}$$

b) The vector A^1 is said to be the *vector* projection of A onto B. Make an arrow drawing that explains this terminology.

c) How far is A from the line $X = sB$? [*Hint:* First look at your drawing.]

10 Let $A^1 = [1, 1, 2]'$, $A^2 = [-1, 2, 1]'$, and $B = [1, -2, 0]'$.

a) Solve the following system for B^1, B^2, s, and t.

$$\begin{cases} B = B^1 + B^2, \\ B^1 = sA^1 + tA^2, \\ B^2 \cdot A^1 = 0, \\ B^2 \cdot A^2 = 0. \end{cases}$$

[*Hint:* Eliminate B^1 between Equations 1 and 2; then "multiply" the resulting equation by A^1 and also by A^2.]

b) Find the distance from B to the 2-plane described by $X = sA^1 + tA^2$. [*Hint:* Make an arrow drawing showing the vectors of part (a).]

c) Find the point on $X = sA^1 + tA^2$ that is nearest to B; call this point B^*.

d) Let L be defined on R^{21} by $L(X) = AX$, where the columns of A are A^1 and A^2. Then the range of L is the 2-plane introduced in part (b). Now, the equation $L(X) = B$ has no solution (why?), but the equation $L(X) = B^*$ does. Find the solution to this latter equation.

e) In a certain sense, the solution to $L(X) = B^*$ is the nearest thing to a solution for the solutionless equation $L(X) = B$. Explain.

f) *An exercise in analytic geometry.* The points $(-1, 1)$, $(2, -2)$, and $(\frac{1}{2}, 0)$ are not collinear. Find an equation for the line that seems best—in a certain sense—to accommodate these points. [*Hint:* Substitute the three points into the equation $b + mx = y$, thus producing a linear system in b and m. Then relate this system to the equation $L(X) = B$ of part (e).]

11 Let P be the 2-plane in R^{31} described by $X = A + sB + tC$, where $A = [1, 2, 3]'$, $B = [1, 0, 2]'$, and $C = [0, 1, 1]'$.

a) Find $D \neq \Theta$ such that $D \cdot B = 0$ and $D \cdot C = 0$.
b) It follows that for any X on the given plane we have $D \cdot (X - A) = 0$. Explain.
c) Expand this last equation to get a coordinate equation for P.
d) Obtain a coordinate equation for P directly from the definition of P (by eliminating s and t).

12 Let P be the 3-plane in R^{41} described by $X = A + rB + sC + tD$, where $A = [1, 2, 3, 4]'$, $B = [1, 0, 0, -1]'$, $C = [0, -1, 2, 0]'$, and $D = [1, 1, 0, 1]'$. Find a coordinate equation for P in two ways. [*Hint:* Generalize the scheme laid out in Exercise 11.]

5.2 ABSTRACT EUCLIDEAN SPACES

In the last section we defined the notions of length and angle for any column space, each of these notions being defined in terms of the dot product. In this section we would like to go a step further: we would like to formulate definitions for length and angle that make sense in any vector space—even those of infinite dimension. The first step would seem to be that of generalizing our definition for a *dot product*.

The dot product that we introduced into the column space R^{n1} was defined in terms of the (natural) coordinates of the column vectors. This definition is easily generalized to any finite dimensional vector space by first coordinatizing that space. For example, given an n-dimensional vector space V and an ordered basis $\mathcal{B}$, then we might define a dot product for V by means of the formula

$$X \cdot Y = x_1 y_1 + x_2 y_2 + \cdots + x_n y_n,$$

where the x_i and y_i are the $\mathcal{B}$-coordinates of X and Y.

Example 1. Let

$$\mathcal{B} = \left[\begin{bmatrix} 1 & 0 \\ 0 & 0 \end{bmatrix}, \begin{bmatrix} 0 & 0 \\ 1 & 0 \end{bmatrix}, \begin{bmatrix} 0 & 1 \\ 0 & 0 \end{bmatrix}, \begin{bmatrix} 0 & 0 \\ 0 & 1 \end{bmatrix} \right].$$

Then $\mathcal{B}$ is an ordered basis for R^{22}. We define a dot product for R^{22} by

$$X \cdot Y = x_1 y_1 + x_2 y_2 + x_3 y_3 + x_4 y_4,$$

where the x_i and y_i are the $\mathcal{B}$-coordinates of X and Y, or, equivalently,

$$X \cdot Y = X'_\mathcal{B} Y_\mathcal{B}.$$

For example,

$$\begin{bmatrix} 1 & 2 \\ 7 & 3 \end{bmatrix} \cdot \begin{bmatrix} 3 & -1 \\ 2 & 4 \end{bmatrix} = \begin{bmatrix} 1 \\ 7 \\ 2 \\ 3 \end{bmatrix}_\mathcal{B}' \cdot \begin{bmatrix} 3 \\ 2 \\ -1 \\ 4 \end{bmatrix}_\mathcal{B}$$
$$= (1)(3) + (7)(2) + (2)(-1) + (3)(4)$$
$$= 27.$$

The dot on the left denotes the dot product; that on the right denotes ordinary matrix multiplication.

It is not difficult to verify that the properties listed in Theorem 5.1 continue to hold for this new dot product. It follows, in a sense to be made more precise in the next section, that the structure consisting of R^{22} along with the dot product based on $\mathcal{B}$ is *isomorphic* to the structure consisting of R^{41} along with the *standard* dot product. □

Example 1 suggests the possibility of defining different dot products for the same vector space: we have only to base our definitions on different coordinate systems. In other words, where $\mathcal{B}$ and $\mathcal{C}$ are both ordered bases for V, we have the option—why such an option should be exercised is for the moment beside the point—of defining a dot product for V by either of the following formulas:

$$X \cdot Y = X'_\mathcal{B} Y_\mathcal{B} \quad \text{or} \quad X \cdot Y = X'_\mathcal{C} Y_\mathcal{C}$$

This observation is not awfully exciting; however, the fact that our basic coordinate formula gives rise to many different dot products for the same vector space does suggest that we should be even bolder: perhaps there are other dot products based on more exotic coordinate formulas.

Example 2. We introduce a "dot product" into the space R^{21} by the formula

$$\begin{bmatrix} x_1 \\ x_2 \end{bmatrix} \cdot \begin{bmatrix} y_1 \\ y_2 \end{bmatrix} = x_1 y_1 + 2 x_1 y_2 + 2 x_2 y_1 + 5 x_2 y_2.$$

Thus, for example,

$$\begin{bmatrix} 3 \\ 1 \end{bmatrix} \cdot \begin{bmatrix} -2 \\ 5 \end{bmatrix} = (3)(-2) + 2(3)(5) + 2(1)(-2) + 5(1)(5)$$
$$= 45. \quad \square$$

The reader should react a bit at this point: Wait a minute! Are we not getting rather far from our original ideas. This thing called in Example 2 a "dot product" is—like our original dot product—a function that assigns to ordered pairs of vectors a number. But is it fair to call it a dot product if there are no geometric implications? In other words, would this dot product serve as the basis for definitions of length and angle?

An answer to this question is inspired by the discussion following Example 1 of Section 5.1. There we noted that the notions of length and angle had been introduced into column spaces in such a manner that the notions of minimal distance and orthogonality were interrelated: the shortest distance from a point to a line coincides with the distance as measured along a perpendicular to that line. The basis for this observation was found in certain computations; those computations were possible because the dot product used there had certain properties: the properties listed under Theorem 5.1. Thus, it seems appropriate to conclude that the dot product of Example 1 will have analogous geometric consequences if it has the four properties of Theorem 5.1. It does. We leave this verification to the reader as it is perfectly straightforward.

We are now prepared to introduce an *abstract dot product*. We define it not by a formula involving the coordinates of vectors, but rather by the properties—those of Theorem 5.1—that seem to make a dot product "work." These properties are held in common by all of the dot products we have considered so far; thus, the definition that follows may be viewed as having been *abstracted* from the various examples considered.

Definition 5.4. Let V be a vector space. Then a *dot product for V* (also called an *inner product*) is a function that assigns to each ordered pair of vectors chosen from V a real numerical value. In addition, it must satisfy the following conditions:

1. $X \cdot X \geq 0$, with equality holding only for the case $X = \Theta$.
2. $X \cdot Y = Y \cdot X$.
3. $(cX) \cdot Y = c(X \cdot Y)$.
4. $X \cdot (Y + Z) = X \cdot Y + X \cdot Z$.

The structure consisting of a vector space along with a dot product is called a *Euclidean space*. By the *basis (dimension)* of a certain Euclidean space we mean the basis (dimension) of the associated vector space.

Example 3. The fact that we have defined the notion of a dot product in coordinate-free terms means that we may contemplate the existence of infinite dimension Euclidean spaces. Let V be the vector space consisting of *all* polynomial functions. We introduce a dot product for V by the formula
$$p \cdot q = \int_0^1 p(t)q(t)\, dt.$$
For example,
$$(t^2) \cdot (2t - 3) = \int_0^1 t^2(2t - 3)\, dt = -\tfrac{1}{2}.$$

5.2 ABSTRACT EUCLIDEAN SPACES

It is instructive to verify that this dot product has the four properties required of it:

1. $p \cdot p = \int_0^1 [p(t)]^2 \, dt.$

A basic theorem of calculus states that the integral of a nonnegative function is necessarily nonnegative. Thus, $p \cdot p \geq 0$. A somewhat deeper theorem states that the integral of a continuous nonnegative function can be zero only if the function is zero throughout the interval. Thus, $p \cdot p = 0$ is possible only if $p = \Theta$.

2. Since
$$\int_0^1 p(t)q(t) \, dt = \int_0^1 q(t)p(t) \, dt,$$
we have
$$p \cdot q = q \cdot p.$$

3. A basic theorem about integrals states that
$$\int_a^b cF(t) \, dt = c \int_a^b F(t) \, dt.$$
Thus, we may write
$$(cp) \cdot q = \int_0^1 cp(t)q(t) \, dt = c \int_0^1 p(t)q(t) \, dt = c(p \cdot q).$$

4. A basic theorem about integrals states that
$$\int_a^b [F(t) + G(t)] \, dt = \int_a^b F(t) \, dt + \int_a^b G(t) \, dt.$$
Thus, we may write
$$p \cdot (q + r) = \int_0^1 p(t)[q(t) + r(t)] \, dt$$
$$= \int_0^1 p(t)q(t) \, dt + \int_0^1 p(t)r(t) \, dt$$
$$= (p \cdot q) + (p \cdot r). \quad \square$$

The next step is to define for arbitrary Euclidean spaces the notions of length and angle. Length is easy, for Property 1 of Definition 5.1 guarantees that the formula $|X| = \sqrt{X \cdot X}$ makes sense for every X and that the only vector whose length is 0 is the zero vector. On the other hand, the idea of an angle between vectors is not so immediate. We want to define the angle between nonzero vectors X and Y as the number ϕ

between 0 and π inclusive that satisfies the equation

$$\cos \phi = \frac{X \cdot Y}{|X|\,|Y|}.$$

But an unsettling possibility suggests itself: Suppose that this fraction—for certain choices of X and Y—were to take on a value that exceeded 1 in absolute value. Then there would be no ϕ satisfying this angle-defining equation; consequently, there would be no angle between that particular pair of vectors. Happily, such a possibility is precluded by the following theorem. Thus, we may apply the definitions of length, angle, and orthogonality given in Definition 5.1 to any Euclidean space.

Theorem 5.2. *Cauchy-Schwarz inequality.* In any Euclidean space we have

$$(X \cdot Y)^2 \leq |X|^2 |Y|^2$$

for every choice of X and Y. (Or, equivalently, $|X \cdot Y| \leq |X|\,|Y|$.)

Proof. There are two cases:

1. Suppose that $X = \Theta$. We must then prove that $(\Theta \cdot Y)^2 \leq |\Theta|^2 |Y|^2$ for every choice of Y. Now $|\Theta|^2 = \Theta \cdot \Theta = 0$. (Why?) Therefore, $|\Theta|^2 |Y|^2 = 0$. Since the number zero times any vector gives the zero vector, we may write $\Theta \cdot Y = (0Y) \cdot Y = 0(Y \cdot Y) = 0$. Therefore, $(\Theta \cdot Y)^2 = 0$. The desired result follows since $0 \leq 0$.

2. Suppose that $X \neq \Theta$ and that Y is arbitrary. We then let $a = X \cdot X$, $b = X \cdot Y$, and $c = Y \cdot Y$, noting that $a > 0$. Now, consider the following argument:

$$\begin{aligned} 0 &\leq (bX - aY) \cdot (bX - aY) \\ &= b^2(X \cdot X) - 2ab(X \cdot Y) + a^2(Y \cdot Y) \\ &= ab^2 - 2ab^2 + a^2 c \\ &= a(-b^2 + ac). \end{aligned}$$

It follows, since $a > 0$, that $b^2 \leq ac$. That is,

$$(X \cdot Y)^2 \leq (X \cdot X)(Y \cdot Y).$$

Corollary. If X and Y are both different from Θ, then

$$-1 \leq \frac{X \cdot Y}{|X|\,|Y|} \leq 1.$$

Example 4. Let $E = (R^{21}, \cdot\,)$, where

$$X \cdot Y = X' \begin{bmatrix} 1 & 1 \\ 1 & 2 \end{bmatrix} Y.$$

We first verify that E, a structure consisting of a vector space along with a dot product, is a Euclidean space:

1. $X \cdot X = X' \begin{bmatrix} 1 & 1 \\ 1 & 2 \end{bmatrix} X.$

   Thus, recalling Section 1.7, $X \cdot X$ is a quadratic form. We introduce a new coordinate system with the formula

   $$X_{\mathfrak{N}} = \begin{bmatrix} 1 & -1 \\ 0 & 1 \end{bmatrix} X_{\mathfrak{B}}.$$

   Then we may write

   $$\begin{aligned} X \cdot X &= X'_{\mathfrak{N}} \begin{bmatrix} 1 & 1 \\ 1 & 2 \end{bmatrix} X_{\mathfrak{N}} \\ &= X'_{\mathfrak{B}} \begin{bmatrix} 1 & 0 \\ -1 & 1 \end{bmatrix} \begin{bmatrix} 1 & 1 \\ 1 & 2 \end{bmatrix} \begin{bmatrix} 1 & -1 \\ 0 & 1 \end{bmatrix} X_{\mathfrak{B}} \\ &= X'_{\mathfrak{B}} \begin{bmatrix} 1 & 0 \\ 0 & 1 \end{bmatrix} X_{\mathfrak{B}} \\ &= b_1^2 + b_2^2 \geq 0. \end{aligned}$$

2. Since $X \cdot Y$ is a 1 by 1 matrix, we may write

   $$\begin{aligned} X \cdot Y = (X \cdot Y)' &= \left\{ X' \begin{bmatrix} 1 & 1 \\ 1 & 2 \end{bmatrix} Y \right\}' \\ &= Y' \begin{bmatrix} 1 & 1 \\ 1 & 2 \end{bmatrix} X \\ &= Y \cdot X. \end{aligned}$$

We leave Properties 3 and 4 to the reader; they follow directly from the fact that our dot product is defined as a product of matrices.

It is instructive to "see" the Euclidean space E, that is, to picture the natural coordinate system for E. Our sketch begins with the choosing of arrows for N^1 and N^2. This choice—now that our vectors have length and direction—is no longer quite so arbitrary. We must first make some computations:

$$N^1 \cdot N^1 = \begin{bmatrix} 1 \\ 0 \end{bmatrix}' \begin{bmatrix} 1 & 1 \\ 1 & 2 \end{bmatrix} \begin{bmatrix} 1 \\ 0 \end{bmatrix} = 1.$$

$$N^2 \cdot N^2 = \begin{bmatrix} 0 \\ 1 \end{bmatrix}' \begin{bmatrix} 1 & 1 \\ 1 & 2 \end{bmatrix} \begin{bmatrix} 0 \\ 1 \end{bmatrix} = 2.$$

$$N^1 \cdot N^2 = \begin{bmatrix} 1 \\ 0 \end{bmatrix}' \begin{bmatrix} 1 & 1 \\ 1 & 2 \end{bmatrix} \begin{bmatrix} 0 \\ 1 \end{bmatrix} = 1.$$

248 EUCLIDEAN SPACES

Thus, $|N^1| = 1$, $|N^2| = \sqrt{2}$, and for the angle ϕ between N^1 and N^2 we have

$$\cos \phi = \frac{N^1 \cdot N^2}{|N^1| \, |N^2|} = \frac{1}{\sqrt{2}}.$$

Thus, $\phi = 45°$. Now, if we agree to the convention that N^1 shall always point to the right, and that ϕ shall be measured in a counterclockwise manner from N^1, then we are led to the sketch in Fig. 5.6.

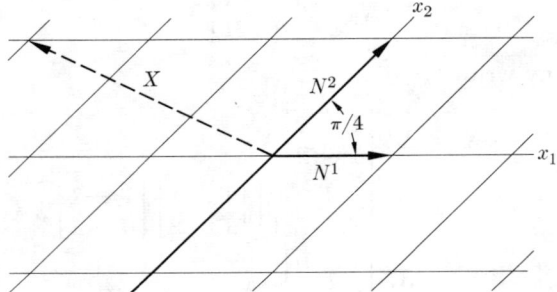

Figure 5.6

As an exercise, the reader should determine the length of the arrow X shown on Fig. 5.6: first, as the side of a certain triangle whose other two sides and the included angle are known, and second, as a vector in the Euclidean space E.

EXERCISE SET 5.2

1 Let E be the Euclidean space of Example 3. Let $p(t) = \sqrt{6}\, t - \sqrt{6}$, $q(t) = 5t^2 - 6t$, $r = p + q$, and $s = p - q$.
  a) Show that $|p| = |q|$.
  b) Show that the angle between p and r is the same as that between q and r.
  c) Show that r and s are orthogonal.
  d) Make an arrow drawing that would lead us to expect (b) and (c) as a consequence of (a).

2 Let $E = (R^{21}, \cdot)$, where the dot product is defined by

$$X \cdot Y = X' \begin{bmatrix} 1 & -\sqrt{3} \\ -\sqrt{3} & 4 \end{bmatrix} X.$$

Verify that E is indeed a Euclidean space, and make a sketch of the natural coordinate system for E. [*Hint:* Follow the pattern laid down in Example 4.]

3 Let f be defined on pairs of column vectors by the formula $f(X, Y) = X'AY$. For which of the A's given below is f a dot product? For any A that is not, indicate (by means of a counterexample) a property of Definition 5.4 that is violated.

a) $\begin{bmatrix} 3 & 0 & 0 \\ 0 & 2 & 0 \\ 0 & 0 & 1 \end{bmatrix}$ b) $\begin{bmatrix} 3 & 0 & 0 \\ 0 & 2 & 0 \\ 0 & 2 & 1 \end{bmatrix}$ c) $\begin{bmatrix} 3 & 0 & 0 \\ 0 & 2 & 2 \\ 0 & 2 & 1 \end{bmatrix}$ d) $\begin{bmatrix} 3 & 0 & 0 \\ 0 & 2 & 2 \\ 0 & 2 & 3 \end{bmatrix}$

4. Recall Theorem 5.2. It was shown in the proof of that theorem that our "inequality" reduces to equality in the case where $X = \Theta$. If $X \neq \Theta$, then we have equality if and only if Y is a multiple of X. Prove this latter statement.

Exercises 5 through 8 deal with propositions that are valid in any Euclidean space.

5. a) Prove: If $A \cdot X = 0$ for all X, then $A = \Theta$.
   b) Prove: If $A \cdot X = B \cdot X$ for all X, then $A = B$.

6. Prove: $X \cdot Y = 0$ if and only if $|X + Y|^2 = |X|^2 + |Y|^2$. [Hint: $|X + Y|^2 = (X + Y) \cdot (X + Y)$.]

7. a) Prove: $|X + Y|^2 + |X - Y|^2 = 2|X|^2 + 2|Y|^2$.
   b) This theorem is called the *parallelogram law*. Why?

8. a) Prove: $|X + Y| \leq |X| + |Y|$. [Hint: Work first with the square of each side.]
   b) Under what circumstances will the equality part hold?
   c) This theorem is called the *triangle inequality*. Why?

9. Let E be the Euclidean space consisting of the vector space of all functions that are continuous on the interval $0 \leq t \leq 2\pi$ along with the following dot product:

$$f \cdot g = \int_0^{2\pi} f(t)g(t)\, dt.$$

   a) Show that $f(t) = \sin t$ is orthogonal to $g(t) = \cos t$.
   b) Let $f(t) = \sin t$ and $g(t) = \cos t$. Verify that $|f + g| < |f| + |g|$.

10. Let $E = (R^{21}, \cdot)$, where $X \cdot Y = X'Y$, and let L_α be the linear transformation from E into E defined by

$$L_\alpha(X) = \begin{bmatrix} \cos \alpha & -\sin \alpha \\ \sin \alpha & \cos \alpha \end{bmatrix} X,$$

where α is between 0 and π inclusive.
   a) Compare $|X|$ with $|L_\alpha(X)|$.
   b) Find the angle between X and $L_\alpha(X)$.
   c) Give a geometric description of the transformation L_α, that is, interpreting E as a plane of arrows, tell what L_α does to the arrow depicting X.

11. *An alternative proof of the Cauchy-Schwarz inequality.*
    a) Prove: There is no real number t such that $(tA + B) \cdot (tA + B) < 0$.
    b) Prove: The equation $(A \cdot A)t^2 + 2(A \cdot B)t + (B \cdot B) = 0$ does not have distinct real roots.

c) Consider the quadratic-formula solution of the equation in part (b). What can be said about the quantity under the square-root sign?

12 a) Let $E = (R^{n1}, \cdot)$, where $X \cdot Y = X'Y$. Let L be the linear transformation from E into E defined by $L(X) = AX$. Now prove: If A is symmetric, then for every X and Y in E we have $L(X) \cdot Y = X \cdot L(Y)$. And conversely.

b) Let E be the Euclidean space of Example 3. Let L be the linear transformation from E into E defined by $L[p(t)] = p(1-t)$. Now show that $L[p(t)] \cdot q(t) = p(t) \cdot L[q(t)]$.

c) Let L be a linear transformation from E into E, where E is any Euclidean space. Define the phrase *L is a symmetric transformation* in such a way that it makes sense even if E is infinite dimensional.

d) Returning to part (b), it is instructive to compare the graph of $p(t)$ with that of $L[p(t)]$.

13 Read Appendix E.

5.3 ISOMORPHIC EUCLIDEAN SPACES

We have already seen that it is possible to have two Euclidean spaces both incorporating the same vector space, their distinctiveness arising out of the use of different dot products. It is natural to wonder just how different such Euclidean spaces really are.

Example 1. We consider two Euclidean spaces incorporating the vector space R^{21}:

$$E = (R^{21}, \cdot), \quad \text{where} \quad X \cdot Y = X' \begin{bmatrix} 1 & 0 \\ 0 & 1 \end{bmatrix} Y$$

and

$$F = (R^{21}, \odot), \quad \text{where} \quad X \odot Y = X' \begin{bmatrix} 1 & \frac{1}{2} \\ \frac{1}{2} & 1 \end{bmatrix} Y.$$

We next introduce a one-to-one linear transformation from E onto F:

$$L(X) = AX, \quad \text{where} \quad A = \begin{bmatrix} 1 & -1/\sqrt{3} \\ 0 & 2/\sqrt{3} \end{bmatrix}.$$

(The A used here has not been chosen at random; its origin is considered in Exercise 3.) We want now to compare the lengths of vectors in E with the lengths of their images (under L) in F. We want also to compare the angle between two vectors in E with that between their images in F. We begin with a numerical illustration.

Let
$$X = \begin{bmatrix} 1 \\ 2 \end{bmatrix} \quad \text{and} \quad Y = \begin{bmatrix} 1 \\ 0 \end{bmatrix}.$$

Then
$$|X|_E = \sqrt{5} \quad \text{and} \quad |Y|_E = 1.$$

(Since the formulas for length and angle are different in E and F, we shall add a distinguishing subscript to our notation.) For the angle between X and Y we have

$$\cos \phi_E = \frac{X \cdot Y}{(\sqrt{5})(1)} = \frac{1}{\sqrt{5}}.$$

We next turn to the images of X and Y:

$$L(X) = \begin{bmatrix} 1 & -1/\sqrt{3} \\ 0 & 2/\sqrt{3} \end{bmatrix} \begin{bmatrix} 1 \\ 2 \end{bmatrix} = \begin{bmatrix} (1 - 2/\sqrt{3}) \\ 4/\sqrt{2} \end{bmatrix}.$$

Similarly, $L(Y) = [1, 2]'$. To find the lengths of these vectors (in F) and the angle between them, we first compute certain dot products:

$$L(X) \cdot L(X) = \begin{bmatrix} (1 - 2/\sqrt{3}) \\ 4/\sqrt{3} \end{bmatrix}' \begin{bmatrix} 1 & \frac{1}{2} \\ \frac{1}{2} & 1 \end{bmatrix} \begin{bmatrix} (1 - 2/\sqrt{3}) \\ 4/\sqrt{3} \end{bmatrix} = 5.$$

Similarly, $L(X) \cdot L(Y) = 1$, and $L(Y) \cdot L(Y) = 1$. It then follows that

$$|L(X)|_F = \sqrt{5}, \qquad |L(Y)|_F = 1, \qquad \text{and} \qquad \cos \phi_F = \frac{1}{\sqrt{5}}.$$

These are the same results as we obtained for X and Y in the space E.

To show that the results just obtained are not exceptional we have only to show, for any X and Y in E, that

$$L(X) \odot L(Y) = X \cdot Y.$$

For if the dot products of the two spaces correspond in this way, then so also must the length and angle functions. A simple calculation is all that is required:

$$L(X) \odot L(Y) = (AX)' \begin{bmatrix} 1 & \frac{1}{2} \\ \frac{1}{2} & 1 \end{bmatrix} (AY)$$

$$= X' \left(\begin{bmatrix} 1 & 0 \\ -1/\sqrt{3} & 2/\sqrt{3} \end{bmatrix} \begin{bmatrix} 1 & \frac{1}{2} \\ \frac{1}{2} & 1 \end{bmatrix} \begin{bmatrix} 1 & -1/\sqrt{3} \\ 0 & 2/\sqrt{3} \end{bmatrix} \right) Y$$

$$= X' \begin{bmatrix} 1 & 0 \\ 0 & 1 \end{bmatrix} Y = X \cdot Y.$$

Further understanding of the relationship between E and F is obtained by sketching the "coordinate planes" that arise on choosing the natural basis in each space. (Recall Example 4, Section 5.2.) For E we have the customary picture of a *rectangular* coordinate system (see Fig. 5.7), but for F (see Fig. 5.8) we have an *oblique* coordinate system. This follows from the fact that the E-angle between N^1 and N^2 is 90° while the corresponding F-angle—the reader should verify this—is 60°. Suppose now that

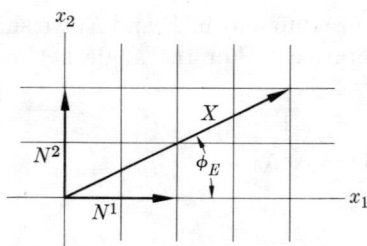

Figure 5.7 **Figure 5.8**

we take an arbitrary X from E, say $X = [2, 1]'$, and place it on our sketch of E. Then $L(X)$ will appear on the sketch of F unchanged in length and direction. This follows from the fact that

$$|L(X)|_F = |X|_E$$

and

$$\cos \phi_F = \frac{L(X) \odot N^1}{|L(X)|_F |N^1|_F}$$
$$= \frac{L(X) \odot L(N^1)}{|L(X)|_F |L(N^1)|_F}$$
$$= \frac{X \cdot N^1}{|X|_E |N^1|_E}$$
$$= \cos \phi_E.$$

Our sketches lead us to the following somewhat colorful description of the relationship between E and F: If we sketch our arrows on the coordinate plane corresponding to E (Fig. 5.7), then questions of length and angle can be resolved by using the *usual* dot product. If we sketch our arrows on the coordinate plane corresponding to F (Fig. 5.8), then notions of length and angle can still be studied in terms of the coordinates "collected" by our arrows, but we must now employ an *unusual* dot product. It follows that Euclidean plane geometry can be studied analytically on either coordinate plane. In this sense, then, E and F are indistinguishable.

Definition 5.5. Let $E = (V, \cdot)$ and $F = (W, \odot)$ be Euclidean spaces of the same dimension. Let L be a vector space isomorphism from V onto W which has the additional property that

$$X \cdot Y = L(X) \odot L(Y).$$

Then we say that E is *isomorphic* to F.

The remainder of this section is devoted to proving three theorems. Each is important in its own right; however, the last one—after our experience with Example 1—seems especially significant. It declares that *all* Euclidean spaces of the same dimension are isomorphic.

Theorem 5.3. Let V be an n-dimensional vector space, let B be a basis for V, and let A be an nth-order, symmetric, positive definite matrix. Finally, let f be defined by

$$f(X,Y) = X'_\mathcal{B} A Y_\mathcal{B}.$$

Then f is a dot product for V, or, equivalently, (V, f) is a Euclidean space.

Proof. We have the four conditions of Definition 5.4 to verify. Conditions 3 and 4 are satisfied as a direct consequence of f being defined as a matrix product, that is, they follow from the laws of matrix arithmetic. Thus, we look in detail only at the first two conditions:

1. We have $f(X, X) = X'_\mathcal{B} A X_\mathcal{B}$, a quadratic form. Since A is positive definite, $f(X, X)$ is a positive definite quadratic form. This means that $f(X, X) \geq 0$, with equality holding only for $X = \Theta$.
2. Strictly speaking, $f(X, Y)$ is a first-order matrix. As such it is equal to its transpose. Thus, using here the hypothesis that A is symmetric, we may write

$$\begin{aligned} f(X, Y)' = (X'_\mathcal{B} A Y_\mathcal{B})' &= Y'_\mathcal{B} A' X_\mathcal{B} \\ &= Y'_\mathcal{B} A X_\mathcal{B} \\ &= f(Y, X). \end{aligned}$$

Theorem 5.4. Given an n-dimensional Euclidean space $(V, \cdot)$ and a basis $\mathcal{B}$, there exists a unique, positive definite, symmetric matrix A such that

$$X \cdot Y = X'_\mathcal{B} A Y_\mathcal{B}.$$

Moreover, $A = [B^i \cdot B^j]_{nn}$.

Proof. We seek a formula for $X \cdot Y$ in terms of the $\mathcal{B}$-coordinates of X and Y. It is natural, then, to start by expressing X and Y in terms of the basis vectors. The remainder of the proof consists of a sequence of calculations exploiting the arithmetic properties of the dot product. We shall carry out the details for the case $n = 2$ only; the general case—using summation symbols—follows the same pattern:

$$\begin{aligned} X \cdot Y &= (x_1 B^1 + x_2 B^2) \cdot (y_1 B^1 + y_2 B^2) \\ &= (x_1 B^1) \cdot (y_1 B^1 + y_2 B^2) + (x_2 B^2) \cdot (y_1 B^1 + y_2 B^2) \\ &= x_1 [y_1 (B^1 \cdot B^1) + y_2 (B^1 \cdot B^2)] + x_2 [y_1 (B^2 \cdot B^1) + y_2 (B^2 \cdot B^2)] \\ &= [x_1, x_2]_\mathcal{B} \begin{bmatrix} (B^1 \cdot B^1) & (B^1 \cdot B^2) \\ (B^2 \cdot B^1) & (B^2 \cdot B^2) \end{bmatrix} \begin{bmatrix} y_1 \\ y_2 \end{bmatrix}_\mathcal{B}. \end{aligned}$$

That the resulting matrix is unique follows from the fact that our construction allows no alternative possibility; that it is positive definite follows from a consideration of $X \cdot X$.

Theorem 5.5. All Euclidean spaces of the same dimension are isomorphic.

Proof. Let $E = (V, \cdot)$ and $F = (W, \odot)$ be any two Euclidean spaces of dimension n. Let $\mathcal{E}$ and $\mathcal{F}$ be ordered bases for E and F, respectively. With respect to these bases, the dot products of E and F may be represented by symmetric, positive matrices, say B and C. Now recall the work of Section 1.7. Since B and C are symmetric, there exist nonsingular matrices P and Q such that

$$P'BP = \begin{bmatrix} b_1 & & & \\ & b_2 & & \\ & & \ddots & \\ & & & b_n \end{bmatrix} \quad \text{and} \quad Q'CQ = \begin{bmatrix} c_1 & & & \\ & c_2 & & \\ & & \ddots & \\ & & & c_n \end{bmatrix}.$$

And since B and C are both positive definite, the diagonal entries (in both cases) are all positive numbers. If we now let

$$R = \begin{bmatrix} \sqrt{b_1/c_1} & & & \\ & \sqrt{b_2/c_2} & & \\ & & \ddots & \\ & & & \sqrt{b_n/c_n} \end{bmatrix},$$

then we have

$$R'Q'CQR = \begin{bmatrix} b_1 & & & \\ & b_2 & & \\ & & \ddots & \\ & & & b_n \end{bmatrix} = P'BP.$$

It follows that

$$[(P^{-1})'R'Q']C[QRP^{-1}] = B.$$

Now, let L be the linear transformation from E into F having $A = QRP^{-1}$ as its $\mathcal{FE}$-representation. Since A is nonsingular, L is one-to-one. Thus, L is a vector space isomorphism from V onto W. But also, we have

$$\begin{aligned} L(X) \odot L(Y) &= \{L(X)_\mathcal{F}\}'C\{L(Y)_\mathcal{F}\} \\ &= (AX_\mathcal{E})'C(AY_\mathcal{E}) \\ &= X'_\mathcal{E}(A'CA)Y_\mathcal{E} \\ &= X'_\mathcal{E}BY_\mathcal{E} \\ &= X \cdot Y. \end{aligned}$$

EXERCISE SET 5.3

1 Let f be defined in R^{51} by the formula $f(X, Y) = X'AY$, where

$$A = \begin{bmatrix} 1 & 0 & 1 & 0 & 1 \\ 0 & 2 & 0 & 0 & 0 \\ 1 & 0 & 3 & 0 & 3 \\ 0 & 0 & 0 & 4 & 0 \\ 1 & 0 & 3 & 0 & 5 \end{bmatrix}.$$

a) Prove that f is a dot product. (Recall Theorem 5.3.)
b) Find the lengths of the natural basis vectors in the Euclidean space (R^{51}, f).

2 Write out a proof of Theorem 5.4 for the special case $n = 3$.

3 Recall Example 1. Find an isomorphism from E to F by following the pattern laid down in the proof of Theorem 5.5, choosing $\mathcal{E} = \mathcal{F} = \mathfrak{N}$. Compare your result with the isomorphism employed in Example 1.

4 Let E be a Euclidean space incorporating R^{21}. A formula for the dot product is not given. However, we are told that $X = [-1, 0]'$ has length 1, that $X^2 = [1, -1]'$ has length $\sqrt{3}$, and that the angle between X^1 and X^2 is 90°.

a) Find $|N^1|$ and $|N^2|$. [*Hint:* $N^1 = -X^1$ and $N^2 = -X^1 - X^2$.]
b) Find the angle between N^1 and N^2.
c) Choosing $\mathfrak{N}$ as a basis for E, sketch the resulting coordinate plane.
d) Find the natural coordinate formula for the dot product of E. (That is, find A such that $X \cdot Y = X'AY$.)
e) Find the $\mathcal{B}$-label formula for the dot product of E, where $\mathcal{B} = [X^1, X^2]$.

5 *An alternative point of view with respect to Example 1.* Let $E = (R^{21}, \cdot\,)$, where $X \cdot Y = X'Y$. Let a $\mathcal{B}$-coordinate system be introduced into E by the formula

$$X_\mathcal{B} = \begin{bmatrix} 1 & -1/\sqrt{3} \\ 0 & 2/\sqrt{3} \end{bmatrix} X_\mathfrak{N}.$$

a) Find the $\mathcal{B}$-label formula for the dot product of E.
b) Let $X = [1, 1]'$ and $Y = [0, -1]'$. Find the angle between X and Y in two ways: first, by using the $\mathfrak{N}$-labels of these vectors; and second, by using their $\mathcal{B}$-labels.

6 Let $E = (P^2, \cdot\,)$, where

$$p \cdot q = \int_0^1 p(t) q(t)\, dt,$$

and let $\mathcal{B} = [t^2, t, 1]$.

a) Find the $\mathcal{B}$-label formula for the dot product of E.
b) Find a basis $\mathcal{C}$ such that $p \cdot q = p'_\mathcal{C} q_\mathcal{C}$.
c) Find the length of $p(t) = t^2 + 2t + 3$ in three different ways: using p directly, using its $\mathcal{B}$-label, and using its $\mathcal{C}$-label.

7 Let $E = (R^{21}, \cdot\,)$, where $X \cdot Y = X'Y$.
   a) Sketch the coordinate plane based on $\mathfrak{N}$.
   b) Place on this sketch arrows depicting $B^1 = [\tfrac{1}{2}, \sqrt{3}/2]'$ and $B^2 = [-\sqrt{3}/2, \tfrac{1}{2}]'$.
   c) Find the $\mathfrak{B}$-label formula for the dot product of E.
   d) Assuming that your result obtained under part (c) is not a coincidence (and is correct), make a conjecture.

5.4 ORTHONORMAL BASES

To motivate the new ideas of this section it will be convenient to work with a specific Euclidean space. Thus, let us take $E = (R^{31}, \cdot\,)$, where the dot product is the standard one ($X \cdot Y = X'Y$), and let F be the subspace of E spanned by $B^1 = [1, 2, 1]'$ and $B^2 = [0, 1, 1]'$. Then, choosing $\mathfrak{B} = [B^1, B^2]$ as a basis for F, we may identify each vector in F with its $\mathfrak{B}$-column, a 2 by 1 column. For example, if

$$X = \begin{bmatrix} 2 \\ 7 \\ 5 \end{bmatrix},$$

Then, as the reader may verify,

$$X_{\mathfrak{B}} = \begin{bmatrix} 2 \\ 3 \end{bmatrix}.$$

Suppose that we are required to find the lengths of various vectors in F. Then we are prompted to make the necessary calculations using their $\mathfrak{B}$-columns, for then only two coordinates are involved. Thus, for the X given above we might write

$$|X| = \begin{bmatrix} 2 \\ 3 \end{bmatrix}_{\mathfrak{B}} \cdot \begin{bmatrix} 2 \\ 3 \end{bmatrix}_{\mathfrak{B}} = \sqrt{(2)(2) + (3)(3)} = \sqrt{13}.$$

However, in doing so we would have obtained an incorrect result, for we cannot—without some special authority—use a two-coordinate abbreviation of the dot product formula given for E (and hence for F) as the $\mathfrak{B}$-coordinate formula for that dot product. The correct result, working directly with X, is given by

$$|X| = \sqrt{2^2 + 7^2 + 5^2} = \sqrt{78}.$$

Now we can certainly find $|X|$ using its $\mathfrak{B}$-column. However, to do so we must first have the correct $\mathfrak{B}$-coordinate formula for our dot product. Appealing to Theorem 5.4, we obtain the result

$$X \cdot Y = \begin{bmatrix} x_1 \\ x_2 \end{bmatrix}'_{\mathfrak{B}} \begin{bmatrix} 6 & 3 \\ 3 & 2 \end{bmatrix} \begin{bmatrix} y_1 \\ y_2 \end{bmatrix}_{\mathfrak{B}}$$
$$= 6x_1y_1 + 6x_1y_2 + 2x_2y_2.$$

Returning to
$$X = \begin{bmatrix} 2 \\ 3 \end{bmatrix}_{\mathcal{B}},$$
we see that
$$\begin{aligned} |X| &= \sqrt{\begin{bmatrix} 2 \\ 3 \end{bmatrix}_{\mathcal{B}} \cdot \begin{bmatrix} 2 \\ 3 \end{bmatrix}_{\mathcal{B}}} \\ &= \sqrt{6(2)(2) + 6(2)(3) + 2(3)(3)} \\ &= \sqrt{78}; \end{aligned}$$

the correct result. Unfortunately, the complexity of this $\mathcal{B}$-coordinate formula for our dot product negates the seeming advantage of there being only two coordinates.

There are many ways of coordinatizing F. We happened—it seemed a reasonable thing to do—to introduce a coordinate system that was based on the two vectors that defined F. Perhaps there is a way of coordinatizing F that will lead to a simple, two-coordinate formula for our dot product. Perhaps it can even be done in such a way that the formula is $x_1 y_1 + x_2 y_2$. It can. The trick is this: We must use a basis whose members are mutually orthogonal and are all of length 1.

Now how can such a basis be found? We start with $\mathcal{B} = [B^1, B^2]$ and construct $\mathcal{C} = [C^1, C^2]$ as follows:

1. Let
$$C^1 = B^1 = \begin{bmatrix} 1 \\ 2 \\ 1 \end{bmatrix}.$$

2. Determine a scalar b_1 such that the vector $(b_1 B^1 + B^2)$ is orthogonal to B^1. We do this by requiring that
$$B^1 \cdot (b_1 B^1 + B^2) = 0.$$

Solving for b_1 leads us to
$$b_1 = -\frac{B^1 \cdot B^2}{B^1 \cdot B^1} = -\frac{1}{2}.$$

Now take
$$\begin{aligned} C^2 &= b_1 B^1 + B^2 \\ &= -\frac{1}{2} \begin{bmatrix} 1 \\ 2 \\ 1 \end{bmatrix} + \begin{bmatrix} 0 \\ 1 \\ 1 \end{bmatrix} = \begin{bmatrix} -\frac{1}{2} \\ 0 \\ \frac{1}{2} \end{bmatrix}. \end{aligned}$$

Consider the ordered pair $\mathcal{C} = [C^1, C^2]$. Since C^1 and C^2 are linearly independent and since both are in F (why?), we are assured that $\mathcal{C}$ is a basis for F. Moreover, we have in $\mathcal{C}$ a basis with a distinguishing feature: its members are orthogonal. But we want our basis to have one other property: we want the length of each vector to be 1. To accomplish this we multiply each C^i by the reciprocal of its length, taking

$$D^1 = \frac{1}{|C^1|}C^1 = \frac{1}{\sqrt{6}}\begin{bmatrix} 1 \\ 2 \\ 1 \end{bmatrix} = \begin{bmatrix} 1/\sqrt{6} \\ 2/\sqrt{6} \\ 1/\sqrt{6} \end{bmatrix}$$

and

$$D^2 = \frac{1}{|C^2|}C^2 = \sqrt{2}\begin{bmatrix} -\frac{1}{2} \\ 0 \\ \frac{1}{2} \end{bmatrix} = \begin{bmatrix} -1/\sqrt{2} \\ 0 \\ 1/\sqrt{2} \end{bmatrix}.$$

We have thus constructed $\mathfrak{D} = [D^1, D^2]$ as an ordered basis with the desired special properties.

We return again to the vector $X = [2, 7, 5]'$ whose length is $\sqrt{78}$. A now familiar calculation produces the $\mathfrak{D}$-column for X:

$$X_{\mathfrak{D}} = \begin{bmatrix} 7\sqrt{6}/2 \\ 3\sqrt{2}/2 \end{bmatrix}.$$

Finally—and this is the point of our discussion—we may compute the length of X from its $\mathfrak{D}$-column, using the standard dot product formula

$$|X| = \sqrt{(7\sqrt{6}/2)^2 + (3\sqrt{2}/2)^2} = \sqrt{78}.$$

Definition 5.6. Let $S = \{X^1, X^2, \ldots, X^k\}$ be a set of vectors chosen from some Euclidean space $E = (V, \cdot)$. Suppose that S has the following special properties:
1. $X^i \cdot X^j = 1$ if $i = j$.
2. $X^i \cdot X^j = 0$ if $i \neq j$.

(We could also write, more compactly, $X^i \cdot X^j = \delta_{ij}$.) Then we say that *S is an orthonormal set in E*. If only the second condition is met, we say that *S is an orthogonal set in E*.

Theorem 5.6. Let $E = (V, \cdot)$ be an Euclidean space of dimension n. Let $\mathfrak{B}$ be an ordered basis whose members form an orthonormal set. Then for any X and Y in E we have

$$X \cdot Y = X'_{\mathfrak{B}} Y_{\mathfrak{B}},$$

or, equivalently, $X \cdot Y = x_1 y_1 + \cdots + x_n y_n$, where the x_i and the y_i are the $\mathfrak{B}$-coordinates of X and Y. Conversely, if the $\mathfrak{B}$-coordinate formula for the dot product is $X'_{\mathfrak{B}} Y_{\mathfrak{B}}$, then $\mathfrak{B}$ is an *orthonormal basis*.

Proof. Theorem 5.4 tells us that

$$X \cdot Y = X'_{\mathfrak{B}} A Y_{\mathfrak{B}},$$

5.4 ORTHONORMAL BASES

where $A = [B^i \cdot B^j]_{nn}$. But $B^i \cdot B^j = \delta_{ij}$. (Why?) Thus, $A = I$. Conversely, if $A = I$, then $B^i \cdot B^j = \delta_{ij}$, from which it follows that $\mathcal{B}$ is an orthonormal basis. (It should be observed that if we relax our hypothesis, requiring only that $\mathcal{B}$ be an orthogonal basis, then we can conclude only that the matrix of the dot product will be a diagonal matrix. And conversely.) □

One question remains: Given a finite-dimensional Euclidean space, does there always exist for it an orthonormal basis? The answer is "yes." We shall show how to find one in the example that follows; that the algorithm used in this example always works will be proved immediately after.

Example 1. Let $E = (R^{31}, \cdot)$, where

$$X \cdot Y = \begin{bmatrix} x_1 \\ x_2 \\ x_3 \end{bmatrix}' \begin{bmatrix} 1 & 1 & 0 \\ 1 & 2 & 1 \\ 0 & 1 & 3 \end{bmatrix} \begin{bmatrix} x_1 \\ x_2 \\ x_3 \end{bmatrix}.$$

Find an orthonormal basis for E.

Solution. We begin by choosing an arbitrary basis for E. A natural choice is $\{N^1, N^2, N^3\}$. We now derive from this set an orthonormal set; the algorithm that we use is called the *Gram-Schmidt* process.

1. Let $B^1 = N^1$.
2. Determine b_1 such that

$$(b_1 B^1 + N^2) \cdot B^1 = 0.$$

Solving this equation, we obtain

$$b_1 = -\frac{B^1 \cdot N^2}{B^1 \cdot B^1}.$$

Now

$$B^1 \cdot N^2 = \begin{bmatrix} 1 \\ 0 \\ 0 \end{bmatrix}' \begin{bmatrix} 1 & 1 & 0 \\ 1 & 2 & 1 \\ 0 & 1 & 3 \end{bmatrix} \begin{bmatrix} 0 \\ 1 \\ 0 \end{bmatrix} = 1,$$

and

$$B^1 \cdot B^1 = \begin{bmatrix} 1 \\ 0 \\ 0 \end{bmatrix}' \begin{bmatrix} 1 & 1 & 0 \\ 1 & 2 & 1 \\ 0 & 1 & 3 \end{bmatrix} \begin{bmatrix} 1 \\ 0 \\ 0 \end{bmatrix} = 1.$$

Thus, $b_1 = -1$. We then take

$$B^2 = b_1 B^1 + N^2 = \begin{bmatrix} -1 \\ 1 \\ 0 \end{bmatrix}.$$

[It should be observed at this point that B^1 and B^2 span the same space as that spanned by N^1 and N^2 (why?) and that B^1 and B^2 are orthogonal (why?).]

3. Determine b_1 (a new b_1) and b_2 such that

$$\begin{cases} (b_1 B^1 + b_2 B^2 + N^3) \cdot B^1 = 0, \\ (b_1 B^1 + b_2 B^2 + N^3) \cdot B^2 = 0. \end{cases}$$

Using appropriate properties of the dot product, we are led to

$$\begin{cases} b_1(B^1 \cdot B^1) + b_2(B^2 \cdot B^1) + (N^3 \cdot B^1) = 0, \\ b_1(B^1 \cdot B^2) + b_2(B^2 \cdot B^2) + (N^3 \cdot B^2) = 0. \end{cases}$$

Calculating the necessary dot products—some are already available from part 2—we get finally

$$\begin{cases} b_1 = 0, \\ b_2 = -1. \end{cases}$$

We then take

$$B^3 = b_1 B^1 + b_2 B^2 + N^3 = \begin{bmatrix} 1 \\ -1 \\ 1 \end{bmatrix}.$$

The reader may now check to see that the set $\{B^1, B^2, B^3\}$ is orthogonal. (The method that we have used above has a simple geometrical interpretation. See Fig. 5.9.) To make it orthonormal, we *normalize* each vector; that is, we replace each B^i by a multiple having length 1. Little effort is required in the present case, for $|B^1| = |B^2| = 1$ and $|B^3| = \sqrt{2}$. Thus, as an orthonormal basis for E, we may take

$$\mathcal{C} = \left\{ \begin{bmatrix} 1 \\ 0 \\ 0 \end{bmatrix}, \begin{bmatrix} -1 \\ 1 \\ 0 \end{bmatrix}, \begin{bmatrix} 1/\sqrt{2} \\ -1/\sqrt{2} \\ 1/\sqrt{2} \end{bmatrix} \right\}.$$

(That $\mathcal{C}$ is linearly independent is a consequence of its orthogonality; see Exercise 6.)

Theorem 5.7. Let $S = \{B^1, B^2, \ldots, B^k\}$, $k < n$, be an orthonormal set of vectors in an n-dimensional Euclidean space. Then there exists a vector B^{k+1} such that $S^* = \{B^1, B^2, \ldots, B^k, B^{k+1}\}$ is orthonormal.

Proof. Take C as any vector not in the space spanned by S. (How do we know there is such a C?) We then determine scalars $b_1, b_2, \ldots, b_k$ such that

$$B^* = b_1 B^1 + b_2 B^2 + \cdots + b_k B^k + C$$

is orthogonal to each member of S. This is easily done. To find each b_i

5.4 ORTHONORMAL BASES

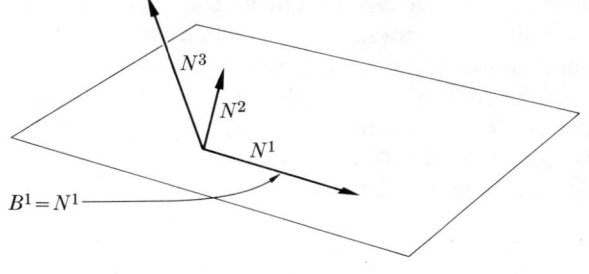

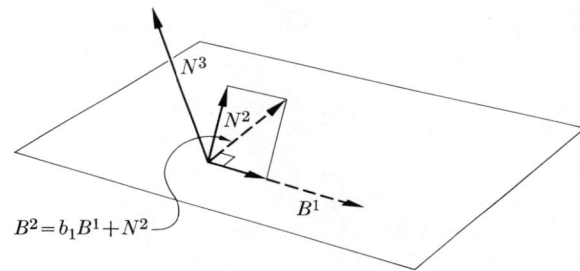

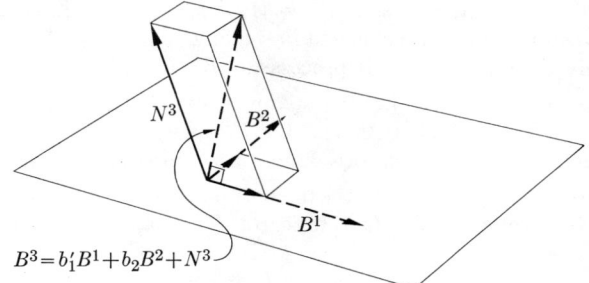

Figure 5.9

we simply multiply each side of the above equation by B^i, producing the equation

$$0 = b_i + (B^i \cdot C).$$

Thus, the desired B^* is obtained by taking $b_i = -(B^i \cdot C)$ for $i = 1, 2, \ldots, k$. Finally, take

$$B^{k+1} = \frac{1}{|B^*|} B^*.$$

Corollary. Every n-dimensional Euclidean space has an orthonormal basis.

Proof. Take B^1 as any vector of length 1. Then $S_1 = \{B^1\}$ is an orthonormal set. By our theorem there exists a vector B^2 such that $S_2 = \{B^1, B^2\}$ is orthonormal. A second application of our theorem produces a vector B^3 such that $S_3 = \{B^1, B^2, B^3\}$ is orthonormal, and so on, until we have the orthonormal set $S_n = \{B^1, B^2, \ldots, B^n\}$. Now, it is easily proved (see Exercise 6) that a set of orthogonal vectors is necessarily linearly independent. Thus S_n is a basis.

EXERCISE SET 5.4

1. Return to the set $\{B^1, B^2, B^3\}$ of Example 1 and verify that it is orthogonal.
2. Let $E = (P^1, \cdot)$, where

$$p \cdot q = \int_0^1 p(t)q(t)\, dt.$$

   a) Verify that $\mathcal{B} = [1, 2\sqrt{3}\, t - \sqrt{3}]$ is an orthonormal basis for E.
   b) Find the $\mathcal{B}$-column for $X = 2t - 2$.
   c) Find $|X|$ both directly and by using its $\mathcal{B}$-column.
   d) Use the Gram-Schmidt process (the method of Example 1) to derive an orthonormal basis from the basis $[1, t]$.

3. Let $E = (R^{31}, \cdot)$, where the dot product is the standard one. Find an orthogonal basis that contains $B^1 = [1, 1, 1]'$. [*Hint:* First adjoin to B^1 any two vectors that will result in a basis for E, say $B^2 = N^1$ and $B^3 = N^2$. Then apply the Gram-Schmidt process to $\{B^1, B^2, B^3\}$.]

4. Let V be the subspace of R^{51} spanned by $[1, 1, 0, 0, 0]'$, $[0, 1, 1, 0, 0]'$, $[0, 0, 0, 1, -2]'$, and $[0, 0, 0, 0, 1]'$. Let $E = (V, \cdot)$, where the dot product is the standard one. Find an orthogonal basis for E.

5. Given an orthonormal basis $\mathcal{B}$, the $\mathcal{B}$-coordinates of a given vector are readily obtained by exploiting the fact that $B_i \cdot B_j = \delta_{ij}$. For example, we first write

$$X = b_1 B^1 + \cdots + b_i B^i + \cdots + b_n B^n.$$

To find b_i we "multiply" both sides by B^i obtaining $B^i \cdot X = b_i$. Try this technique in the following situation: $E = (R^{31}, \cdot)$, the dot product being the standard one:

$$\mathcal{B} = \left[\begin{bmatrix} \frac{2}{3} \\ -\frac{2}{3} \\ \frac{1}{3} \end{bmatrix}, \begin{bmatrix} \frac{2}{3} \\ \frac{1}{3} \\ -\frac{2}{3} \end{bmatrix}, \begin{bmatrix} \frac{1}{3} \\ \frac{2}{3} \\ \frac{2}{3} \end{bmatrix} \right] \quad \text{and} \quad X = \begin{bmatrix} 6 \\ -3 \\ 12 \end{bmatrix}.$$

6. **Theorem 5.8.** If the set $\{X^1, X^2, \ldots, X^k\}$ is orthogonal, then it is necessarily linearly independent.

   Prove this theorem.

7 We have used the phrase *subspace of a Euclidean space* without defining it. Give a formal definition for this phrase.

8 Prove: If X is orthogonal to each of $X^1, X^2, \ldots, X^k$, then X is orthogonal to every vector in the space spanned by these vectors.

9 Recall the proof of Theorem 5.5. Suppose that the bases $\mathcal{E}$ and $\mathcal{F}$ chosen there had been orthonormal. Follow the proof through making this assumption, noting the remarkable simplification that occurs. (This illustrates the technical advantage that often results on using orthonormal bases in theoretical arguments.)

10 *An alternative to the Gram-Schmidt process.* Let $E = (P^2, \cdot)$, where

$$p \cdot q = \int_0^1 p(t)q(t) \, dt.$$

The problem is to find an orthogonal basis for E; consider the following method:

a) Take any basis for E, say $\mathcal{B} = [1, t, t^2]$, and find the matrix A such that
$$X \cdot Y = X'_\mathcal{B} A Y_\mathcal{B}.$$

b) Find (recall Section 1.6) a matrix C such that CAC' is diagonal. (The entries along the diagonal are unimportant; thus, row multiplication to avoid fractions may be freely used.)

c) Let $\mathcal{D}$ be the basis for the coordinate system introduced by the formula $X_\mathcal{B} = C' X_\mathcal{D}$. Find the matrix A^* such that
$$X \cdot Y = X'_\mathcal{D} A^* X_\mathcal{D}.$$

d) By Theorem 5.6 (the converse part) the basis $\mathcal{D}$ must be orthogonal. Find it.

e) Inspired by your activities above, write a "cookbook procedure" for deriving an orthogonal basis from a given basis. As a check, try it out on Exercise 4.

11 Let $E = (P^3, \cdot)$, where
$$p \cdot q = \int_{-1}^1 p(t)q(t) \, dt.$$

Use the method of Exercise 10 to find an orthogonal basis for E.

5.5 THE CHARACTERISTIC VALUE PROBLEM FOR SYMMETRIC MATRICES

We return to the diagonalization problem of Section 4.7: Given a square matrix A, find a nonsingular matrix B such that $B^{-1}AB$ is diagonal. In this section we shall restrict our attention to the case in which A is symmetric. Under these special circumstances a B can always be found; indeed, a B of a very special nature can be found. (Since the reader has already had some experience with this diagonalization problem, we shall be more terse here, moving directly through the theory without further discussion.)

Theorem 5.9. Let A be a symmetric matrix of order n, and let X and Y be characteristic vectors corresponding to two different characteristic values, say λ and μ. Then X and Y are orthogonal. [Reference to orthogonality without mention of a specific Euclidean space always refers to the space $E = (R^{n1}, \cdot \,)$, where the dot product is the standard one.]

Proof. Some proofs are especially easy to follow because they are computational in nature; however, it is frequently the case that such proofs are marred by "inspired beginnings." This is such a proof:

$$\begin{aligned}(\lambda - \mu)(X \cdot Y) &= (\lambda - \mu)X'Y \\ &= \lambda X'Y - \mu X'Y \\ &= (\lambda X)'Y - X'(\mu Y) \\ &= (AX)'Y - X'(AY) \\ &= X'A'Y - X'AY \\ &= 0. \quad \text{(Why?)}\end{aligned}$$

Finally, since $\lambda - \mu \neq 0$, it follows that

$$X \cdot Y = 0.$$

Theorem 5.10. If A is a symmetric matrix of order n, then its characteristic values are all real.

Proof. For this proof we are obliged to work with matrices having complex number entries and with scalars that are complex numbers. We need first some notation: Let $\bar{z}$ denote the complex conjugate of z; and let $\bar{B}$ denote the matrix obtained from B by replacing each entry with its complex conjugate. We leave it to the reader to prove the following elementary facts:

1. For any scalar z and matrix B, we have

$$\overline{zB} = \bar{z}\,\bar{B}.$$

2. For any matrix produce BC, we have

$$\overline{BC} = \bar{B}\,\bar{C}.$$

3. If the column X is not the zero column, then $X'\bar{X} > 0$.

When we accept these preliminaries, the proof itself is quite similar to that given for the preceding theorem.

5.5 CHARACTERISTIC VALUES FOR SYMMETRIC MATRICES

Let λ be a characteristic value of A and let X be an associated characteristic vector. To show that λ is real we have only to show that $\bar{\lambda} = \lambda$.

$$\begin{aligned}(\lambda - \bar{\lambda})X'\overline{X} &= \lambda X'\overline{X} - \bar{\lambda}X'\overline{X} \\ &= (\lambda X)'\overline{X} - X'(\bar{\lambda}\overline{X}) \\ &= (AX)'\overline{X} - X'(\overline{AX}) \\ &= X'A'\overline{X} - X'\overline{A}\,\overline{X} \\ &= X'A\overline{X} - X'A\overline{X} \quad \text{(Why?)} \\ &= 0.\end{aligned}$$

Finally, since $X'\overline{X} \neq 0$, it follows that $\lambda = \bar{\lambda}$.

Example 1. Let

$$A = \begin{bmatrix} 2 & 0 & 0 \\ 0 & 3 & 2 \\ 0 & 2 & 3 \end{bmatrix}.$$

Then

$$|A - \lambda I| = -(\lambda - 1)(\lambda - 2)(\lambda - 5).$$

Thus, the characteristic values of A are—the order having no special significance—$\lambda_1 = 1$, $\lambda_2 = 2$, and $\lambda_3 = 5$. And, leaving the computations to the reader, we find that the corresponding spaces of characteristic vectors are given by

$$X^1 = \begin{bmatrix} 0 \\ -r \\ r \end{bmatrix}, \quad X^2 = \begin{bmatrix} s \\ 0 \\ 0 \end{bmatrix}, \quad X^3 = \begin{bmatrix} 0 \\ t \\ t \end{bmatrix}.$$

It should be observed—this is the promise of Theorem 5.9—that no matter how r, s, and t are chosen, we obtain an orthogonal set of vectors. If we take any such set as the columns of B (in the given order) we will have the result

$$B^{-1}AB = \begin{bmatrix} 1 & 0 & 0 \\ 0 & 2 & 0 \\ 0 & 0 & 5 \end{bmatrix}.$$

(The fact that B is nonsingular is guaranteed by Theorem 5.8.)

Of special interest, however, is the matrix B that results on choosing for its columns an orthonormal set of characteristic vectors. To get such a set we may choose $r = s = t = 1$ and then divide each of the resulting vectors by its length. Thus, we take

$$B^1 = \frac{1}{\sqrt{2}}\begin{bmatrix} 0 \\ -1 \\ 1 \end{bmatrix} = \begin{bmatrix} 0 \\ -1/\sqrt{2} \\ 1/\sqrt{2} \end{bmatrix}, \quad B^2 = \frac{1}{1}\begin{bmatrix} 1 \\ 0 \\ 0 \end{bmatrix} = \begin{bmatrix} 1 \\ 0 \\ 0 \end{bmatrix},$$

and
$$B^3 = \frac{1}{\sqrt{2}}\begin{bmatrix}0\\1\\1\end{bmatrix} = \begin{bmatrix}0\\1/\sqrt{2}\\1/\sqrt{2}\end{bmatrix}.$$
Then
$$B = \begin{bmatrix}0 & 1 & 0\\ -1/\sqrt{2} & 0 & 1/\sqrt{2}\\ 1/\sqrt{2} & 0 & 1/\sqrt{2}\end{bmatrix}.$$
A straightforward calculation produces
$$B^{-1} = \begin{bmatrix}0 & -1/\sqrt{2} & 1/\sqrt{2}\\ 1 & 0 & 0\\ 0 & 1/\sqrt{2} & 1/\sqrt{2}\end{bmatrix}.$$
But this is a surprising result: $B^{-1} = B'$. Thus, in B we have found a matrix that simultaneously solves the two diagonalization problems that we have considered:

$$B^{-1}AB = \text{diagonal matrix} \quad \text{(Section 4.7)}$$
$$B'AB = \text{diagonal matrix} \quad \text{(Section 1.6)}.$$

Definition 5.7. Let B be an nth-order matrix whose columns form an orthonormal set. Then B is said to be an *orthogonal matrix*. (It would seem more appropriate to call B an *orthonormal matrix*, but we must adhere to common usage.)

Theorem 5.11. If B is orthogonal, then $B^{-1} = B'$. And conversely.

Proof

1. Using a column notation for B, we write $B = [B_1, B_2, \ldots, B_n]$.

   Then, using a row notation for B', we have
   $$B' = \begin{bmatrix}B'_1\\B'_2\\\vdots\\B'_n\end{bmatrix}.$$

   It follows that $B'B = [B'_i B_j]_{nn}$. And, by our hypothesis, $B'_i B_j = B_i \cdot B_j = \delta_{ij}$. Thus, $B'B = I$, which means that $B' = B^{-1}$.

2. The argument is reversible: If $B^{-1} = B'$, then $B'B = I$. It follows that $B_i \cdot B_j = \delta_{ij}$. Thus the columns of B form an orthonormal set.

Corollary. If B is orthogonal, then its rows form an orthonormal set.

Proof. Let $C = B'$. Then, using the "forward part" of our theorem, we have
$$C'C = BB' = BB^{-1} = I.$$

It follows that $C^{-1} = C'$. But this means, using the "backward part" of our theorem, that C is orthogonal. Thus, by definition, the columns of C form an orthonormal set. But the columns of C are the rows of B. □

Let us pause to summarize our activities. We are interested in A, a symmetric matrix of order n. If A has n distinct characteristic values, then (Theorem 5.9) there are n mutually orthogonal characteristic vectors; moreover (Theorem 5.8) such a set of vectors is necessarily linearly independent. It follows that there is a B such that $B^{-1}AB$ is diagonal. Furthermore, if we first "reduce" each characteristic vector to length 1, then the diagonalizing matrix will be orthogonal. But what if A has fewer than n characteristic values? The answer to this question is given in the following theorem; its somewhat difficult proof will be omitted.

Theorem 5.12. Let A be a symmetric matrix of order n. Suppose that its characteristic equation can be written in the form

$$(\lambda - \lambda_1)^{n_1}(\lambda - \lambda_2)^{n_2} \cdots (\lambda - \lambda_k)^{n_k} = 0.$$

Then there is an orthogonal matrix B such that

$$B^{-1}AB = \begin{bmatrix} D(\lambda_1) & & & \\ & D(\lambda_2) & & \\ & & \ddots & \\ & & & D(\lambda_k) \end{bmatrix},$$

where $D(\lambda_i)$, $i = 1, 2, \ldots, k$, is a scalar matrix with diagonal entry λ_i and of order n_i.

Example 2. Let

$$A = \begin{bmatrix} 1 & -4 & 2 \\ -4 & 1 & -2 \\ 2 & -2 & -2 \end{bmatrix}.$$

Find an orthogonal matrix B such that $B^{-1}AB$ is diagonal.

Solution

1. The characteristic equation for A—after some effort—takes the form

$$(\lambda + 3)^2(\lambda - 6) = 0.$$

We order our characteristic values as follows:

$$\lambda_1 = -3, \quad \lambda_2 = 6.$$

Then, using the notation of Theorem 5.12, we have

$$n_1 = 2 \quad \text{and} \quad n_2 = 1.$$

Thus, we are anticipating a B such that

$$B^{-1}AB = \begin{bmatrix} -3 & 0 & 0 \\ 0 & -3 & 0 \\ 0 & 0 & 6 \end{bmatrix}.$$

2. We next find a maximal independent set of characteristic vectors: First, corresponding to $\lambda_1 = -3$, we have the equation

$$\begin{bmatrix} 4 & -4 & 2 \\ -4 & 4 & -2 \\ 2 & -2 & 1 \end{bmatrix} X = \Theta.$$

The solution space is given by

$$X = s \begin{bmatrix} 1 \\ 1 \\ 0 \end{bmatrix} + t \begin{bmatrix} -\frac{1}{2} \\ 0 \\ 1 \end{bmatrix}.$$

We take

$$X^1 = \begin{bmatrix} 1 \\ 1 \\ 0 \end{bmatrix} \quad \text{and} \quad X^2 = \begin{bmatrix} -1 \\ 0 \\ 2 \end{bmatrix}.$$

Secondly, corresponding to $\lambda_2 = 6$, we have the equation

$$\begin{bmatrix} -5 & -4 & 2 \\ -4 & -5 & -2 \\ 2 & -2 & -8 \end{bmatrix} X = \Theta.$$

The solution space is given by

$$X = t \begin{bmatrix} 2 \\ -2 \\ 1 \end{bmatrix}.$$

We take

$$X^3 = \begin{bmatrix} 2 \\ -2 \\ 1 \end{bmatrix}.$$

3. The matrix having X^1, X^2, and X^3 as its columns will diagonalize A. However, we are seeking an orthogonal matrix that plays this role. Now, we have at this point—and in general—a set of independent characteristic vectors corresponding to each λ_i. Each such set serves as a basis for a certain vector space. Using the Gram-Schmidt process we can exchange each of these bases for one that is orthonormal. And—this follows from Theorem 5.9—the vectors of one such basis are orthogonal to those of any other. Thus, the totality of these character-

istic vectors constitutes an orthonormal set. It follows that these vectors—taken in the proper order—give us the desired matrix B.

In the present example, our application of the Gram-Schmidt process is an easy one: From $\{X^1, X^2\}$ we obtain $\{B^1, B^2\}$, where

$$B^1 = \begin{bmatrix} \sqrt{2}/2 \\ \sqrt{2}/2 \\ 0 \end{bmatrix} \quad \text{and} \quad B^2 = \begin{bmatrix} -\sqrt{2}/6 \\ \sqrt{2}/6 \\ 2\sqrt{2}/3 \end{bmatrix},$$

and from $\{X^3\}$ we obtain $\{B^3\}$, where

$$B^3 = \begin{bmatrix} \frac{2}{3} \\ -\frac{2}{3} \\ \frac{1}{3} \end{bmatrix}.$$

Finally, then, we take $B = [B^1, B^2, B^3]$.

EXERCISE SET 5.5

1 Find an orthogonal matrix B such that $B^{-1}AB = C$, where A is the first matrix given and C is the second one.

a) $\begin{bmatrix} 1 & 2 \\ 2 & 1 \end{bmatrix}, \begin{bmatrix} 3 & 0 \\ 0 & -1 \end{bmatrix}$

b) $\begin{bmatrix} 0 & 3 \\ 3 & 0 \end{bmatrix}, \begin{bmatrix} 3 & 0 \\ 0 & -3 \end{bmatrix}$

c) $\begin{bmatrix} 0 & 2 & 1 \\ 2 & 0 & 1 \\ 1 & 1 & 3 \end{bmatrix}, \begin{bmatrix} 1 & 0 & 0 \\ 0 & -2 & 0 \\ 0 & 0 & 4 \end{bmatrix}$

d) $\begin{bmatrix} 0 & 1 & 1 \\ 1 & 0 & 1 \\ 1 & 1 & 0 \end{bmatrix}, \begin{bmatrix} -1 & 0 & 0 \\ 0 & -1 & 0 \\ 0 & 0 & 2 \end{bmatrix}$

e) $\begin{bmatrix} 1 & 0 & 0 & 0 \\ 0 & 1 & 0 & 0 \\ 0 & 0 & 2 & 1 \\ 0 & 0 & 1 & 2 \end{bmatrix}, \begin{bmatrix} 1 & 0 & 0 & 0 \\ 0 & 1 & 0 & 0 \\ 0 & 0 & 1 & 0 \\ 0 & 0 & 0 & 3 \end{bmatrix}$

2 a) Prove: The product of two orthogonal matrices is orthogonal.
  b) Prove: If A is orthogonal, then $|A| = \pm 1$.
  c) Prove: A symmetric matrix A is positive definite if and only if its characteristic values are all positive.

3 Let E be a Euclidean space and suppose that two coordinate systems for E are related by the equation $X_{\mathfrak{E}} = AX_{\mathfrak{B}}$, where A is orthogonal. Prove: If $\mathfrak{B}$ is orthonormal, then so is $\mathfrak{C}$. [*Hint:* Use Theorem 5.6.]

4 Let $E = (R^{21}, \cdot)$, where the dot product is the standard one. Suppose that a basis $\mathfrak{B}$ is introduced by

$$\mathfrak{B} = \begin{bmatrix} 1/\sqrt{2} & 1/\sqrt{2} \\ -1/\sqrt{2} & 1/\sqrt{2} \end{bmatrix} \mathfrak{N}.$$

a) Show that $\mathfrak{B}$ is orthonormal.
b) Make an arrow drawing showing N^1, N^2, B^1, and B^2.
c) Find the matrix that takes $\mathfrak{N}$-columns into $\mathfrak{B}$-columns.

5 Let $\mathcal{B}$ and $\mathcal{C}$ be bases for a vector space V. Prove: If the matrix A that transforms $\mathcal{B}$ into $\mathcal{C}$ is orthogonal, then A is also the matrix that transforms $\mathcal{B}$-columns into $\mathcal{C}$-columns.

6 Let $A_\alpha = \begin{bmatrix} \cos\alpha & \sin\alpha \\ -\sin\alpha & \cos\alpha \end{bmatrix}$ and $B_\alpha = \begin{bmatrix} \cos\alpha & \sin\alpha \\ \sin\alpha & -\cos\alpha \end{bmatrix}$.

    a) Show that both A_α and B_α are orthogonal.
    b) Show that $|A_\alpha| = 1$ and $|B_\alpha| = -1$.
    c) Let $E = (R^{21}, \cdot)$, where the dot product is the standard one, and let $\mathcal{C}$ be introduced by the equation $\mathcal{C} = A_\alpha \mathcal{N}$. Find A^1 and A^2 and then make an arrow drawing showing N^1, N^2, A^1, and A^2. (Take $0 < \alpha < \pi/2$.)
    d) Same as (c), where the new basis is given by $\mathcal{B} = B_\alpha \mathcal{N}$.
    e) Prove: Every orthogonal, 2 by 2 matrix is of the form A_α or B_α.

7 Let $\begin{bmatrix} \frac{2}{3} & \frac{2}{3} & \frac{1}{3} \\ -\frac{2}{3} & \frac{1}{3} & \frac{2}{3} \\ \frac{1}{3} & -\frac{2}{3} & \frac{2}{3} \end{bmatrix}$.

    a) Verify that A is orthogonal.
    b) Show that the characteristic values of A all have absolute value 1.
    c) Prove: If A is an orthogonal matrix, then each of its characteristic values has absolute value 1.

5.6 APPLICATIONS TO GEOMETRY

Problems in plane geometry are often most effectively solved by "placing" the given figure on the natural coordinate plane of some Euclidean space. For then the points, lines, and curves of the figure may be translated into coordinate pairs and coordinate equations from which—by algebraic means—various distances, angles, etc., can be computed. This is the method of *analytic geometry*. (This method is also applicable, of course, to solid geometry.) The Euclidean space chosen for this purpose is generally the space $E = (R^{21}, \cdot)$, where the dot product is the standard one. However, as we shall show in the example that follows (Solution 2) other Euclidean spaces may also be appropriate.

Example 1. The major and minor axes of an ellipse have lengths 6 and 4, respectively. A line passes through the center of the ellipse in such a way as to bisect the right angle formed by the axes of the ellipse. The ellipse cuts from this line a certain line segment; how long is this line segment? (See Fig. 5.10.) There are, of course, two such line segments, but they have the same length.

Solution 1. We identify the plane of plane geometry with the natural coordinate plane of $E = (R^{21}, \cdot)$, where the dot product is the standard one. We next place the line ellipse configuration on our coordinate plane, positioning it to our advantage: we make the axes of the ellipse coincide

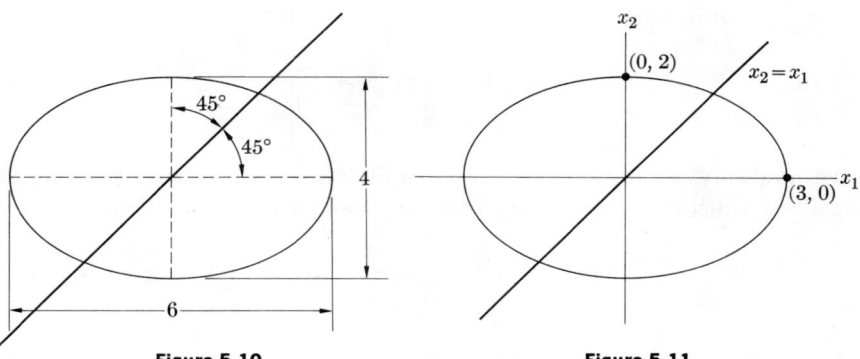

Figure 5.10 **Figure 5.11**

with those of our coordinate system, the larger one falling on the x_1-axis. (See Fig. 5.11.)

The line of our configuration now has the equation $x_2 = x_1$, that is, the pairs (x_1, x_2) that satisfy this equation may be identified with the vectors $[x_1, x_2]'$ in E whose "tips" make up the line.

To find the equation of the ellipse we must first recall two basic geometric facts about an ellipse: (1) The distance from the center of an ellipse to either of its foci is given by $\sqrt{a^2 - b^2}$, where a is the length of the semimajor axis and b is the length of the semiminor axis. Thus, in the present case, the *focal distance* is $\sqrt{3^2 - 2^2} = \sqrt{5}$. (2) The sum of the distances from the foci to any point on the ellipse is equal to the length of the major axis. In the present case, 6.

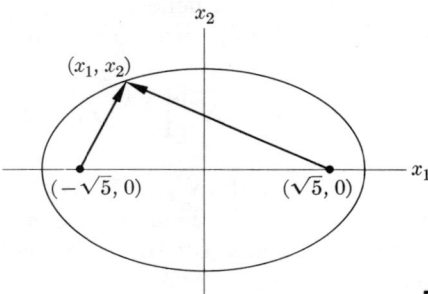

Figure 5.12

Now consider Fig. 5.12. Vectors from the two foci to an arbitrarily chosen point on the ellipse are given by

$$A = \begin{bmatrix} x_1 + \sqrt{5} \\ x_2 \end{bmatrix} \quad \text{and} \quad B = \begin{bmatrix} x_1 - \sqrt{5} \\ x_2 \end{bmatrix}.$$

Thus, our ellipse has the equation $|A| + |B| = 6$, or,

$$\left|\begin{bmatrix} x_1 + \sqrt{5} \\ x_2 \end{bmatrix}\right| + \left|\begin{bmatrix} x_1 - \sqrt{5} \\ x_2 \end{bmatrix}\right| = 6.$$

We now solve simultaneously the line equation with that of the ellipse. A direct substitution, $x_1 = x_2$, produces a single equation in x_2:

$$\left|\begin{bmatrix} x_2 + \sqrt{5} \\ x_2 \end{bmatrix}\right| + \left|\begin{bmatrix} x_2 - \sqrt{5} \\ x_2 \end{bmatrix}\right| = 6,$$

or, recalling the formula for length in this Euclidean space, we have

$$\sqrt{(x_2 + \sqrt{5})^2 + x_2^2} + \sqrt{(x_2 - \sqrt{5})^2 + x_2^2} = 6.$$

We leave it to the reader to show that $x_1 = x_2 = \pm 6/\sqrt{13}$. Thus, the "vectors of intersection" are

$$\begin{bmatrix} 6/\sqrt{13} \\ 6/\sqrt{13} \end{bmatrix} \quad \text{and} \quad \begin{bmatrix} -6/\sqrt{13} \\ -6/\sqrt{13} \end{bmatrix}.$$

The length of the line segment in question (see Fig. 5.11) is twice the length of either of these vectors. Thus, our final answer is

$$2\sqrt{36/13 + 36/13} = 12\sqrt{26}/13.$$

Solution 2. This time—trying to exploit the relationship between the line and the ellipse—we choose a Euclidean space in which the vectors N^1 and N^2 make an angle of $45°$. Such a choice is provided by $E = (R^{21}, \cdot)$, where

$$X \cdot Y = X' \begin{bmatrix} 1 & 1 \\ 1 & 2 \end{bmatrix} Y.$$

An easy computation shows that $|N^1| = 1$, $|N^2| = \sqrt{2}$, and that the angle between these vectors is $45°$. Thus, the natural-coordinate system for E may be viewed as in Fig. 5.13; the placement of our line ellipse configuration is then immediate.

Since our line coincides with the x_2-axis, its equation is now $x_1 = 0$. And the ellipse—the reasoning of Solution 1 is still appropriate since $|N^1| = 1$—is given by

$$\left|\begin{bmatrix} x_1 + \sqrt{5} \\ x_2 \end{bmatrix}\right| + \left|\begin{bmatrix} x_1 - \sqrt{5} \\ x_2 \end{bmatrix}\right| = 6.$$

5.6 APPLICATIONS TO GEOMETRY

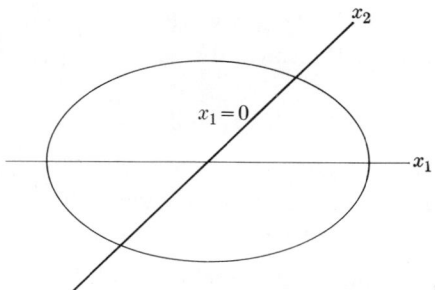

Figure 5.13

Substituting for x_1 we obtain a single equation in x_2:

$$\left|\begin{bmatrix}\sqrt{5}\\x_2\end{bmatrix}\right| + \left|\begin{bmatrix}-\sqrt{5}\\x_2\end{bmatrix}\right| = 6.$$

The next step, because we are working in a "strange" Euclidean space, is not so immediate. We must first derive a length formula for E:

$$|Y| = \sqrt{Y \cdot Y} = \sqrt{y_1^2 + 2y_1 y_2 + 2y_2^2}.$$

Using this formula, we may rewrite our equation in x_2 as

$$\sqrt{2x_2^2 + 2\sqrt{5}x_2 + 5} + \sqrt{2x_2^2 - 2\sqrt{5}x_2 + 5} = 6.$$

This equation is identical with that met in Solution 1; thus, $x_2 = \pm 6/\sqrt{13}$. It follows that the intersection vectors are

$$\begin{bmatrix}0\\6/\sqrt{13}\end{bmatrix} \quad \text{and} \quad \begin{bmatrix}0\\-6/\sqrt{13}\end{bmatrix}.$$

Again the length of the line segment in question is twice the length of either of these vectors. Thus, we must make the following computation:

$$2 \cdot \left|\begin{bmatrix}0\\6/\sqrt{13}\end{bmatrix}\right| = 2\sqrt{2 \cdot \frac{36}{13}} = \frac{12\sqrt{26}}{13}. \quad \square$$

In Example 1 the problem is posed without reference to a Euclidean space. Our solutions involved picking a Euclidean space and then employing the natural coordinate system for that space. The subsequent steps of the solutions—the arithmetic—differed, but the answers agreed. Now in other applications of coordinate systems it may be that the original problem is already posed with respect to some coordinate system. It then may be that a change in the coordinate system will simplify the problem.

Example 2. Let $E = (R^{21}, \cdot\,)$, where the dot product is the standard one. A certain curve on the natural-coordinate plane determined by E has the equation
$$3x_1^2 + 4x_1x_2 + 3x_2^2 = 3.$$
That is, if we identify each pair (x_1, x_2) which satisfies this equation with the tip of the arrow depicting the vector $[x_1, x_2]'$, then a certain curve will arise. Sketch this curve.

Solution 1. We can simplify our sketching problem by introducing a new coordinate system. First, we rewrite the given equation in matrix form:
$$X'_{\mathfrak{N}} A X_{\mathfrak{N}} = 3, \qquad \text{where} \qquad A = \begin{bmatrix} 3 & 2 \\ 2 & 3 \end{bmatrix}.$$
Then we seek a coordinate system in which the matrix equation of the curve will employ—in the place of A—a diagonal matrix.

Now, by the method of Section 1.6, we can find a matrix B such that $B'AB$ is diagonal. For example, we may take
$$B = \begin{bmatrix} 2 & -2 \\ 0 & 3 \end{bmatrix}.$$
Then, as the reader may verify,
$$B'AB = \begin{bmatrix} 2 & 0 \\ -2 & 3 \end{bmatrix} \begin{bmatrix} 3 & 2 \\ 2 & 3 \end{bmatrix} \begin{bmatrix} 2 & -2 \\ 0 & 3 \end{bmatrix} = \begin{bmatrix} 12 & 0 \\ 0 & 15 \end{bmatrix}.$$
Thus, we are led to introduce a new coordinate system with the equation $X_{\mathfrak{N}} = BX_{\mathfrak{C}}$. Substitution into the natural-coordinate equation of our curve leads us directly to its $\mathfrak{C}$-coordinate equation:
$$(BX_{\mathfrak{C}})'A(BX_{\mathfrak{C}}) = 3,$$
$$X'_{\mathfrak{C}}(B'AB)X_{\mathfrak{C}} = 3,$$
$$12u_1^2 + 15u_2^2 = 3,$$
where we have let $X_{\mathfrak{C}} = [u_1, u_2]'$.

Our plan is to produce the desired sketch from this new (and simpler) coordinate equation. Thus, we must first superimpose upon the $x_1 x_2$-coordinate system (based on $\mathfrak{N}$) the new $u_1 u_2$-coordinate system (based upon $\mathfrak{C}$). And for this purpose we need to know C. This is an old problem for us: Since $X_{\mathfrak{N}} = BX_{\mathfrak{C}}$, then $\mathfrak{C} = B'\mathfrak{N}$. Thus,
$$\mathfrak{C} = \begin{bmatrix} 2 & 0 \\ -2 & 3 \end{bmatrix} \mathfrak{N} = \begin{bmatrix} 2N^1 \\ -2N^1 + 3N^2 \end{bmatrix}.$$
That is, $C^1 = 2N^1$ and $C^2 = -2N^1 + 3N^2$. In Fig. 5.14 the two coordinate systems are compared.

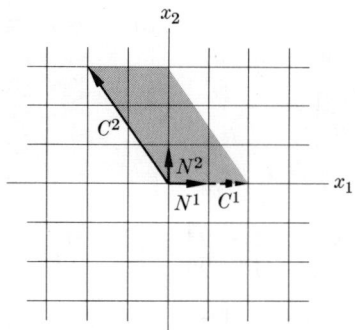
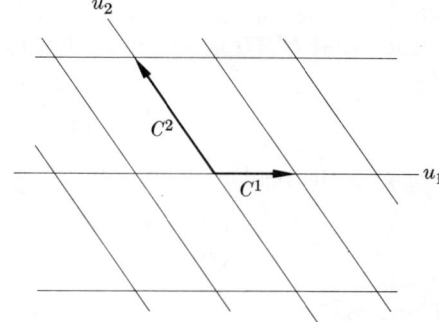

Figure 5.14

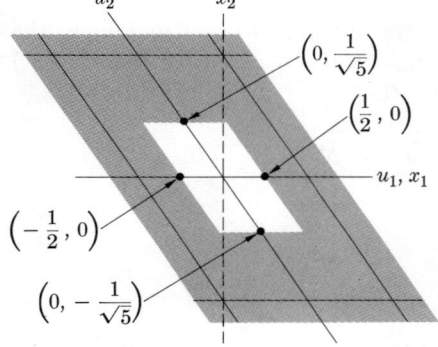
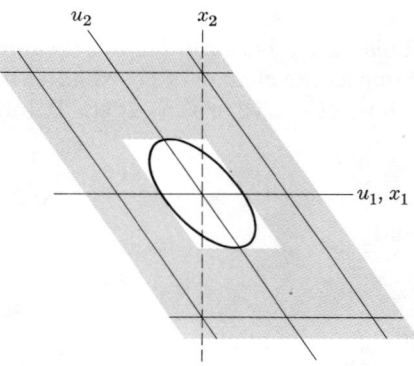

Figure 5.15 **Figure 5.16**

We return now to the equation $12u_1^2 + 15u_2^2 = 3$. Two observations are immediate: If (u_1, u_2) satisfies this equation, then

1. $u_1^2 \leq \frac{1}{4}$; hence $-\frac{1}{2} \leq u_1 \leq \frac{1}{2}$.
2. $u_2^2 \leq \frac{1}{5}$; hence $-\dfrac{1}{\sqrt{5}} \leq u_2 \leq \dfrac{1}{\sqrt{5}}$.

It then follows that the graph of $12u_1^2 + 15u_2^2 = 3$ is restricted to the unshaded region shown in Fig. 5.15. Finally, after locating some specific points on the curve, we obtain the sketch shown in Fig. 5.16. The curve is an ellipse.

Solution 2. As before we begin by putting the given equation in matrix form:

$$X'_{\mathfrak{R}} A X_{\mathfrak{R}} = 3, \quad \text{where} \quad A = \begin{bmatrix} 3 & 2 \\ 2 & 3 \end{bmatrix}.$$

Now, by the method of Section 5.5, we can find an orthogonal matrix B such that $B'AB$ is diagonal. For example, we may take

$$B = \begin{bmatrix} 1/\sqrt{2} & -1/\sqrt{2} \\ 1/\sqrt{2} & 1/\sqrt{2} \end{bmatrix}.$$

Then, as the reader may verify,

$$B'AB = \begin{bmatrix} 1/\sqrt{2} & 1/\sqrt{2} \\ -1/\sqrt{2} & 1/\sqrt{2} \end{bmatrix} \begin{bmatrix} 3 & 2 \\ 2 & 3 \end{bmatrix} \begin{bmatrix} 1/\sqrt{2} & -1/\sqrt{2} \\ 1/\sqrt{2} & 1/\sqrt{2} \end{bmatrix} = \begin{bmatrix} 5 & 0 \\ 0 & 1 \end{bmatrix}.$$

Thus, on introducing a new coordinate system with the equation $X_\mathfrak{N} = BX_\mathfrak{e}$, we obtain the new coordinate equation

$$5u_1^2 + u_2^2 = 3.$$

This time, because B is orthogonal, the change-of-basis matrix is the same as the change-of-coordinates matrix. (Recall Exercise 5, Section 5.5.) Thus, $\mathfrak{N} = B\mathfrak{C}$ or $\mathfrak{C} = B'\mathfrak{N}$. It follows that

$$C^1 = \frac{1}{\sqrt{2}} N^1 + \frac{1}{\sqrt{2}} N^2$$

and

$$C^2 = \frac{-1}{\sqrt{2}} N^1 + \frac{1}{\sqrt{2}} N^2.$$

In Fig. 5.17 the old and new coordinate systems are compared. It should be observed that the new basis is also orthornormal (recall Exercise 3, Section 5.5) and that the new coordinate system may be visualized as having been obtained from the original one through a "rotation of axes."

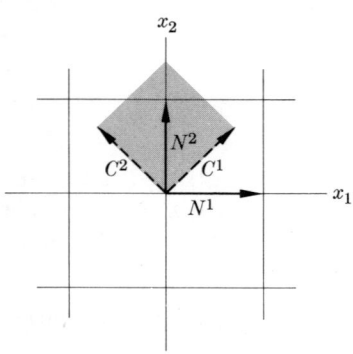 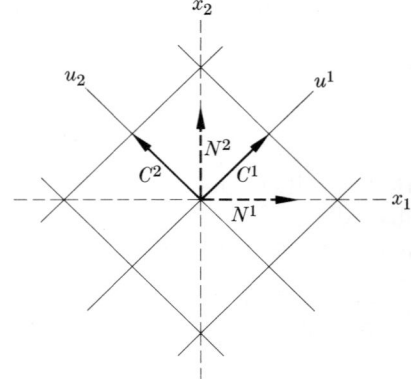

Figure 5.17

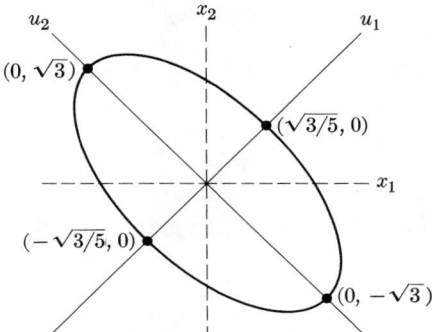

Figure 5.18

Returning to the equation $5u_1^2 + u_2^2 = 3$, we proceed exactly as in Solution 1 and obtain finally the sketch of Fig. 5.18. The fact that the basis for the $u_1 u_2$-coordinate system is orthornormal has—with respect to our ellipse—two important consequences: (1) The ellipse is symmetric with respect to both the u_1- and u_2-axes. (2) The dimensions of the ellipse may be obtained directly from its u_1- and u_2-intercepts:

Length of minor axis: $\sqrt{3/5} - (-\sqrt{3/5}) = 2\sqrt{3/5}$.
Length of major axis: $\sqrt{3} - (-\sqrt{3}) = 2\sqrt{3}$.

EXERCISE SET 5.6

1 Given two concentric circles of radius 1 and 2, respectively, a line is drawn tangent to the inner one. The outer circle cuts from this tangent a chord. The problem is to find the length of this chord.
   a) Solve this problem using the space E of Example 1, Solution 1.
   b) Same, using the space E of Example 1, Solution 2.

2 Let E be as in Example 2. Find the lengths of the major and minor axes of the ellipse whose natural coordinate equation is

$$7x_1^2 - 2\sqrt{3}\, x_1 x_2 + 5x_2^2 = 8.$$

[*Hint:* Find the characteristic values of

$$\begin{bmatrix} 7 & -\sqrt{3} \\ -\sqrt{3} & 5 \end{bmatrix};$$

then you are very nearly finished.]

3 Let E be as in Example 2. Sketch the curve whose natural-coordinate equation is

$$x_1^2 + 4x_1 x_2 + y_1^2 = 3.$$

4 Let $E = (R^{31}, \cdot)$, where the dot product is the standard one. If an ellipsoid with semiaxes having lengths a, b, and c is appropriately placed in a coordinate system for E having an orthonormal basis, then its equation in that coordinate system will be

$$\frac{u_1^2}{a^2} + \frac{u_2^2}{b^2} + \frac{u_3^2}{c^2} = 1.$$

Now, a certain surface S in E has the natural-coordinate equation

$$3x_1^2 - 2x_1x_2 + 5x_2^2 - 2x_2x_3 + 3x_3^2 + 2x_1x_3 = 6.$$

Prove that S is an ellipsoid and find the lengths of its three axes.

5 Return to Solution 1 of Example 2 and complete the following:
  a) Find the $\mathcal{C}$-coordinate equation of the line whose natural-coordinate equation is $x_2 = x_1$.
  b) Solve the equation obtained in part (a) simultaneously with the $\mathcal{C}$-coordinate equation of the ellipse.
  c) Find the distance between the two points found in part (b). (Your answer should be the length of the minor axis of the ellipse as found at the end of Solution 2.)

6 Return to the ellipse of Example 2 and find the lengths of the major and minor axes without introducing a new coordinate system. [*Hint:* Find the points of intersection of the ellipse with the line $x_2 = tx_1$, express the distance between these points as a function of t, and then find the maximum and minimum values achieved by this function.]

7 *A geometric interpretation for a characteristic vector.* Let $E = (R^{31}, \cdot)$, where the dot product is the standard one, and let L be defined on E by

$$L(X) = \begin{bmatrix} 2 & 0 & 0 \\ 0 & 3 & 0 \\ 0 & 0 & 6 \end{bmatrix} X.$$

  a) Sketch a coordinate system for E and show (as arrows) the vectors $A = [1, 1, 1]'$ and $L(A)$. Show $B = [0, 0, 1]'$ and $L(B)$. Show other such pairs. Finally, describe in geometric terms what happens to the unit cube of vectors embraced by N^1, N^2, and N^3.
  b) Consider next the transformation defined by

$$L^*(X) = \begin{bmatrix} 3 & -1 & 1 \\ -1 & 5 & -1 \\ 1 & -1 & 3 \end{bmatrix} X.$$

Show that the action of L^* on a vector is *similar* to that of L in the sense that there is a unit cube of vectors that is transformed by L^* in the same way that the natural unit cube is transformed by L.

APPENDIX A

RELATIONS

In Section 1.8, a considerable effort was made to motivate the notion of an *equivalence relation*. The definition that evolved from that discussion described an equivalence relation as a *relation* that had certain special properties. (Recall Definition 1.22.) However, nowhere in that section did we say precisely what mathematicians mean when they speak of a relation. That oversight will be attended to here.

First of all, there is the natural, intuitive approach: A *relation* is simply a scheme for relating the elements of some set. But such a "definition" leaves us vulnerable to the question: What is a *scheme for relating elements*? The point is this: We must somehow find a characterization of *relation* which does not depend on the verb "relate." How can this be done? The answer is best motivated by considering some examples.

In dealing with real numbers we frequently have occasion to employ the relation b *is less than* a, written $b < a$. This relation has a familiar geometric interpretation; representing real numbers as points on a line, the relation $b < a$ corresponds to the *point* b being to the left of the *point* a. But—and this is our concern here—there is an alternative way of visualizing the less-than relation.

Consider the usual xy-plane, and sketch on it the line $y = x$. This line divides the plane into an upper region and a lower one (see Fig. A.1).

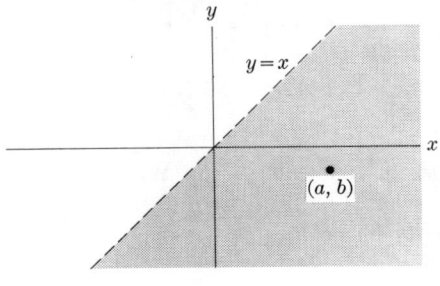

Figure A.1

Consider a point (a, b) that is situated in the lower region. What relation exists between its coordinates? Answer: $b < a$. Conversely, if $b < a$, then the point (a, b) must lie below the line $y = x$. In other words, $b < a$ if and only if the point (a, b) lies below the line $y = x$. Thus, the shaded region in Fig. A.1 also serves to depict the less-than relation. But this shaded region is the picture of a set of ordered pairs of real numbers. This suggests—ever so slightly, perhaps—that we might simply define a *relation in a set S* as a set of ordered pairs with entries chosen from S. Before committing ourselves to this definition, let us try it out with another example.

This time we begin with the set of ordered pairs and then see if we are led to something that we would intuitively accept as a relation. Letting i, j, and k denote positive integers, we introduce the set

$$R = \{(i, j): \text{there exists a number } k \text{ such that } i = kj\}.$$

For example, the pair $(8, 2)$ is in R because $8 = 4 \cdot 2$. Plotting some of the points in R leads to the picture given in Fig. A.2. A picture of all of R can be achieved mentally: Take the point $(1, 1)$ and every *lattice point* to its right; take the point $(2, 2)$ and every second lattice to its right; take the point $(3, 3)$ and every third lattice point to its right; and so on.

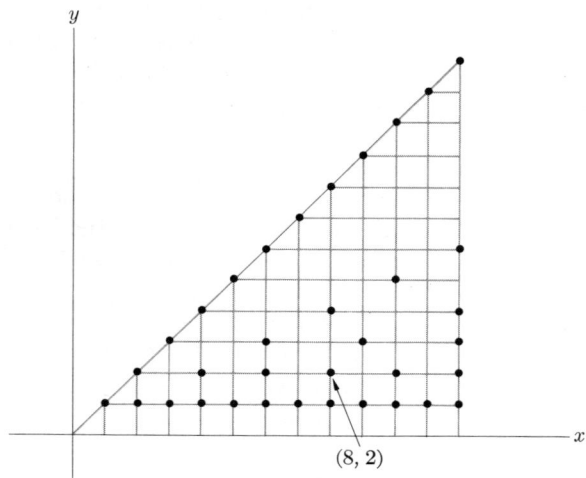

Figure A.2

Now, take (p, q) as any pair in R. What special relation exists between p and q? Answer: The number q divides evenly into p. Thus, we can define the *divisibility relation of positive integers* as follows: we say that q *divides* p if (p, q) is in R.

The point is this: Corresponding to every situation in which mathematicians say, intuitively, that the objects of a certain set are related in such and such a manner, there exists a set of ordered pairs of those objects having the property that a pair is in the set if and only if the second coordinate of the pair is in the prescribed relationship to the first. Thus, the intuitive point of view may be reversed, that is, the set of pairs may be regarded as coming first and the actual relating of the objects coming second.

Definition. Let S be a given set and let R be a set of ordered pairs having entries chosen from S. Then R is said to be a *relation in S*. If (a, b) is in S, then we say that *b is R-related to a*. We write $b \, R \, a$.

It is interesting to observe that this definition makes sense only if the concept of an *ordered pair* is satisfactorily defined. What is an *ordered pair*, really? People that ask such awkward questions are responsible for a branch of mathematics called *foundations of mathematics*. The interested reader should consider the following texts:

1. Fraenkel, Abraham, *Set Theory and Logic*, Addison-Wesley, Reading, Mass., 1966.

2. Halmos, P. R., *Naive Set Theory*, Van Nostrand, Princeton, N.J., 1950.

3. Stoll, Robert R., *Sets, Logic and Axiomatic Theories*, W. H. Freeman, San Francisco, Calif., 1961.

APPENDIX B

WHAT IS A MATRIX, REALLY?

In Section 1.1, we find the following definition: An *m by n matrix* is a rectangular array of numbers having m rows and n columns. This definition is the traditional one; however, it does not hold up under the standards of rigor imposed by modern mathematics.

The primary objection is that numbers have no physical existence: they are mathematical entities. Thus, it makes no sense to think of numbers as being *arranged* in a rectangular array. However, suppose that we replace the word "numbers" in our definition by "numerals"; for numerals—being marks used to denote numbers—can be arranged in rectangular arrays. Then our definition would define a matrix as a rectangular array of symbols, and thus a matrix would be itself simply a symbol. This is one point of view: a matrix is not a mathematical entity at all, but rather, it is a symbol that stands for some mathematical entity. For what mathematical entity? Well, as a result of Theorem 3.9, a matrix can be viewed as a convenient symbol for representing a linear transformation between two finite-dimensional vector spaces.

The foregoing point of view does not allow for the existence of matrices as mathematical entities in their own right. Thus, a discussion of the arithmetic of matrices (recall Section 1.1) becomes meaningless in the absence of the notion of a linear transformation. However, as the reader is well aware, this text begins with a study of matrix arithmetic and does not introduce linear transformations until the third chapter. Thus, in this text a matrix is apparently being regarded as being more than just a symbol, but rather, as a mathematical entity that exists in its own right. How can such a point of view be justified? A suitable answer is found in the following definition of a matrix.

Definition. An *m by n matrix* is a function α whose domain is a set of the form

$$I_{mn} = \{(i,j): i = 1, 2, \ldots, m; j = 1, 2, \ldots, n\}.$$

A *real (complex) matrix* is a matrix whose range is a set of real (complex) numbers.

This definition needs amplification. Most disturbing, perhaps, is the idea that a matrix is a function. How can this be? Well, let us try to portray graphically this special kind of function that is called a matrix. It will help to work with a particular example.

Let α be defined on the domain

$$I_{34} = \{(i,j): i = 1, 2, 3; j = 1, 2, 3, 4\}$$

by the formula $\alpha(i,j) = i + 2j$. Thus, for example, $\alpha(2,3) = 8$ but $\alpha(2,5)$ does not exist. Now, the domain of α, the set I_{34}, may easily be visualized as a set of points on a plane. One way to accomplish this is as follows: Take a rectangular coordinate system in which the positive y-axis is to the right and the positive x-axis is downward. The "plot" of I_{34} is then as in Fig. B.1. Thus, the domain of α may be visualized as a rectangular array of points.

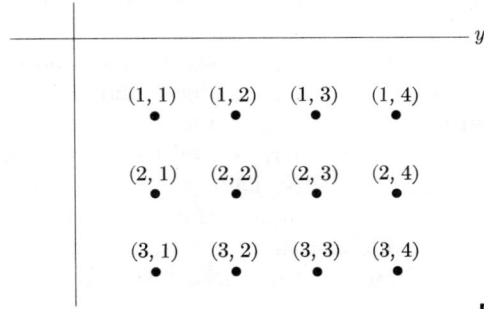

Figure B.1

What about the range of α? Well, the range of α is (in this case) a set of positive integers. Would it not be reasonable to seek a picture that shows each number in the range of α alongside the domain element (or elements) that produced it? One way of doing such a thing would be to erect at each point (i,j) in the domain of α a "flag" bearing the number $\alpha(i,j)$. See Fig. B.2. In this picture we have a vivid portrayal of the function α. We see its domain (a rectangular array of points); we see its

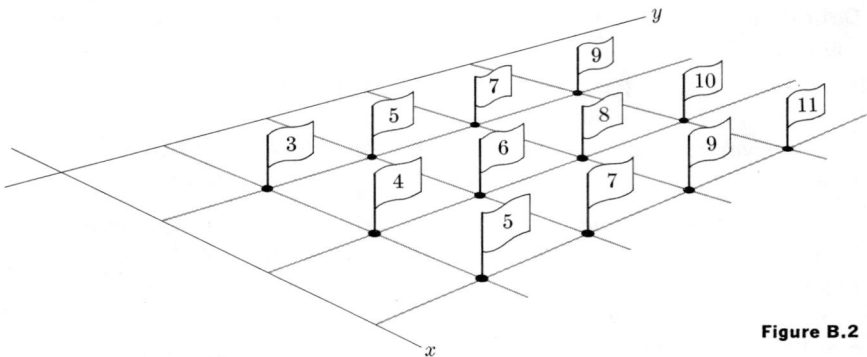

Figure B.2

range (the set of all numbers appearing on the flags); and we see which numbers are assigned by α to which points. But it is a rather complicated picture to draw; is there not a simpler picture that will serve us just as well? Why of course:

$$\begin{bmatrix} 3 & 5 & 7 & 9 \\ 4 & 6 & 8 & 10 \\ 5 & 7 & 9 & 11 \end{bmatrix}.$$

The point is this: That which is usually called a matrix (a rectangular array of numerals) is now seen to be just a graphical symbol for portraying the matrix that is defined here, the matrix that is defined as a certain kind of function. In other words, we use "matrix arrays" as symbols for "matrix functions."

Finally, it is of some interest to note that the notion of equality for matrices requires no special definition; for functions are equal when their domains are identical and the rules for assigning mates are identical. It follows that the rectangular arrays used to denote equal matrices must be of the same shape and must have corresponding entries the same. Similarly, the notions of matrix addition and numerical multiplication are special instances of the addition and numerical multiplication of functions. But matrix multiplication is not a special instance of function multiplication.

APPENDIX C

THE TRADITIONAL APPROACH TO THE DETERMINANT FUNCTION

There are a number of ways of defining the determinant function and developing its basic properties, none of which is particularly easy. The definition chosen in this text was chosen for pedagogical reasons: the idea that the theory of the determinant function could be developed using proofs by mathematical induction was attractive to this author. However, this inductive approach is not the traditional one; thus, so that the reader be better versed on this matter, we shall consider here the traditional definition.

It will be convenient to talk about the determinant of a square matrix of low order, say

$$\begin{bmatrix} a_{11} & a_{12} & a_{13} & a_{14} \\ a_{21} & a_{22} & a_{23} & a_{24} \\ a_{31} & a_{32} & a_{33} & a_{34} \\ a_{41} & a_{42} & a_{43} & a_{44} \end{bmatrix}.$$

First, consider an ordered quadruple of the form

$$[(1, j_1), (2, j_2), (3, j_3), (4, j_4)],$$

where the numbers j_1, j_2, j_3, and j_4 are just the numbers 1, 2, 3, and 4, but not necessarily in their natural order. In other words, (j_1, j_2, j_3, j_4) is one of the *permutations* of $(1, 2, 3, 4)$. Such a quadruple will be called a *path through* A. For example, $P = [(1, 4), (2, 2), (3, 1), (4, 3)]$ is a path through A. The choice of the word "path" is inspired by the realization that a quadruple like P may be viewed as a set of instructions for "walking through A" (see Fig. C.1).

Next comes the hard part. Take an arbitrary path through A, say

$$[(1, j_1), (2, j_2), (3, j_3), (4, j_4)],$$

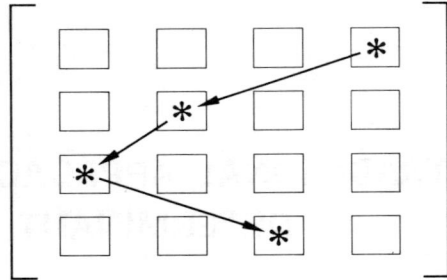

Figure C.1

and look at the permutation (j_1, j_2, j_3, j_4). We engage in a certain curious counting process: (1) How many of the numbers in the set $\{j_2, j_3, j_4\}$ are less than j_1? (2) How many of the numbers in the set $\{j_3, j_4\}$ are less than j_2? (3) How many of the numbers in the set $\{j_4\}$ are less than j_3? If the answers to these questions are totaled and the sum is even, then the path is said to be an *even path*. Otherwise, it is an *odd path*. For the path

$$P = [(1, 4), (2, 2), (3, 1), (4, 3)]$$

the answers to the three questions are, respectively, 3, 1, and 0. Thus, P is an even path.

Finally, with respect to each path through A there is a certain product, the product of the entries in A met along that path. For example, taking P as above, the product *associated with* P is $a_{14}a_{22}a_{31}a_{43}$. We now have all the necessary preliminaries for giving the following definition:

Definition. The *determinant function*, Δ, is defined for any square matrix by the following three-part rule:

1. Consider all of the paths through A, and for each path determine the associated product.

2. To each of the products determined under part 1—there will be $n!$ of these if A is nth-order—prefix a plus sign or a minus sign according as the corresponding path is even or odd.

3. The sum of the terms resulting under part 2 is the value of $\Delta(A)$.

A more formal rule for Δ is the following:

$$\Delta(A) = \sum (\pm a_{1j_1} a_{2j_2} \cdots a_{nj_n}),$$

where the summation is extended over all permutations $(j_1, j_2, \ldots, j_n)$ and the plus or minus sign is used according as a particular permutation is even or odd.

An instructive exercise: starting with the definition of Δ given in Section 4.2, use mathematical induction to prove the above formula.

The reader who would like to see how the theory is developed using this formula should consider the following texts:

1. Shilov, Georgi F., *Introduction to the Theory of Linear Spaces*, Prentice-Hall, Englewood Cliffs, N.J., 1961.

2. Marcus, Marvin, and Mine, Henryk, *Introduction to Linear Algebra*, Macmillan Company, New York, 1965.

APPENDIX D

A GEOMETRIC INTERPRETATION FOR THE DETERMINANT FUNCTION

With respect to a rectangular coordinate system, consider the parallelogram "embraced" by the pair of arrows $A = [a_1, a_2]$ and $B = [b_1, b_2]$ (see Fig. D.1). We wish to determine the area of this parallelogram. It is more convenient to determine first the square of this area:

$$\begin{aligned}
\text{area}^2(P) &= |A|^2 h^2 \\
&= |A|^2 |B|^2 \sin^2 \phi \\
&= |A|^2 |B|^2 (1 - \cos^2 \phi) \\
&= |A|\,|B|^2 - (A \cdot B)^2 \\
&= (a_1 b_2 - a_2 b_1)^2.
\end{aligned}$$

Thus,
$$\text{area}(P) = \pm \begin{vmatrix} a_1 & a_2 \\ b_1 & b_2 \end{vmatrix},$$

the plus sign prevailing when B is counterclockwise of A, as in Fig. D.1. Thus, the 2 by 2 determinant may be viewed (in absolute value) as the area of a certain parallelogram.

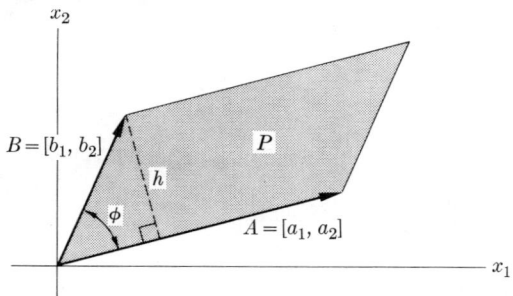

Figure D.1

DETERMINANT FUNCTION: GEOMETRIC INTERPRETATION

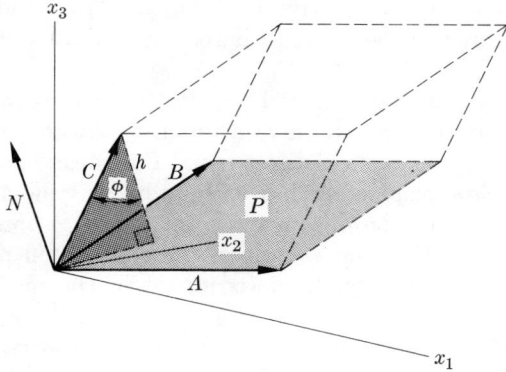

Figure D.2

Consider next the parallelepiped embraced by the triple of arrows $A = [a_1, a_2, a_3]$, $B = [b_1, b_2, b_3]$, and $C = [c_1, c_2, c_3]$ (see Fig. D.2). We wish to determine the volume of this parallelepiped. First, we make two preliminary calculations:

1. Let P be the parallelogram indicated in Fig. D.2. Then, following a computation exactly like that used earlier, we obtain

$$\text{area}^2(P) = \begin{vmatrix} a_2 & a_3 \\ b_2 & b_3 \end{vmatrix}^2 + \begin{vmatrix} a_1 & a_3 \\ b_1 & b_3 \end{vmatrix}^2 + \begin{vmatrix} a_1 & a_2 \\ b_1 & b_2 \end{vmatrix}^2.$$

2. Let $N = [a_2 b_3 - a_3 b_2, a_3 b_1 - a_1 b_3, a_1 b_2 - a_2 b_1]$. Then N is perpendicular to the plane of P (see Fig. D.2.) This follows from a straightforward calculation which shows that $A \cdot N = 0$ and $B \cdot N = 0$. Now, the altitude of the parallelepiped (from the base P) is given by

$$h = |C| \cos \phi = \frac{C \cdot N}{|N|}.$$

Finally, noting that $\text{area}^2(P) = |N|^2$, we compute the volume of our parallelepiped:

$$(\text{volume})^2 = h^2 \, \text{area}^2(P) = \left\{ \frac{C \cdot N}{|N|} \right\}^2 |N|^2 = (C \cdot N)^2.$$

Thus,

$$\text{volume} = \pm C \cdot N$$

$$= \pm [c_1, c_2, c_3] \cdot \left[\begin{vmatrix} a_2 & a_3 \\ b_2 & b_3 \end{vmatrix}, -\begin{vmatrix} a_1 & a_3 \\ b_1 & b_3 \end{vmatrix}, \begin{vmatrix} a_1 & a_2 \\ b_1 & b_2 \end{vmatrix} \right]$$

$$= \pm \begin{vmatrix} a_1 & a_2 & a_3 \\ b_1 & b_2 & b_3 \\ c_1 & c_2 & c_3 \end{vmatrix},$$

the sign depending on the relative positions of A, B, and C. Thus, the 3 by 3 determinant may be viewed (in absolute value) as the volume of a certain parallelepiped.

The results given above may be generalized. We cannot give the precise details here, but we can give a rough description of that generalization. First, working in an n-dimensional Euclidean space, we consider the generalized parallelepiped embraced by n linearly independent vectors $A^1, A^2, \ldots, A^n$. A notion of volume is defined for n-dimensional solids, and then it is shown that the volume of the generalized parallelepiped is given (in absolute value) by the determinant of the matrix whose rows are $A^1, A^2, \ldots, A^n$.

It is interesting to consider some of the basic properties of the determinant function with respect to this volume interpretation; we consider the third-order case:

1. If the given determinant has a zero row, then one of the three arrows embracing our parallelepiped is the zero arrow. It follows that the parallelepiped collapses to a parallelogram, and the *volume* of a parallelogram is surely zero.

2. If one row of a determinant is multiplied by $c \neq 0$, then one of the arrows is lengthened by a factor of $|c|$. This evidently increases the volume of the parallelepiped by a factor of $|c|$.

3. If two rows of a determinant are interchanged, then the arrows embracing our parallelepiped are simply reordered; the volume—hence the absolute value of the determinant—stays the same.

4. Adding a multiple of one row of a determinant to another row has the effect of *shearing* the parallelepiped embraced by the original three arrows. To see this consider Fig. D.3. Since the resulting parallelepiped has the same base and altitude as the original one, its volume is unchanged.

The reader who finds this brief discussion intelligible and interesting should look into the following text to find a more complete discussion:

Birkhoff, G., and MacLane, S., *A Survey of Modern Algebra*, Third edition, Macmillan Company, New York, 1965.

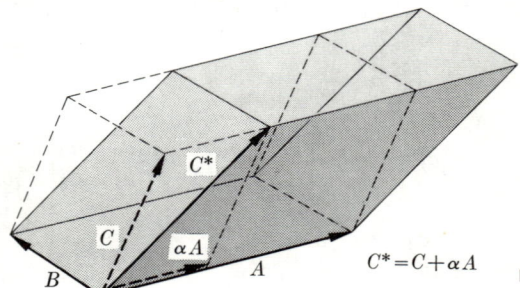

Figure D.3

APPENDIX E

AN APPLICATION FOR LINEAR TRANSFORMATIONS

The notion of a *tangent line* first appears in connection with the circle: the line tangent to a circle at a point P is defined to be the line through P that is perpendicular to the radius at P. This definition is extended, early in a first course in calculus, to include the case of a line tangent to the graph of an equation of the form $y = f(x)$, where f is a differentiable function. That definition is as follows: the line tangent to the graph of $y = f(x)$ at the point $(a, f(a))$ is given by the equation

$$y = f'(a)(x - a) + f(a)$$

(see Fig. E.1). Indeed, the desire to generalize the notion of a tangent line might well be regarded as the primary motivation for the "invention" of the derivative.

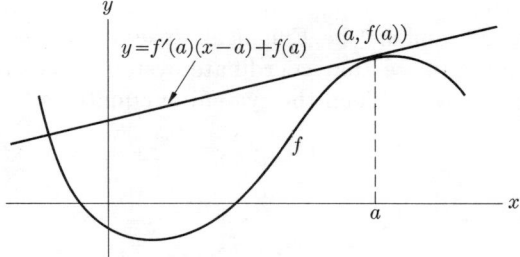

Figure E.1

In more advanced courses in calculus the notion of tangency is generalized even further: one considers the case of a line tangent to a curve that twists through three-dimensional space and also the case of a plane tangent to a curved surface. Ultimately, the notion of tangency is generalized to the point of abstraction, that is, an abstract concept of tangency evolves that is independent of any pictorial considerations. It is this ultimate notion of tangency that we wish to discuss at this point.

First, let us return to the case of a line tangent to $y = f(x)$ at the point $(a, f(a))$: the line $y = f'(a)(x - a) + f(a)$. Looking at pictures of such tangent lines leads to the following observation: If g is the function defined by

$$g(x) = f'(a)(x - a) + f(a),$$

then for values of x near to a the functions g and f produce values that are very nearly the same. For example, let $f(x) = x^3$ and let

$$g(x) = f'(1)(x - 1) + f(1)$$
$$= 3(x - 1) + 1.$$

If we evaluate each of these functions near $x = 1$, say at $x = 1.01$, then we get values that differ only in the fourth decimal place:

$$f(1.01) = 1.030301 \quad \text{and} \quad g(1.01) = 1.03.$$

In other words, the function g that gives rise to the line tangent to f at $(a, f(a))$ evidently will serve to approximate f for values of x near to a. Thus, the geometric notion of a tangent line may be tied to the analytic (numerical) notion of "good approximating function." It is this analytic counterpart of tangency that is most easily generalized, rather than tangency itself.

Now, it would take us too deeply into the study of calculus to say precisely what we mean above by the phrase "good approximating function." However, it is possible here simply to state the formula for this "good approximating function" in the general case and to consider an example.

Definition. Let E^n and E^m be Euclidean spaces of dimensions n and m, respectively, and suppose that coordinate systems have been introduced for each of these spaces. Then the system of equations

$$\begin{cases} y_1 = f^1(x_1, x_2, \ldots, x_n), \\ y_2 = f^2(x_1, x_2, \ldots, x_n), \\ \vdots \\ y_m = f^m(x_1, x_2, \ldots, x_n), \end{cases}$$

serves to associate with the coordinates of some X in E^n the coordinates of some Y in E^m. Thus, this system may be viewed as the coordinate manifestation of some vector function F whose domain is some subset of E^n and whose range is some subset of E^m (see Fig. E.2.) Next, taking

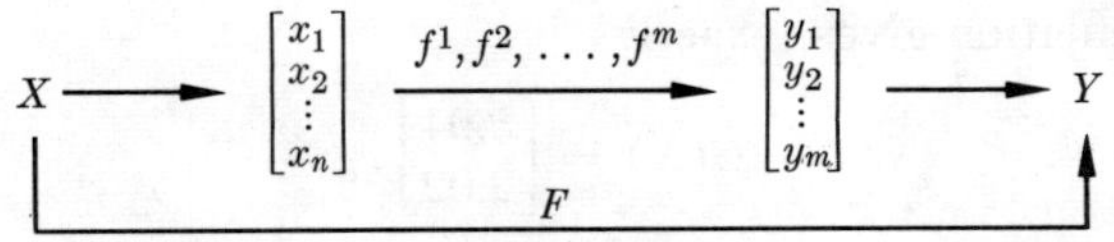

AN APPLICATION FOR LINEAR TRANSFORMATIONS

$$X \longrightarrow \begin{bmatrix} x_1 \\ x_2 \\ \vdots \\ x_n \end{bmatrix} \xrightarrow{f^1, f^2, \ldots, f^m} \begin{bmatrix} y_1 \\ y_2 \\ \vdots \\ y_m \end{bmatrix} \longrightarrow Y$$

$$F$$

Figure E.2

any A in the domain of F, we let G be defined by

$$G(X) = [f_j^i(A)]_{mn} \cdot (X - A) + F(A),$$

where

$$f_j^i(A) = \left. \frac{\partial f^i}{\partial x_j} \right|_{(a_1, a_2, \ldots, a_n)}.$$

Then G is said to be the *linear approximation of F at A*.

Example. Let $E^3 = (R^{31}, \cdot)$ and $E^2 = (R^{21}, \cdot)$, where the dot products are the standard ones, and suppose that in each space we have elected to use the natural-coordinate system. We then define F from E^3 into E^2 by the following natural-coordinate formula:

$$\begin{cases} y_1 = f^1(x_1, x_2, x_3) = x_1 x_2^2 + 3x_3, \\ y_2 = f^2(x_1, x_2, x_3) = 2x_1^3 - 5x_2 x_3. \end{cases}$$

We shall determine the linear approximation of F at $A = [1, 0, 2]$. There are two items that must be computed: $F(A)$ and $[f_j^i(A)]_{23}$.

1. $F(A) = \begin{bmatrix} f^1(A) \\ f^2(A) \end{bmatrix} = \begin{bmatrix} (1 \cdot 0^2) + (3 \cdot 2) \\ (2 \cdot 1^3) - 5(0 \cdot 2) \end{bmatrix} = \begin{bmatrix} 6 \\ 2 \end{bmatrix}.$

2. $f_1^1(A) = x_2^2 |_{(1,0,2)} = 0.$ $f_1^2(A) = 6x_1^2 |_{(1,0,2)} = 6.$

    $f_2^1(A) = 2x_1 x_2 |_{(1,0,2)} = 0.$ $f_2^2(A) = -5x_3 |_{(1,0,2)} = -10.$

    $f_3^1(A) = 3 |_{(1,0,2)} = 3.$ $f_3^2(A) = -5x_2 |_{(1,0,2)} = 0.$

Thus,

$$G(X) = \begin{bmatrix} 0 & 0 & 3 \\ 6 & -10 & 0 \end{bmatrix} (X - A) + \begin{bmatrix} 6 \\ 2 \end{bmatrix}.$$

As a test—to justify the naming of G as a linear approximation of F at A—let us evaluate $F(X)$ and $G(X)$ for some X near to A, say $X = [1.02, 0.01, 1.98]'$. (Note that the phrase "X near to A" makes sense because we are working in a Euclidean space. How far is this X from A?) A tiresome calculation gives (exactly)

$$F(X) = \begin{bmatrix} 5.940102 \\ 2.023416 \end{bmatrix}.$$

A simple calculation gives (exactly)

$$G(X) = \begin{bmatrix} 5.94 \\ 2.02 \end{bmatrix}.$$

Thus, G does appear to approximate F, at least for X near to A.

Finally, it should be noted that the formula for G can be written as

$$G(X) = \begin{bmatrix} 0 & 0 & 3 \\ 6 & -10 & 0 \end{bmatrix} X - \begin{bmatrix} 0 & 0 & 3 \\ 6 & -10 & 0 \end{bmatrix} A + \begin{bmatrix} 6 \\ 2 \end{bmatrix}$$

$$= \begin{bmatrix} 0 & 0 & 3 \\ 6 & -10 & 0 \end{bmatrix} X + \begin{bmatrix} 0 \\ -4 \end{bmatrix}$$

Thus, G may be viewed as the composition of two transformations: a linear transformation followed by a *translation*. Such a transformation is said to be an *affine* transformation. ▯

The reader that finds this brief discourse both intelligible and interesting should look into the following undergraduate texts:

1. Hummel, James A., *Introduction to Vector Functions*, Addison-Wesley, Reading, Mass., 1967.

2. Crowell, R. H., and Williams, R. E., *Calculus of Vector Functions*, Revised edition, Prentice-Hall, Englewood Cliffs, N.J., 1968.

ANSWERS TO SELECTED EXERCISES

SECTION 1.1

1 a) $\begin{bmatrix} -2 & 0 & 2 \\ -3 & 0 & 3 \\ -4 & 0 & 4 \end{bmatrix}$ b) $\begin{bmatrix} 0 & 1 & 2 & 3 & 4 \\ 1 & 0 & 1 & 2 & 3 \\ 2 & 1 & 0 & 1 & 2 \end{bmatrix}$

2 $AB = \begin{bmatrix} 4 & 0 & -4 & -4 & -4 \\ 6 & 0 & -6 & -6 & -6 \\ 8 & 0 & -8 & -8 & -8 \end{bmatrix}$

$A(AB) = \begin{bmatrix} 8 & 0 & -8 & -8 & -8 \\ 12 & 0 & -12 & -12 & -12 \\ 16 & 0 & -16 & -16 & -16 \end{bmatrix}$

$AB + 2B = \begin{bmatrix} 4 & 2 & 0 & 2 & 4 \\ 8 & 0 & -4 & -2 & 0 \\ 12 & 2 & -8 & -6 & -4 \end{bmatrix}$

3 $A(2A + B) = 2A^2 + AB = \begin{bmatrix} 15 & 18 & 21 \\ 30 & 36 & 42 \\ 45 & 54 & 63 \end{bmatrix}$

4 a) $\begin{bmatrix} 1 & 0 & 0 & 0 \\ 0 & 1 & 0 & 0 \\ 0 & 0 & 1 & 0 \end{bmatrix}$ b) $\begin{bmatrix} 1 & 0 & 0 & 0 \\ 0 & 2 & 0 & 0 \\ 0 & 0 & 3 & 0 \\ 0 & 0 & 0 & 4 \end{bmatrix}$ c) $\begin{bmatrix} 0 & 5 & 0 \\ 0 & 5 & 0 \\ 0 & 5 & 0 \end{bmatrix}$

6 a) $\begin{bmatrix} 0 & -1 & -2 \\ 1 & 0 & -1 \\ 2 & 1 & 0 \end{bmatrix}$

7 a) $AB = \begin{bmatrix} 19 & 22 \\ 43 & 50 \end{bmatrix}, BA = \begin{bmatrix} 23 & 34 \\ 31 & 46 \end{bmatrix}$
b) $AB = \Theta$, but $A \neq \Theta$ and $B \neq \Theta$.

296 ANSWERS TO SELECTED EXERCISES

c) $A^2 = A$, but $A \neq I$ and $A \neq \Theta$.

d) $AB = [14]$, $BA = \begin{bmatrix} 0 & 0 & 0 \\ 6 & 18 & -24 \\ 1 & 3 & 4 \end{bmatrix}$

8 a) $D = A + 2B + 3C$
b) $D = (A + B)C$
c) $D = (CA)B$

9 a) $B = [\delta_{pj}]_{1m}$
b) $C = [\delta_{iq}]_{n1}$

SECTION 1.2

1 One illustration:

$$(A + B) + C = \begin{bmatrix} 6 & 5 \\ 5 & 5 \end{bmatrix} + \begin{bmatrix} 1 & 2 \\ 3 & 0 \end{bmatrix} = \begin{bmatrix} 7 & 7 \\ 8 & 5 \end{bmatrix}$$

$$A + (B + C) = \begin{bmatrix} 1 & 2 \\ 3 & 4 \end{bmatrix} + \begin{bmatrix} 6 & 5 \\ 5 & 1 \end{bmatrix} = \begin{bmatrix} 7 & 7 \\ 8 & 5 \end{bmatrix}$$

2 a) $A = \begin{bmatrix} 4 & 1 & \frac{3}{2} \\ 3 & 2 & \frac{7}{2} \\ 2 & \frac{1}{2} & \frac{5}{2} \end{bmatrix}$ b) $A = \begin{bmatrix} -17 & 0 \\ 0 & -17 \end{bmatrix}$

3 a) $A = \begin{bmatrix} \frac{8}{7} & \frac{12}{7} \\ \frac{16}{7} & \frac{20}{7} \end{bmatrix}$, $B = \begin{bmatrix} -\frac{9}{7} & -\frac{10}{7} \\ -\frac{11}{7} & -\frac{12}{7} \end{bmatrix}$

b) $A = \begin{bmatrix} -1 & -\frac{3}{2} \\ -1 & -1 \end{bmatrix}$, $B = \begin{bmatrix} 4 & -7 \\ -9 & -2 \end{bmatrix}$

4 a) $A = \begin{bmatrix} 5 & -2 \\ -2 & 1 \end{bmatrix}$ b) $B = A$ c) $C = \begin{bmatrix} 1 & 18 \\ 1 & -7 \end{bmatrix}$

5 b) Same as answer to part (a).
c) $A = \Theta_{22}$
d) Disproved by first sentence in exercise.

6 a) $d_{ij} = \sum_{k=1}^{n} a_{ik}(b_{kj} + c_{kj})$

e) $A(B + C) = AB + AC$

7 a) $A = \begin{bmatrix} 2 & 0 & 0 \\ 0 & 5 & 0 \\ 0 & 0 & 7 \end{bmatrix}$ b) $B = \begin{bmatrix} \frac{1}{2} & 0 & 0 \\ 0 & \frac{1}{5} & 0 \\ 0 & 0 & \frac{1}{7} \end{bmatrix}$

c) $AC = \begin{bmatrix} 2 & 6 & 4 \\ 20 & 0 & 30 \\ 49 & 7 & 7 \end{bmatrix}$, $CA = \begin{bmatrix} 2 & 15 & 14 \\ 8 & 0 & 42 \\ 14 & 5 & 7 \end{bmatrix}$

In $[d_j \, \delta_{ij}]C$, the first row of C is multiplied by d_1, the second by d_2, and so on. In $C[d_j \, \delta_{ij}]$, the first column of C is multiplied by d_1, the second by d_2, and so on.

ANSWERS TO SELECTED EXERCISES   297

d) Let $C = AB$. Then

$$c_{ij} = \sum_{k=1}^{n} a_{ik}b_{kj} = \sum_{k=1}^{n} (d_k\delta_{ik})(e_j\delta_{kj}) = d_je_j\delta_{ij}.$$

8 a) $A = \begin{bmatrix} 1 & 2 & 3 \\ 2 & 4 & 6 \\ 3 & 6 & 9 \end{bmatrix}$, $B = \begin{bmatrix} -2 & 3 & -4 \\ 3 & 4 & 5 \\ -4 & 5 & -6 \end{bmatrix}$

b) $A + B = \begin{bmatrix} -1 & 5 & -1 \\ 5 & 8 & 11 \\ -1 & 11 & 3 \end{bmatrix}$, $AB = \begin{bmatrix} -8 & 26 & -12 \\ -16 & 52 & -24 \\ -24 & 87 & 36 \end{bmatrix}$

c) No. Let $C = A + B$. Then

$$c_{ji} = a_{ji} + b_{ji} = a_{ij} + b_{ij} = c_{ij}.$$

Therefore, C is symmetric.

9 a) $T = \begin{bmatrix} 1 & 0 & 0 \\ 1 & 1 & 0 \\ 1 & 1 & 1 \end{bmatrix}$ b) $A = \begin{bmatrix} 3 & 0 & 0 \\ 4 & 6 & 0 \\ 5 & 7 & 9 \end{bmatrix}$ c) $B = \begin{bmatrix} 1 & 0 & 0 \\ 4 & 8 & 0 \\ 9 & 18 & 27 \end{bmatrix}$

d) $A + B = \begin{bmatrix} 4 & 0 & 0 \\ 8 & 14 & 0 \\ 14 & 25 & 36 \end{bmatrix}$, $AB = \begin{bmatrix} 3 & 0 & 0 \\ 28 & 48 & 0 \\ 114 & 218 & 243 \end{bmatrix}$

e) Sums and products of lower triangular matrices are themselves lower triangular.

f) In Definition 1.10, replace "$i < j$" by "$i > j$" and "lower" by "upper."

SECTION 1.3

1 a) $\begin{bmatrix} 1 & 0 & 0 & 0 \\ 0 & 0 & 0 & 1 \\ 0 & 0 & 1 & 0 \\ 0 & 1 & 0 & 0 \end{bmatrix}$ b) $\begin{bmatrix} 1 & 0 & 0 & 0 \\ 0 & 1 & 0 & 0 \\ 0 & 0 & 2 & 0 \\ 0 & 0 & 0 & 1 \end{bmatrix}$ c) $\begin{bmatrix} 1 & 0 & 0 & 0 \\ 0 & 1 & 0 & -3 \\ 0 & 0 & 1 & 0 \\ 0 & 0 & 0 & 1 \end{bmatrix}$

2 $E_1A = \begin{bmatrix} 1 & 2 \\ 1 & 3 \\ 2 & 0 \\ 4 & -3 \end{bmatrix}$, $E_2A = \begin{bmatrix} 1 & 2 \\ 4 & -3 \\ 4 & 0 \\ 1 & 3 \end{bmatrix}$, $E_3A = \begin{bmatrix} 1 & 2 \\ 1 & -12 \\ 2 & 0 \\ 1 & 3 \end{bmatrix}$

3 a) $\begin{bmatrix} 1 & 0 & 0 \\ 0 & 1 & 0 \\ 0 & 0 & -13 \end{bmatrix}$ b) $\begin{bmatrix} 0 & 0 & 1 \\ 0 & 1 & 0 \\ 1 & 0 & 0 \end{bmatrix}$ c) $\begin{bmatrix} 1 & 0 & 0 \\ 0 & 1 & 0 \\ 0 & 5 & 1 \end{bmatrix}$

4 a) $AE_1 = \begin{bmatrix} 2 & 1 & 26 \\ 3 & 0 & -65 \end{bmatrix}$, $AE_2 = \begin{bmatrix} -2 & 1 & 2 \\ 5 & 0 & 3 \end{bmatrix}$,

$AE_3 = \begin{bmatrix} 2 & -9 & -2 \\ 3 & 25 & 5 \end{bmatrix}$

b) Do unto the columns of the identity matrix that which you would have the resulting elementary matrix do (as a postmultiplier) unto the columns of A.

c) $E = \begin{bmatrix} 1 & 0 & 0 \\ 0 & 1 & 0 \\ 4 & 0 & 1 \end{bmatrix}$

6 a) $\begin{bmatrix} 1 & 0 & 0 \\ 0 & \frac{1}{2} & 0 \\ 0 & 0 & \frac{1}{3} \end{bmatrix}$ b) $\begin{bmatrix} 1 & 0 & 0 \\ 0 & 1 & -1 \\ 0 & 0 & 1 \end{bmatrix}$ c) $\begin{bmatrix} 1 & 0 & 0 \\ 0 & 0 & 1 \\ 0 & 1 & 0 \end{bmatrix}$

7 $E_1 = [i \leftrightarrow j]_n$, $E_1^{-1} = [i \leftrightarrow j]_n$
$E_2 = [i \to (c)i]_n$, $E_2^{-1} = [i \to (1/c)i]_n$
$E_3 = [i \to i + (c)k]_n$, $E_3^{-1} = [i \to i + (-c)k]_n$

8 a) $B = IB = (CA)B = C(AB) = CI = C$. Thus, $B = C$.
b) $B = BI = B(AC) = (BA)C = IC = C$. Thus, $B = C$.
Returning to our hypotheses, we can write $CA = I$ and $AC = I$. It follows that $C = A^{-1}$.

9 Assume that A is nonsingular. Then A^{-1} exists and we can write
$$AB = \Theta, \quad A^{-1}(AB) = A^{-1}(\Theta),$$
$$(A^{-1}A)B = \Theta, \quad IB = \Theta, \quad B = \Theta.$$
But $B \neq \Theta$. Therefore, A is singular.

SECTION 1.4

1 a) $A^{-1} = \begin{bmatrix} -5 & 2 \\ 3 & -1 \end{bmatrix}$, $B^{-1} = \begin{bmatrix} -1 & -1 \\ -1 & -2 \end{bmatrix}$, $(AB)^{-1} = \begin{bmatrix} 2 & -1 \\ -1 & 0 \end{bmatrix}$.
Then $(AB)^{-1} = B^{-1}A^{-1}$.

b) To prove that $(AB)^{-1} = B^{-1}A^{-1}$, we must show that
$$(AB)(B^{-1}A^{-1}) = (B^{-1}A^{-1})(AB) = I.$$
We do here just one of the necessary calculations:
$$(AB)(B^{-1}A^{-1}) = A[B(B^{-1}A^{-1})]$$
$$= A[(BB^{-1})A^{-1}]$$
$$= A[IA^{-1}] = AA^{-1} = I.$$

2 a) $A = \begin{bmatrix} 2 & 3 & 4 \\ 3 & 6 & 10 \\ 4 & 10 & 20 \end{bmatrix}$, $A^{-1} = \begin{bmatrix} 5 & -5 & \frac{3}{2} \\ -5 & 6 & -2 \\ \frac{3}{2} & -2 & \frac{3}{4} \end{bmatrix}$

b) $B = \begin{bmatrix} 1 & 2 & 3 & 4 \\ 0 & 1 & 1 & 2 \\ 0 & 0 & 1 & 1 \\ 0 & 0 & 0 & 1 \end{bmatrix}$, $B^{-1} = \begin{bmatrix} 1 & -2 & -1 & 1 \\ 0 & 1 & -1 & -1 \\ 0 & 0 & 1 & -1 \\ 0 & 0 & 0 & 1 \end{bmatrix}$

c) Conjecture: The inverse of a symmetric (triangular) matrix is also symmetric (triangular).

3 a) $A^{-1} = \begin{bmatrix} -3 & -1 & 2 \\ 6 & 3 & -4 \\ -1 & 1 & 1 \end{bmatrix}$, $X = \begin{bmatrix} 17 \\ -34 \\ 7 \end{bmatrix}$

b) $A^{-1} = \frac{1}{2} \begin{bmatrix} 1 & -1 & 1 & -1 \\ 1 & 1 & -1 & 1 \\ -1 & 1 & 1 & -1 \\ 1 & -1 & 1 & 1 \end{bmatrix}$, $X = \begin{bmatrix} -1 \\ 2 \\ 0 \\ 3 \end{bmatrix}$

4 $A^{-1} = \begin{bmatrix} -3 & 6 & -1 \\ -1 & 3 & 1 \\ 2 & -4 & 1 \end{bmatrix}$, $X = [17, -34, 7]$

5 $\begin{bmatrix} 1 & -2 & 3 \\ 2 & 7 & -1 \\ 4 & 3 & 5 \end{bmatrix} \stackrel{\text{(row)}}{=} \begin{bmatrix} 1 & -2 & 3 \\ 0 & 11 & -7 \\ 0 & 11 & -7 \end{bmatrix} \stackrel{\text{(row)}}{=} \begin{bmatrix} 1 & -2 & 3 \\ 0 & 11 & -7 \\ 0 & 0 & 0 \end{bmatrix}$

6 In Definition 1.13, replace "$(E_n \cdots E_2 E_1)A$" by "$A(E_1 E_2 \cdots E_n)$" and "row" by "column."

$\begin{bmatrix} 1 & 2 & 1 \\ 5 & 7 & 5 \\ 9 & 3 & -1 \end{bmatrix} \stackrel{\text{(col)}}{=} \begin{bmatrix} 1 & 0 & 0 \\ 5 & -3 & 0 \\ 9 & -5 & -10 \end{bmatrix} \stackrel{\text{(col)}}{=} \begin{bmatrix} 1 & 0 & 0 \\ 2 & 3 & 0 \\ 4 & 5 & 10 \end{bmatrix}$

7 a) Let $C = AB$. Then for $j = 1, 2, \ldots, n$ we can write

$$c_{kj} = \sum_{h=1}^{n} a_{kh} b_{hj}$$

$$= \sum_{h=1}^{n} 0 \cdot b_{hj} = 0.$$

b) By part (a), there is no matrix B such that $AB = I$ (for I has no zero rows).

c) There exist elementary matrices $E^1, \ldots, E^k$ such that $A^* = E^k \cdots E^1 A$ has a zero row. Now each elementary matrix has an inverse. Suppose also that A^{-1} exists. Then it would follow from Theorem 1.17 that A^* would have an inverse. By part (b) this is impossible. (Note that our inverse-finding algorithm fails only when we run into a row of zeros.)

SECTION 1.5

1 a) $\begin{bmatrix} 1 & 2 & 0 & 3 & -1 \\ 4 & 8 & 1 & 14 & -3 \\ 3 & 6 & 1 & 11 & -2 \end{bmatrix} \stackrel{\text{(row)}}{=} \begin{bmatrix} 1 & 2 & 0 & 3 & -1 \\ 0 & 0 & 1 & 2 & 1 \\ 0 & 0 & 0 & 0 & 0 \end{bmatrix}$

c) $X = \begin{bmatrix} (-2s - 3t - 1) \\ s \\ (-2t + 1) \\ t \end{bmatrix}$

2 a) $X = \begin{bmatrix} -8 \\ -12 \\ 12 \end{bmatrix}$ b) $X = \begin{bmatrix} (-s + 4) \\ -s \\ s \end{bmatrix}$ c) $X = \begin{bmatrix} (-2s - 3t + 4) \\ s \\ t \end{bmatrix}$

3 a) $X = \begin{bmatrix} 2 \\ 3 \\ 4 \end{bmatrix}$ b) No solutions. c) $X = \begin{bmatrix} 2 \\ 3 \\ s \end{bmatrix}$

4 a) No solutions. b) $X = \begin{bmatrix} 2 \\ 3 \\ 4 \end{bmatrix}$ c) $X = \begin{bmatrix} (-2s+4) \\ (-3s+5) \\ s \end{bmatrix}$

5 a) $X = \begin{bmatrix} (-2s+5) \\ (-3s+6) \\ (-4s+7) \\ s \end{bmatrix}$ b) No solutions.

c) $X = \begin{bmatrix} (-2s-3t+4) \\ s \\ t \\ 5 \end{bmatrix}$

6 Suppose that the augmented matrix of such a system is reduced to row echelon form. Because of our second hypothesis there will be no row of the form $[0, 0, \ldots, 0, 1]$. Thus, the system has solutions. There will be a single solution only if every variable in the reduced system is a beginner variable. This situation occurs when the coefficient matrix of the reduced system is the identity matrix. In the present case, because of hypothesis 1, this matrix is not even square.

7 a) $X = \begin{bmatrix} (-2s+4) \\ (s-2) \\ (-s+1) \\ s \end{bmatrix}$ b) $X = \begin{bmatrix} (\frac{5}{2}s + \frac{1}{2}t + \frac{3}{2}) \\ (-\frac{5}{2}s - \frac{3}{2}t + \frac{1}{2}) \\ s \\ t \end{bmatrix}$

c) In the solution to part (b), take any s and t such that $s \neq -t + 1$.

SECTION 1.6

1 $B'A' = \begin{bmatrix} 4 & 3 & 2 \\ 0 & 1 & 2 \\ 1 & 3 & 5 \\ 2 & 1 & -1 \end{bmatrix} \begin{bmatrix} 1 & 2 \\ 7 & 0 \\ 3 & -4 \end{bmatrix} = \begin{bmatrix} 31 & 0 \\ 13 & -8 \\ 37 & -18 \\ 6 & 8 \end{bmatrix} = (AB)'$

2 Let $B = AA'$. Then

$$B' = (AA')' = (A')'A' \quad \text{(Theorem 1.11)}$$
$$= AA' = B.$$

It follows, by the converse part of Theorem 1.12, that B is symmetric.

3 a) $I = I' = (AA^{-1})' = (A^{-1})'A'$. But $A' = A$. Therefore, $I = (A^{-1})'A$. It follows that $(A^{-1})' = A^{-1}$.

b) $AA^{-1} = I$, $(AA^{-1})' = I'$, $(A^{-1})'A' = I$. Similarly, starting with $A^{-1}A = I$, we obtain $A'(A^{-1})' = I$. It follows that $(A^{-1})' = (A')^{-1}$.

4 Since this problem does not have a unique solution, the reader's solution may be different from the one given. In such an event the reader should simply check to see that BAB' is diagonal.

ANSWERS TO SELECTED EXERCISES   301

a) $\begin{bmatrix} 1 & 0 \\ -7 & 1 \end{bmatrix}$ b) $\begin{bmatrix} 1 & 0 & 0 \\ -2 & 1 & 0 \\ 1 & -2 & 3 \end{bmatrix}$ c) $\begin{bmatrix} 1 & 0 & 1 \\ -1 & 0 & 1 \\ 3 & -5 & 2 \end{bmatrix}$

d) $\begin{bmatrix} 1 & 0 & 0 \\ 5 & -4 & 0 \\ 10 & -11 & 15 \end{bmatrix}$ e) $\begin{bmatrix} 0 & 0 & 1 \\ 0 & 5 & -4 \\ 1 & 6 & -5 \end{bmatrix}$ f) $\begin{bmatrix} 1 & 1 & 0 \\ 1 & -1 & 0 \\ 3 & 2 & -1 \end{bmatrix}$

g) $\begin{bmatrix} 1 & 0 & 0 & 0 & 0 \\ -1 & 1 & 0 & 0 & 0 \\ -1 & 1 & 0 & 1 & 0 \\ 1 & 1 & -2 & 1 & 0 \\ 0 & 0 & 0 & 0 & 1 \end{bmatrix}$

5 b) Let B be any matrix such that BAB' is diagonal; and consider $C = EB$, where E is a "row multiplier." Then

$$CAC' = (EB)A(EB)' = E(BAB')E' = EDE'.$$

Now EDE' is diagonal. (Why?) Thus, every such E gives rise to a C such that CAC' is diagonal.

SECTION 1.7

1 $f(X) = X'AX$, where A is

a) $\begin{bmatrix} 1 & \frac{3}{2} & 0 \\ \frac{3}{2} & 5 & -1 \\ 0 & -1 & -1 \end{bmatrix}$ b) $\begin{bmatrix} 3 & \frac{5}{2} \\ \frac{5}{2} & 2 \end{bmatrix}$ c) $\begin{bmatrix} 1 & 3 & \frac{9}{2} \\ 3 & 5 & \frac{3}{2} \\ \frac{9}{2} & \frac{3}{2} & 0 \end{bmatrix}$

2 a) By the method of Section 1.6, we find that

$$BAB' = \begin{bmatrix} 1 & 0 & 0 \\ 0 & 2 & 0 \\ 0 & 0 & 2 \end{bmatrix}, \text{ where } B = \begin{bmatrix} 1 & 0 & 0 \\ -2 & 1 & 0 \\ -4 & -1 & 2 \end{bmatrix}.$$

It follows, taking $C = B'$, that $g(U) = f(CU) = U_1^2 + 2U_2^2 + 2U_3^2$.

b) Since $g(U)$ is a sum of squares, we may conclude that $g(U) \geq 0$ for all U. But also, such a sum can be zero only if each term is zero. Thus, $g(U) = 0$ only when $U = \Theta$. It follows that $g(U)$ is positive definite.

c) Yes. By Theorem 1.14 (or its corollary).

3 Take $B = \begin{bmatrix} 0 & 0 & 1 \\ 0 & 1 & -2 \\ 1 & -2 & 5 \end{bmatrix}$. Then $BAB' = \begin{bmatrix} 1 & 0 & 0 \\ 0 & 1 & 0 \\ 0 & 0 & 1 \end{bmatrix}$.

It follows, by Definition 1.19, that B is positive definite.

4 We note first that $f(x, y) = [x, y]\begin{bmatrix} a & b/2 \\ b/2 & c \end{bmatrix}\begin{bmatrix} x \\ y \end{bmatrix}$. Taking

$$\begin{bmatrix} x \\ y \end{bmatrix} = \begin{bmatrix} 1 & -b/2a \\ 0 & 1 \end{bmatrix}\begin{bmatrix} u \\ v \end{bmatrix},$$

we obtain

$$g(u, v) = au^2 + \left(\frac{4ac - b^2}{4a}\right)v^2.$$

Now suppose that f is positive definite. Then (Theorem 1.14) g is also. Thus, $g(1, 0) = a > 0$. Also, $g(0, 1) > 0$. Thus,

$$4a \cdot g(0, 1) = 4ac - b^2 > 0.$$

On the other hand, suppose that $a > 0$ and $4ac - b^2 > 0$. Then g is positive definite and (Theorem 1.14) f is also.

5 Taking $C = \begin{bmatrix} 1 & -2 & -3 \\ 0 & 1 & 1 \\ 0 & 0 & 1 \end{bmatrix}$, we have

$$g(U) = f(CU) = 2u_1^2 + u_2^2 + 0 \cdot u_3^2.$$

It follows immediately that $g(U) \geq 0$ for all U. Thus, g is positive semidefinite. By Theorem 1.14, f is also.

6 For *negative definite* replace the word "positive" by "negative" in Definition 1.19. For *negative semidefinite*, interchange the words "positive" and "negative" in Definition 1.20.

7 Taking $C = \begin{bmatrix} -1 & 0 & -1 \\ 1 & 1 & -2 \\ 0 & 0 & 1 \end{bmatrix}$, we have $g(U) = f(CU) = -u_1^2 + 4u_2^2$.

SECTION 1.8

1 a) No; 1, 2   b) Yes   c) Yes   d) No; 2
   e) No; 3   f) Yes   g) Yes

2 $S_6 = \{30, 60, 90, 120, 150, 240, \ldots\}$

3 a) $2, 2 \pm 2\pi, 2 \pm 4\pi$
   b) If $a \sim b$, then $\sin a = \sin b$.

4 a) Let $C = I$. Then $CAC' = A$. Thus, $A \sim A$. Next, suppose that $A \sim B$. Then there is a nonsingular matrix C such that $CAC' = B$. It follows that $C^{-1}B(C^{-1})' = A$. Thus, $B \sim A$. Finally, let $A \sim B$ and $B \sim C$. Then there exist nonsingular matrices G and H such that $GAG' = B$ and $HBH' = C$. It follows that $H(GAG')H' = C$. That is,

$$(HG)A(HG)' = C.$$

Moreover, since G and H are both nonsingular, then so is HG. Thus, $A \sim C$.

   b) Let $f(X) = X'AX$. In determining the sign behavior of the quadratic form f, we may replace A by any matrix that is congruent to A.

   c) Using the method of Section 1.6, show that both matrices are congruent to the same diagonal matrix. Then use the fact that congruence is an equivalence relation.

5 a) The function f is defined except where $t^2 - 5t + 6 = 0$. Thus, f is defined except for $t = 2$ and 3. It follows that f is defined in a deleted neighborhood of 2.

b) For all $t < 3$ except $t = 2$ we may write
$$f(t) = (3-t)\frac{(t-2)(t+2)}{(t-2)(t-3)} = -t-2.$$
Letting $g(t) = -t-2$, we have $g \stackrel{2}{=} f$.

c) Since $\lim_{t \to 2} g(t) = -4$, it follows that
$$\lim_{t \to 2} f(t) = -4.$$

SECTION 2.1

1 a) The function f defined by $f(t) = 0$ for all t.
b) Let $g(t) = -f(t)$. Then g is the negative of f.
c) Let $h = f \oplus g$. Then
$$h''(t) - h'(t) - 2h(t) = [f''(t) + g''(t)] - [f'(t) + g'(t)] - 2[f(t) + g(t)]$$
$$= [f''(t) - f'(t) - 2f(t)] + [g''(t) - g'(t) - 2g(t)]$$
$$= 0 + 0 = 0.$$
Thus, h is a solution of $y'' - y' - 2y = 0$. (Similarly with $h = c \odot f$.)

2 Axioms 7 and 8 are violated.

3 Note that $(x, y) \oplus (1, 1) = (x, y)$ for *every* pair (x, y), and that $(1, 1)$ is the only pair filling this role. Thus, $(1, 1)$ plays the role of the *zero vector* in this structure. However, it then happens that not every pair has a *negative*; for example, there is no pair (x, y) such that $(4, 0) \oplus (x, y) = (1, 1)$. Thus, Axiom 4 is violated. Also, Axioms 5 and 6 are violated.

4 Axioms 5 and 6 are violated.

5 a) $c \odot \Theta = (c+0) \odot \Theta = [c \odot \Theta] \oplus [0 \odot \Theta]$, using Axiom 6. Now add $-[c \odot \Theta]$ to both sides.
b) $X = 1 \odot X = (1+0) \odot X = [1 \odot X] \oplus [0 \odot X] = X \oplus [0 \odot X]$, using Axioms 6 and 8. Now add $-X$ to both sides.
c) $[1 \odot X] \oplus [-1 \odot X] = [1 + (-1)] \odot X = 0 \odot X = \Theta$, using Axiom 6 and Exercise 5(a).

6 From *loner* $\oplus$ *loner* = *loner* it follows that $X \oplus$ *loner* = X for every X in V, there being only one element in V. Therefore, *loner* is the zero vector required by Axiom 3. Consider next Axiom 4. In this case the only requirement made by Axiom 4 is that there be a unique X such that *loner* $\oplus X = \Theta$, that is, that *loner* $\oplus X =$ *loner*. Evidently, $X =$ *loner* satisfies this requirement. All other axioms are satisfied in a completely trivial manner: all computations lead to the answer *loner*.

7 A suitable zero vector is given by the function f defined by $f(t) = 0$. The negative of a given f is given by g, where $g(t) = -f(t)$. Since $\oplus$ and $\odot$ are defined in terms of the ordinary addition and multiplication of real numbers, all of the other axioms are fulfilled.

ANSWERS TO SELECTED EXERCISES

SECTION 2.2

1 a) A 3-by-1 matrix is in W if its 3rd-row entry is 0. Now

$$\begin{bmatrix} a \\ b \\ 0 \end{bmatrix} + \begin{bmatrix} \alpha \\ \beta \\ 0 \end{bmatrix} = \begin{bmatrix} (a+\alpha) \\ (b+\beta) \\ 0 \end{bmatrix} \quad \text{and} \quad c \begin{bmatrix} a \\ b \\ 0 \end{bmatrix} = \begin{bmatrix} ac \\ bc \\ 0 \end{bmatrix}.$$

These calculations show that the sum of any two vectors in W is in W, and that every multiple of a vector in W is in W. Thus, by Theorem 2.1, W is a subspace of V.

b) $[2, 2, 4]'$ is in W, but $3[2, 2, 4]' = [6, 6, 12]'$ is not. (Because $12 \neq 6^2$.) Thus, the operation $c \odot X$ cannot always be performed in W. Thus, W is not a subspace of V.

c) Reasoning is as in 1(a). Two calculations are required:

$$\begin{bmatrix} a \\ b \\ (3a+4b) \end{bmatrix} + \begin{bmatrix} \alpha \\ \beta \\ (3\alpha+4\beta) \end{bmatrix} = \begin{bmatrix} (a+\alpha) \\ (b+\beta) \\ 3(a+\alpha)+4(b+\beta) \end{bmatrix}$$

and

$$c \begin{bmatrix} a \\ b \\ (3a+4b) \end{bmatrix} = \begin{bmatrix} ac \\ bc \\ (3ac+4bc) \end{bmatrix}.$$

2 a) W is a subspace; reasoning is as in 1(a).

b) If $p(t)$ is in W, then $(-1) \odot p(t)$ is not. Thus, the negatives of the vectors in W are not in W. It follows that W is not a subspace of V.

c) W is a subspace; reasoning is as in 1(a).

d) W is not a subspace of V for the addition $(2t^2 - 1) \oplus (-t^2 + 2)$ cannot be performed in W. (Why?)

3 a) W is a subspace; reasoning is as in 1(a).

b) Same as (a).

c) W is not a subspace of V for the addition $\begin{bmatrix} 1 & 0 \\ 0 & 0 \end{bmatrix} \oplus \begin{bmatrix} 0 & 0 \\ 0 & 1 \end{bmatrix}$ cannot be performed in W. (Why?)

d) Same as (c). Consider the sum $\begin{bmatrix} 1 & 3 \\ 1 & 3 \end{bmatrix} \oplus \begin{bmatrix} -3 & -1 \\ -3 & -1 \end{bmatrix}$.

4 a) See final paragraph under Example 4.

b) No. (Recall Exercise 6 of Section 2.1.)

c) No. For suppose that X^* is a solution of $AX = B$. Then $A(2X^*) = 2(AX^*) = 2B \neq B$. Thus, the set of solutions of $AX = B$ is not closed with respect to numerical multiplication.

SECTION 2.3

1 a)

$$\text{Id}(X) + \text{Id}(Y) = \begin{bmatrix} 1 \\ 2 \\ 3 \end{bmatrix} + \begin{bmatrix} 7 \\ 4 \\ -2 \end{bmatrix} = \begin{bmatrix} 8 \\ 6 \\ 1 \end{bmatrix}.$$

$$\text{Id}(X + Y) = \text{Id}\left(\begin{bmatrix} 8 & 6 \\ 6 & 1 \end{bmatrix}\right) = \begin{bmatrix} 8 \\ 6 \\ 1 \end{bmatrix}.$$

b)
$$17 \, \text{Id}(X) = 17 \begin{bmatrix} 1 \\ 2 \\ 3 \end{bmatrix} = \begin{bmatrix} 17 \\ 34 \\ 51 \end{bmatrix}.$$

$$\text{Id}(17X) = \text{Id}\left(\begin{bmatrix} 17 & 34 \\ 34 & 51 \end{bmatrix}\right) = \begin{bmatrix} 17 \\ 34 \\ 51 \end{bmatrix}.$$

c) $\begin{bmatrix} 0 & 1 \\ 1 & 2 \end{bmatrix}$

d) Three

2 a) Let $p(t) = (t - 5)(at^2 + bt + c)$. Then $\text{Id}(p) = [a, b, c]'$.
Let $q(t) = (t - 5)(\alpha t^2 + \beta t + \gamma)$. Then $\text{Id}(q) = [\alpha, \beta, \gamma]'$.
Next, we observe that

$$p(t) + q(t) = (t - 5)[(a + \alpha)t^2 + (b + \beta)t + (c + \gamma)].$$

Thus, $\text{Id}(p + q) = \begin{bmatrix} (a + \alpha) \\ (b + \beta) \\ (c + \gamma) \end{bmatrix}$. The desired result follows.

b) $t^3 - 3t^2 - 7t - 15$

c) Three

3 It is evident that Id assigns different columns to different diagonal matrices. Thus, Id is one-to-one. It next must be established that Id has the two properties required of it by Definition 2.4. We consider here only the first property:

$$\text{Id}\left(\begin{bmatrix} a & & \\ & b & \\ & & c \end{bmatrix} + \begin{bmatrix} \alpha & & \\ & \beta & \\ & & \gamma \end{bmatrix}\right) = \text{Id}\left(\begin{bmatrix} (a + \alpha) & & \\ & (b + \beta) & \\ & & (c + \gamma) \end{bmatrix}\right)$$

$$= \begin{bmatrix} (a + \alpha) \\ (b + \beta) \\ (c + \gamma) \end{bmatrix} = \begin{bmatrix} a \\ b \\ c \end{bmatrix} \oplus \begin{bmatrix} \alpha \\ \beta \\ \gamma \end{bmatrix}$$

$$= \text{Id}\left(\begin{bmatrix} a & & \\ & b & \\ & & c \end{bmatrix}\right) \oplus \text{Id}\left(\begin{bmatrix} \alpha & & \\ & \beta & \\ & & \gamma \end{bmatrix}\right).$$

4 a) $\text{Id}(X) = X'$; dimension is three.

b) $\text{Id}\left(\begin{bmatrix} a & b \\ c & d \\ e & f \end{bmatrix}\right) = [a, b, c, d, e, f]'$; six.

c) $\text{Id}([a, a, a, b, b]) = \begin{bmatrix} a \\ b \end{bmatrix}$; two.

d) $\text{Id}(at^5 + bt) = \begin{bmatrix} a \\ b \end{bmatrix}$; two.

e) $\text{Id}(a \cos t + b \sin t) = \begin{bmatrix} a \\ b \end{bmatrix}$; two.

f) $\text{Id}(a) = [a]$; one.

g) $\mathrm{Id}[p(t)] = \mathrm{Id}[(t-2)(\alpha t^2 + \beta t + \gamma)] = \begin{bmatrix} \alpha \\ \beta \\ \gamma \end{bmatrix}$; three.

h) $\mathrm{Id}[p(t)] = \mathrm{Id}[(t-2)(t-3)(\alpha t + \beta)] = \begin{bmatrix} \alpha \\ \beta \end{bmatrix}$; two.

5 a) V is isomorphic to V. If V is isomorphic to W, then W is isomorphic to V. If V is isomorphic to W, and W is isomorphic to U, then V is isomorphic to U.

b) We consider here only the third theorem listed under (a): If V is isomorphic to W, and W to U, then there exist isomorphisms as indicated by the symbolism $V \xrightarrow{\mathrm{Id}_1} W \xrightarrow{\mathrm{Id}_2} U$. Now let Id be defined on V by the formula $\mathrm{Id}(X) = \mathrm{Id}_2[\mathrm{Id}_1(X)]$. Then the range of Id is the space U. For let Y be any vector in U and consider the equation $Y = \mathrm{Id}(X)$, X being the unknown. Equivalently, consider $Y = \mathrm{Id}_2[\mathrm{Id}_1(X)]$. Since isomorphisms have inverses, we may solve for the unique solution $X = \mathrm{Id}_1^{-1}[\mathrm{Id}_2^{-1}(Y)]$. This argument also shows that there is only one X whose Id-mate is Y; thus, Id is a one-to-one function from V onto U. We have yet to prove the arithmetical properties required of an isomorphism. Consider the following:

$$\begin{aligned}
\mathrm{Id}(X^1 + X^2) &= \mathrm{Id}_2[\mathrm{Id}_1(X^1 + X^2)] \\
&= \mathrm{Id}_2[\mathrm{Id}_1(X^1) + \mathrm{Id}_2(X^2)] \\
&= \mathrm{Id}_2[\mathrm{Id}_1(X^1)] + \mathrm{Id}_2[\mathrm{Id}_1(X^2)] \\
&= \mathrm{Id}(X^1) + \mathrm{Id}(X^2).
\end{aligned}$$

(We leave the other property to the reader.)

6 Let X be any vector in V^1 and let $Y = \mathrm{Id}(X)$. Then we may write, using Exercise 5 of Section 2.1,

$$\mathrm{Id}(\Theta_1) = \mathrm{Id}(0 \cdot X) = 0 \cdot \mathrm{Id}(X) = 0 \cdot Y = \Theta_2.$$

SECTION 2.4

1 Let the vectors of S, in the order given, be denoted by X^1, X^2, X^3, and X^4.

a) By inspection, $\begin{bmatrix} 1 & 2 \\ 3 & 4 \end{bmatrix} = 1X^1 + 2X^2 + 3X^3 + 4X^4$.

b) A typical vector in R^{22} is denoted by $X = \begin{bmatrix} \alpha & \beta \\ \gamma & \delta \end{bmatrix}$. By inspection, $X = \alpha X^1 + \beta X^2 + \gamma X^3 + \delta X^4$. This shows that every vector in R^{22} can be expressed as a linear combination of the vectors in S. Thus, S is a spanning set for R^{22}.

c) Set $aX^1 + bX^2 + cX^3 + dX^4 = \Theta$. Equivalently, carrying out the operations on the left side, we have

$$\begin{bmatrix} a & b \\ c & d \end{bmatrix} = \begin{bmatrix} 0 & 0 \\ 0 & 0 \end{bmatrix}.$$

Thus, $a = b = c = d = 0$. This shows that the only Θ-producing combination of vectors in S is the trivial one; it follows that S is linearly independent.

2 Set $a[4, 1, 0] + b[2, 1, 2] + c[2, 0, -2] = [\alpha, \beta, \gamma]$. The augmented matrix of the corresponding system of numerical equations is

$$\begin{bmatrix} 4 & 2 & 2 & \alpha \\ 1 & 1 & 0 & \beta \\ 0 & 2 & -2 & \gamma \end{bmatrix} \overset{(\text{row})}{=} \begin{bmatrix} 1 & 0 & 1 & (\tfrac{1}{2}\alpha - \beta) \\ 0 & 1 & -1 & (-\tfrac{1}{2}\alpha + 2\beta) \\ 0 & 0 & 0 & (\alpha - 4\beta + \gamma) \end{bmatrix}.$$

a) Taking $\alpha = 4$, $\beta = 3$, $\gamma = 8$ produces a system having infinitely many solutions; for example, take $a = -1$, $b = 4$, $c = 0$.
b) Taking $\alpha = 1$, $\beta = 0$, $\gamma = 0$ produces a system having no solution, for then $\alpha - 4\beta + \gamma \neq 0$. Thus, the vector $[1, 0, 0]$ is not "reached" by the vectors in S.
c) Taking $\alpha = \beta = \gamma = 0$ produces a system with solutions other than the trivial one, say $a = 1$, $b = -1$, $c = -1$.

3 Set $a(t) + b(t^2 + 1) + c(2t - 3) + d(2t - 4) = \alpha t^2 + \beta t + \gamma$. Matching off coefficients of like powers of t leads to the system whose augmented matrix is

$$\begin{bmatrix} 0 & 1 & 0 & 0 & \alpha \\ 1 & 0 & 2 & 2 & \beta \\ 0 & 1 & -3 & -4 & \gamma \end{bmatrix} \overset{(\text{row})}{=} \begin{bmatrix} 1 & 0 & 0 & -\tfrac{2}{3} & (-\tfrac{2}{3}\alpha + \beta + \tfrac{2}{3}\gamma) \\ 0 & 1 & 0 & 0 & (\alpha) \\ 0 & 0 & 1 & \tfrac{4}{3} & (\tfrac{1}{3}\alpha - \tfrac{2}{3}\gamma) \end{bmatrix}$$

a) Taking $\alpha = 3$, $\beta = \gamma = 0$ leads to a solution $a = -2$, $b = 3$, $c = 1$.
b) Take $a = -\tfrac{2}{3}\alpha + \beta + \tfrac{2}{3}\gamma$, $b = \alpha$, $c = \tfrac{1}{3}\alpha - \tfrac{1}{3}\gamma$, and $d = 0$. The corresponding linear combination produces $\alpha t^2 + \beta t + \gamma$. Thus, any vector in P^2 can be "reached" using the vectors in S.
c) Taking $\alpha = \beta = \gamma = 0$ permits the nontrivial solution $a = 2$, $b = 0$, $c = -4$, and $d = 3$.

4 a) Let A and B be in W. Then we can write

$$A = \sum_{j=1}^{n} a_j X^j \quad \text{and} \quad B = \sum_{j=1}^{n} b_j X^j.$$

It follows that

$$A + B = \sum_{j=1}^{n} (a_j + b_j) X^j;$$

that is, $A + B$ is a linear combination of the vectors X^j. Thus, $A + B$ is in W. Similarly, cA is in W, provided that A is. Then, by Theorem 2.1, W is a subspace of V.
b) Let the vectors of the given set be X^1, X^2, X^3, X^4, and solve the equation

$$aX^1 + bX^2 + cX^3 + dX^4 = [5, -1, 3, 8]'.$$

One solution is $a = -3$, $b = 1$, $c = 2$, $d = 0$.

5 a) $0[1, 2, 3] + 0[4, 5, 6] + 1[0, 0, 0] = \Theta$.
b) Let $S = \{X^1, \ldots, X^n, \Theta\}$. Then $0X^1 + \cdots + 0X^n + 1\Theta = \Theta$. It follows (by the definition of linear dependence) that S is linearly dependent.

6 a) Let X be linearly dependent upon $\{X^1, \ldots, X^n\}$. This means that there exist numbers $a_1, \ldots, a_n$ such that $X = a_1 X^1 + \cdots + a_n X^n$. It follows that $\Theta = -X + a_1 X^1 + \cdots + a_n X^n$. Since we have here a nontrivial combination of the vectors in $S = \{X, X^1, \ldots, X^n\}$, we have proved that S is linearly dependent.

b) Let $\{X^1, \ldots, X^n\}$ be linearly dependent. Then there exist numbers $a_1, \ldots, a_n$—not all zero—such that

$$a_1 X^1 + \cdots + a_n X^n = \Theta.$$

We may suppose (since the vectors may be taken in any order) that $a_1 \neq 0$. Then we may solve for X^1:

$$X^1 = -\frac{a_2}{a_1} X^2 - \cdots - \frac{a_n}{a_1} X^n.$$

Thus, at least one vector is linearly dependent upon the remaining vectors.

c) $\{X^1, X^2, X^3\}$ is linearly dependent because $2X^1 - X^2 + 0X^3 = \Theta$. X^2 is linearly dependent upon $\{X^1, X^3\}$ because $X^2 = 2X^1 + 0X^3$. X^3 is not linearly dependent upon $\{X^1, X^2\}$ because the equation $X^3 = aX^1 + bX^2$ leads to the augmented matrix

$$\begin{bmatrix} 1 & 2 & 7 \\ 2 & 4 & -11 \\ 3 & 6 & 46 \end{bmatrix} \stackrel{(\text{row})}{=} \begin{bmatrix} 1 & 2 & 7 \\ 0 & 0 & -18 \\ 0 & 0 & 32 \end{bmatrix}.$$

SECTION 2.5

1 The pattern of proof is the same for all three cases; we consider here only the proof of part (b). Let X^1, X^2, X^3, X^4 be the vectors of the given set, and consider the equation

$$aX^1 + bX^2 + cX^3 + dX^4 = \begin{bmatrix} \alpha & \beta \\ \gamma & \delta \end{bmatrix},$$

where the matrix on the right denotes a typical vector in R^{22}. Equivalently, substituting for the X^j and combining the terms on the left side, we have

$$\begin{bmatrix} a & b \\ c & d \end{bmatrix} = \begin{bmatrix} \alpha & \beta \\ \gamma & \delta \end{bmatrix}.$$

Thus, our equation has a unique solution: $a = \alpha$, $b = \beta$, $c = \gamma$, $d = \delta$. That there is a solution for every choice of α, β, γ, and δ, shows that the set $S = \{X^1, X^2, X^3, X^4\}$ is a spanning set for R^{22}. That the solution is unique shows that S is linearly independent. (Explain.) Thus, S is a basis for R^{22}.

2 a) $\left\{ \begin{bmatrix} 1 \\ 0 \\ 0 \\ 0 \end{bmatrix}, \begin{bmatrix} 0 \\ 1 \\ 0 \\ 0 \end{bmatrix}, \begin{bmatrix} 0 \\ 0 \\ 1 \\ 0 \end{bmatrix}, \begin{bmatrix} 0 \\ 0 \\ 0 \\ 1 \end{bmatrix} \right\}$

b) $\{[1, 0, 0, 0], [0, 1, 0, 0], [0, 0, 1, 0], [0, 0, 0, 1]\}$

c) $\left\{ \begin{bmatrix} 1 & 0 & 0 \\ 0 & 0 & 0 \end{bmatrix}, \begin{bmatrix} 0 & 1 & 0 \\ 0 & 0 & 0 \end{bmatrix}, \ldots, \begin{bmatrix} 0 & 0 & 0 \\ 0 & 0 & 1 \end{bmatrix} \right\}$

d) $\{t^4, t^3, t^2, t, 1\}$

3 A proof may be patterned after that of Example 3.

4 b) Since $\begin{bmatrix} 1 & 1 \\ 0 & 0 \end{bmatrix} - \begin{bmatrix} 0 & 1 \\ 0 & 1 \end{bmatrix} + \begin{bmatrix} 0 & 0 \\ 1 & 1 \end{bmatrix} - \begin{bmatrix} 1 & 0 \\ 1 & 0 \end{bmatrix} = \Theta$, the given set S is not linearly independent, hence not a basis for any space.

5 Proceeding as in Example 2, we determine that

$$S = \left\{ \begin{bmatrix} -\frac{3}{2} \\ 4 \\ 1 \\ 0 \\ 0 \end{bmatrix}, \begin{bmatrix} -\frac{1}{2} \\ -1 \\ 0 \\ 1 \\ 0 \end{bmatrix}, \begin{bmatrix} -1 \\ 6 \\ 0 \\ 0 \\ 1 \end{bmatrix} \right\}$$

is a basis for V. It follows, by Theorem 2.10, that V has dimension three.

SECTION 2.6

1 The set $T = \left\{ \begin{bmatrix} 1 \\ 2 \\ 3 \end{bmatrix}, \begin{bmatrix} 4 \\ 7 \\ 11 \end{bmatrix} \right\}$ is obviously linearly independent. (Why?) We try for three vectors: Consider the equation

$$a \begin{bmatrix} 1 \\ 2 \\ 3 \end{bmatrix} + b \begin{bmatrix} 4 \\ 7 \\ 11 \end{bmatrix} + c \begin{bmatrix} 4 \\ 8 \\ 12 \end{bmatrix} = \begin{bmatrix} 0 \\ 0 \\ 0 \end{bmatrix}.$$

The reader should verify that the only solution of this equation is the trivial one. The same result occurs if we replace $[4, 8, 12]'$ by $[3, 5, 8]'$. Thus, there is no linearly independent subset of S that properly contains T. By definition, then, T is a maximal independent subset of S.

2 a) The space R^{31} has dimension 3. By Theorems 2.13 and 2.15, it follows that T is linearly dependent.

b) From part (a) it follows that there is a nontrivial combination of the vectors in T that equals Θ. But then this same combination of the vectors in S will equal Θ.

c) $T = \left\{ \begin{bmatrix} 1 \\ 0 \\ 0 \\ 0 \end{bmatrix}, \begin{bmatrix} 1 \\ 2 \\ 0 \\ 0 \end{bmatrix}, \begin{bmatrix} 1 \\ 2 \\ 3 \\ 0 \end{bmatrix} \right\}$

d) By Theorem 2.12, we may take T as a basis for V.

e) By Theorem 2.10, dim $V = 3$.

3 A basis for R^{15} will have 5 vectors. Noting the location of the zero-entries in X^1, X^2, and X^3, we arrive at an easy choice for X^4 and X^5:

$$X^4 = [0, 0, 1, 0, 0] \quad \text{and} \quad X^5 = [0, 0, 0, 0, 1].$$

4 We complete the argument started in the text: Then there must be an X in V that is not dependent upon the X^i. Consider now the equation

$$a_1 X^1 + \cdots + a_n X^n + aX = \Theta.$$

Evidently, $a = 0$. (Why?) But also, since S is linearly independent, the remaining a_i are all zero. Thus, the set $\{X^1, \ldots, X^n, X\}$ is linearly independent. But this is impossible (Theorems 2.13 and 2.15). Thus, returning to the beginning, S must be a spanning set for V. And since S is an independent set, it must be a basis for V.

5 We complete the argument started in the text: Then a maximal independent subset of S can have at most $(n-1)$ vectors. It follows, Theorem 2.12, that V has a basis containing less than n vectors. Consequently, Theorem 2.10, V has dimension less than n, which contradicts our hypothesis.

6 a) Take the set corresponding to S under the isomorphism Id, that is, the set $T = \{t^3 + 2t + 1, t^3 + t^2 + 1, -t^2 + 2t\}$.
   b) Any two of the vectors in T.

SECTION 2.7

1 a) 4
   b) Let the first nonzero entry of A^j be its kth entry. Then the kth entry of every other A^i is zero. (Why?) It follows that A^j cannot be expressed as a linear combination of the other nonzero rows. But j is arbitrary. Therefore, using the contrapositive form of Theorem 2.8, the set

$$S = \{A^1, \ldots, A^n\}$$

is linearly independent. Finally, since S is a basis for the row space of A (why?), the dimension of that row space (the rank of A) is n.

2 a) As in Example 1, we consider

$$A = \begin{bmatrix} 1 & 2 & 0 & 3 \\ 3 & 6 & 1 & 9 \\ 0 & 1 & 0 & 2 \\ 3 & 4 & 1 & 5 \end{bmatrix} \underset{\text{(row)}}{\equiv} \begin{bmatrix} 1 & 2 & 0 & 3 \\ 0 & 1 & 0 & 2 \\ 0 & 0 & 1 & 0 \\ 0 & 0 & 0 & 0 \end{bmatrix} = B.$$

Evidently, the set $S = \{[1, 2, 0, 3], [0, 1, 0, 2], [0, 0, 1, 0]\}$ is a basis for the row space of B. It follows, Theorem 2.18, that the same set S is a basis for the row space of A. (There are, of course, infinitely many such bases.)
   b) $T = t^3 + 2t^2 + 3, t^2 + 2, t\}$. [*Hint:* Apply Theorem 2.14, defining Id by the formula $\text{Id}([a, b, c, d]) = at^3 + bt^2 + ct + d$].
   c) $U = \left\{ \begin{bmatrix} 1 & 3 \\ 2 & 0 \end{bmatrix}, \begin{bmatrix} 0 & 2 \\ 1 & 0 \end{bmatrix}, \begin{bmatrix} 0 & 0 \\ 0 & 1 \end{bmatrix} \right\}$

3 $\begin{bmatrix} 1 & 0 & 2 & 1 & 5 \\ 0 & 0 & 1 & 1 & 3 \\ 1 & 0 & 3 & 0 & 4 \\ 0 & 0 & 1 & 2 & 5 \\ 0 & 0 & 0 & 1 & 2 \end{bmatrix} \underset{\text{(row)}}{\equiv} \begin{bmatrix} 1 & 0 & 0 & 1 & 3 \\ 0 & 0 & 1 & 0 & 1 \\ 0 & 0 & 0 & 1 & 2 \\ 0 & 0 & 0 & 0 & 0 \\ 0 & 0 & 0 & 0 & 0 \end{bmatrix}$

   a) 3
   b) Any three-vector subset containing A^1 and A^3 is linearly independent; all others are dependent.

ANSWERS TO SELECTED EXERCISES   311

4 a) If B has k rows, then the appearance of a zero row implies that the row rank of B is less than k. This means (Theorem 2.18) that the row rank of A is less than k. It follows (by Theorem 2.13) that the rows of A are linearly dependent.

b) In this case the row rank of B and (by Theorem 2.18) that of A is k. Since the k rows of A span a space of dimension k, those rows (Theorem 2.17) form a basis for that space. But a basis is a linearly independent set.

5 $\begin{bmatrix} 1 & 0 & 3 & 2 \\ 1 & 0 & 2 & 1 \\ 0 & 1 & 0 & 1 \\ 1 & 3 & 3 & 5 \end{bmatrix} \underset{=}{\text{(row)}} \begin{bmatrix} 1 & 0 & 0 & -1 \\ 0 & 1 & 0 & 1 \\ 0 & 0 & 1 & 1 \\ 0 & 0 & 0 & 0 \end{bmatrix}$

a) Dependent

b) Independent. *Hint:* First use the isomorphism defined by

$$\text{Id}\left(\begin{bmatrix} a & b & c \\ d & e & f \end{bmatrix}\right) = [a, b, c, d, e, f].$$

SECTION 2.8

1 a) $\begin{bmatrix} 7+i & 8i \\ 4+4i & -3-10i \end{bmatrix}$ b) $\begin{bmatrix} 0 & 6 & 0 \\ 4 & 0 & 0 \end{bmatrix}$

2 a) $\begin{bmatrix} (2+2t^2+2e^t+2t\sin t) & (2t+2t\cos t) \\ (2\sin t + 2e^t \sin t) & (2t \sin t) \end{bmatrix}$

b) $\begin{bmatrix} 0 & 1 \\ \cos t & 0 \end{bmatrix}$

3 $\left[\begin{bmatrix} 2 & 4 \\ 6 & 8 \end{bmatrix} \begin{bmatrix} 3 & 2 \\ 0 & 5 \end{bmatrix}\right], \left[\begin{bmatrix} 6 & 18 \\ 11 & 29 \end{bmatrix} \begin{bmatrix} 9 & 7 \\ 15 & 10 \end{bmatrix}\right]$
$\left[\begin{bmatrix} 3 & 2 \\ 0 & 5 \end{bmatrix} \begin{bmatrix} 2 & 2 \\ 4 & 2 \end{bmatrix}\right], \left[\begin{bmatrix} 2 & 9 \\ 8 & 17 \end{bmatrix} \begin{bmatrix} 7 & 2 \\ 8 & 7 \end{bmatrix}\right]$

4 $A = \begin{bmatrix} -1 & 1 & 0 \\ 0 & -1 & 1 \\ -1 & -1 & 1 \end{bmatrix}, B = \begin{bmatrix} 0 & 1 & -1 \\ 1 & 1 & -1 \\ 1 & 2 & -1 \end{bmatrix}$

5 a) $\begin{bmatrix} 0 & 0 & 1 \\ 0 & 1 & 0 \\ 1 & 0 & 0 \end{bmatrix}$ b) $\begin{bmatrix} 0 & \frac{5}{16} & \frac{1}{16} \\ 0 & -\frac{1}{16} & \frac{3}{16} \\ \frac{1}{2} & -\frac{5}{32} & -\frac{1}{32} \end{bmatrix}$

SECTION 2.9

1 a) $X_{\mathcal{B}} = \begin{bmatrix} 2 \\ 1 \\ 4 \end{bmatrix}$ b) 5   c) $X = 2t^2 + 10t + 4$

2 a) $X_{\mathcal{B}} = \begin{bmatrix} 6 \\ -2 \end{bmatrix}$ b) $Y = \begin{bmatrix} 1 \\ 3 \\ 5 \end{bmatrix}$

c) $Z_\mathcal{B} = 2X_\mathcal{B} + 3Y_\mathcal{B} = 2\begin{bmatrix} 6 \\ -2 \end{bmatrix} + 3\begin{bmatrix} -1 \\ 2 \end{bmatrix} = \begin{bmatrix} 9 \\ 2 \end{bmatrix}$.

Therefore, $Z = [11, 13, 15]'$.

3 a) 3   b) 7

4 a) $X_\mathcal{B} = \begin{bmatrix} 10 \\ 9 \\ 7 \\ 4 \end{bmatrix}$ b) $X_e = \begin{bmatrix} -1 \\ 2 \\ -5 \\ 3 \end{bmatrix}$

c) $X_\mathcal{B} = \begin{bmatrix} a \\ b \\ c \\ d \end{bmatrix}$, $X_e = \begin{bmatrix} (a-b) \\ (b-c) \\ (c-d) \\ d \end{bmatrix}$

d) $A = \begin{bmatrix} 1 & -1 & 0 & 0 \\ 0 & 1 & -1 & 0 \\ 0 & 0 & 1 & -1 \\ 0 & 0 & 0 & 1 \end{bmatrix}$

5 e) $X_e = [\frac{4}{3}, \frac{1}{3}]'$

SECTION 2.10

1 a) $\mathcal{C} = \left[\begin{bmatrix} 12 \\ 11 \end{bmatrix}, \begin{bmatrix} 5 \\ 5 \end{bmatrix} \right]$ b) $X_e = \begin{bmatrix} -2 \\ 5 \end{bmatrix}$

c) $\begin{bmatrix} 1 & -1 \\ -2 & 3 \end{bmatrix}$ d) $X_e = \begin{bmatrix} 1 & -1 \\ -2 & 3 \end{bmatrix} \begin{bmatrix} -1 \\ 1 \end{bmatrix} = \begin{bmatrix} -2 \\ 5 \end{bmatrix}$

2 a) $X_e = \begin{bmatrix} 0 \\ 1 \\ 1 \end{bmatrix}$ b) $\mathcal{C} = \left[\begin{bmatrix} -1 \\ 0 \\ 0 \end{bmatrix}, \begin{bmatrix} 0 \\ 1 \\ 1 \end{bmatrix}, \begin{bmatrix} 1 \\ 1 \\ 2 \end{bmatrix} \right]'$

3 a) $G(f) = g(f_\mathcal{B}) = x_1^2 + 4x_1 x_2 + 4x_2^2$ b) $\mathcal{C} = [e^t, -2e^t + e^{2t}]$

4 a) $\begin{bmatrix} 1 & 2 \\ 2 & 8 \end{bmatrix}$ b) $\begin{bmatrix} 1 & -1 \\ 0 & \frac{1}{2} \end{bmatrix}$ c) $g\left(\begin{bmatrix} u_1 \\ u_2 \end{bmatrix} \right) = u_1^2 + u_2^2$

d) The quadratic forms corresponding to congruent matrices give rise to surfaces that—provided appropriate coordinate systems are chosen—are *congruent* in the geometric sense.

SECTION 2.11

1 a) A vector formula for L_1 is given by

$$X = \begin{bmatrix} 0 \\ 2 \\ 10 \\ 8 \end{bmatrix} + r \begin{bmatrix} 3 \\ 6 \\ 9 \\ 12 \end{bmatrix}.$$

ANSWERS TO SELECTED EXERCISES   313

For L_1 to be parallel to P it is necessary that the vector giving direction to L_1, $[3, 6, 9, 12]'$, be in the space spanned by $[2, 1, 0, 3]'$ and $[3, 3, 1, 2]'$. That this is not the case follows from the fact that

$$\begin{bmatrix} 3 & 6 & 9 & 12 \\ 2 & 1 & 0 & 3 \\ 3 & 3 & 1 & 2 \end{bmatrix} \stackrel{(\text{row})}{=} \begin{bmatrix} 1 & 2 & 3 & 4 \\ 0 & -3 & -6 & -5 \\ 0 & 0 & -2 & -5 \end{bmatrix}.$$

The desired point of intersection is found by solving the equation

$$\begin{bmatrix} 0 \\ 2 \\ 10 \\ 8 \end{bmatrix} + r \begin{bmatrix} 3 \\ 6 \\ 9 \\ 12 \end{bmatrix} = \begin{bmatrix} 2 \\ 3 \\ 6 \\ 1 \end{bmatrix} + s \begin{bmatrix} 2 \\ 1 \\ 0 \\ 3 \end{bmatrix} + t \begin{bmatrix} 3 \\ 3 \\ 1 \\ 2 \end{bmatrix}.$$

The solution is $r = -2/3$, $s = 1$, $t = -2$. Evaluating the left side at $r = -2/3$ gives the point $[-2, -2, 4, 0]'$.

b) Proceed as in part (a), noting that a vector formula for L_2 is given by

$$X = \begin{bmatrix} 1 \\ 2 \\ 3 \\ 4 \end{bmatrix} + r \begin{bmatrix} 1 \\ 2 \\ 1 \\ -2 \end{bmatrix}.$$

2 a) Solve $X^1 = X$, obtaining $r = 1$, $s = 0$, $t = 1$, $u = 0$. Evaluating X^1 at $r = 1$ gives the point $[3, 4, 4, 5]'$.
  b) Show that the equation $X^2 = X$ has no solution.
  c) A direct calculation reveals that

$$\begin{bmatrix} 3 \\ 6 \\ 6 \\ 1 \end{bmatrix} = 2 \begin{bmatrix} 1 \\ 2 \\ 1 \\ 0 \end{bmatrix} - 1 \begin{bmatrix} 2 \\ 2 \\ 1 \\ 1 \end{bmatrix} + 1 \begin{bmatrix} 3 \\ 4 \\ 5 \\ 2 \end{bmatrix}.$$

Thus, a new formula for L_3 is given by

$$X^3 = \begin{bmatrix} 1 \\ 2 \\ 3 \\ 4 \end{bmatrix} + 2r \begin{bmatrix} 1 \\ 2 \\ 1 \\ 0 \end{bmatrix} - r \begin{bmatrix} 2 \\ 2 \\ 1 \\ 1 \end{bmatrix} + r \begin{bmatrix} 3 \\ 4 \\ 5 \\ 2 \end{bmatrix}.$$

Compare this formula with that for P.

  d) Definition: L is parallel to P if A is in the space spanned by $\{D, E, F\}$. The fact that L_2 and L_3 are parallel to P then follows from

$$\begin{bmatrix} 3 \\ 6 \\ 6 \\ 1 \end{bmatrix} = 2 \begin{bmatrix} 1 \\ 2 \\ 1 \\ 0 \end{bmatrix} - 1 \begin{bmatrix} 2 \\ 2 \\ 1 \\ 1 \end{bmatrix} + 1 \begin{bmatrix} 3 \\ 4 \\ 5 \\ 2 \end{bmatrix}.$$

3 A typical point on P_1 is given by $A = [a + b + 1, 2a + b + 2, 2b + 3]'$. (Why?) We wish to show that this point is also on P_2. To do so we have only to show that the equation $X^2 = A$ has a solution. Then we reverse this process, showing that a typical point P_2 is also on P_1.

4 a) Let $[\alpha, \beta, \gamma]'$ be on the given 2-plane. Then the equation

$$\begin{bmatrix} 1 \\ 2 \\ 3 \end{bmatrix} + s \begin{bmatrix} 1 \\ 0 \\ 2 \end{bmatrix} + t \begin{bmatrix} 0 \\ 1 \\ 1 \end{bmatrix} = \begin{bmatrix} \alpha \\ \beta \\ \gamma \end{bmatrix}$$

has a solution. Now

$$\begin{bmatrix} 1 & 0 & (\alpha - 1) \\ 0 & 1 & (\beta - 2) \\ 2 & 1 & (\gamma - 3) \end{bmatrix} \stackrel{(\text{row})}{\simeq} \begin{bmatrix} 1 & 0 & (\alpha - 1) \\ 0 & 1 & (\beta - 2) \\ 0 & 0 & (-2\alpha - \beta + \gamma + 1) \end{bmatrix}.$$

It follows that $[\alpha, \beta, \gamma]'$ is on the plane if and only if $2\alpha + \beta - \gamma = 1$. Thus, the plane is characterized by the natural coordinate equation $2x_1 + x_2 - x_3 = 1$.

b) If $x_1 = s$ and $x_2 = t$, then $x_3 = \frac{1}{3}(4 - 2s - 5t)$. Thus, the vectors in S are those arising from the formula $X = [s, t, \frac{1}{3}(4 - 2s - 5t)]'$. It follows that

$$X = \begin{bmatrix} 0 \\ 0 \\ 4/3 \end{bmatrix} + s \begin{bmatrix} 1 \\ 0 \\ -2/3 \end{bmatrix} + t \begin{bmatrix} 0 \\ 1 \\ 5/3 \end{bmatrix}.$$

SECTION 3.1

1 a) $\begin{bmatrix} 14 \\ 19 \end{bmatrix}$ **b)** $A = \begin{bmatrix} 2 & 3 \\ -1 & 5 \end{bmatrix}$

b) That L is a linear transformation follows directly from the fact that $L(X)$ may be expressed by the matrix product AX. For then we may write $L(X^1 + X^2) = A(X^1 + X^2) = AX^1 + AX^2 = L(X^1) + L(X^2)$. And similarly with $L(cX)$.

2 Let $X = [1, 1]'$ and $c = 2$. Then $L(cX) = [10, 20]'$, but $cL(X) = [10, 10]'$.

3 a) $L([a, b, c]') = \begin{bmatrix} (3a + 2b + c) \\ (6a + 4b + 2c) \end{bmatrix}$

b) Since $L([a, b, c]') = a\begin{bmatrix} 3 \\ 6 \end{bmatrix} + b\begin{bmatrix} 2 \\ 4 \end{bmatrix} + c\begin{bmatrix} 1 \\ 2 \end{bmatrix}$,

it follows that $\left\{ \begin{bmatrix} 3 \\ 6 \end{bmatrix}, \begin{bmatrix} 2 \\ 4 \end{bmatrix}, \begin{bmatrix} 1 \\ 2 \end{bmatrix} \right\}$ is a spanning set for the range of L.

Evidently, $\left\{ \begin{bmatrix} 1 \\ 2 \end{bmatrix} \right\}$ is also a spanning set, an independent spanning set.

c) Since the range has a basis containing only 1 vector, its dimension is 1.

4 a) $3t^4 + 4t^3$

b) Briefly, $L(p + q) = t^2 D(p + q) = t^2 D(p) + t^2 D(q) = L(p) + L(q)$, and $L(cp) = t^2 D(cp) = t^2[cD(p)] = c[t^2 D(p)] = cL(p)$.

c) $L(at^3 + bt^2 + ct + d) = a(3t^4) + b(2t^3) + c(t^2)$. It follows that $\{3t^4, 2t^3, t^2\}$ is a spanning set for the range of L. (Explain.) This set being linearly independent, we conclude that the dimension of the range of L is 3.

ANSWERS TO SELECTED EXERCISES   315

5 a) $L\left(\begin{bmatrix} 5 & 6 \\ 3 & 7 \end{bmatrix}\right) = 5\begin{bmatrix} 2 \\ 3 \\ 4 \end{bmatrix} + 1\begin{bmatrix} 0 \\ 0 \\ 0 \end{bmatrix} - 3\begin{bmatrix} 1 \\ -1 \\ -2 \end{bmatrix} + 4\begin{bmatrix} 3 \\ 2 \\ 2 \end{bmatrix} = \begin{bmatrix} 19 \\ 26 \\ 34 \end{bmatrix}.$

b) $L\left(\begin{bmatrix} a & b \\ c & d \end{bmatrix}\right) = a\begin{bmatrix} 2 \\ 3 \\ 4 \end{bmatrix} + b\begin{bmatrix} -1 \\ 1 \\ 2 \end{bmatrix} + c\begin{bmatrix} -2 \\ -3 \\ -4 \end{bmatrix} + d\begin{bmatrix} 3 \\ 2 \\ 2 \end{bmatrix}.$

Since

$$\begin{bmatrix} 2 & 3 & 4 \\ -1 & 1 & 2 \\ -2 & -3 & -4 \\ 3 & 2 & 2 \end{bmatrix} \stackrel{\text{(row)}}{=} \begin{bmatrix} 1 & -1 & -2 \\ 0 & 5 & 8 \\ 0 & 0 & 0 \\ 0 & 0 & 0 \end{bmatrix},$$

it follows that $\{[1, -1, -2]', [0, 5, 8]'\}$ is a basis for the range of L.

6 a) An isomorphism between vector spaces is simply a linear transformation that happens to be one-to-one.
  b) Spaces that are isomorphic are of the same dimension. (Why?) Thus, L is not an isomorphism. It follows—recall part (a)—that L is not one-to-one.

7 a) Let X be any vector in V and let $Y = L(X)$. Then we may write
$$L(\Theta_V) = L(0 \cdot X) = 0 \cdot L(X) = 0 \cdot Y = \Theta_W.$$
  b) $L(X - Y) = L[X + (-1)Y] = L(X) + L[(-1)Y]$
  $\qquad\qquad\quad = L(X) + (-1)L(Y) = L(X) - L(Y).$

SECTION 3.2

1 a) $X = [-2s, s]'$. Thus, L is not one-to-one.
  b) $X = \Theta$. Thus, L is one-to-one.
  c) $L[p(t)] = \Theta$ if $p(t)$ is any constant polynomial. Thus, L is not one-to-one.

3 a) $L(a \sin t + b \cos t) = 2a$. Since $2a = \Theta$ when $a = 0$, it follows that the kernel of L consists of all functions of the form $b \cos t$.
  b) The kernel of L is all of V.
  c) $L(a \sin t + b \cos t) = a \cos t - b \sin t$, and $a \cos t - b \sin t = \Theta$ if and only if $a = b = 0$. Thus, the kernel of L consists of the zero-function alone, from which it follows that L is one-to-one. Two formulas for L^{-1} are as follows:

$$L^{-1}[f(t)] = f(3\pi/2) + \int_0^t f(u)\, du;$$
$$L^{-1}[f(t)] = -f'(t).$$

4 a) Domain dimension is 3, range dimension is 2, kernel dimension is 1.
  b) As in part (a): 4, 1, and 3.
  c) As in part (a): 2, 1, and 1.

5 Let X^1 and X^2 be in K (the kernel of L). Then we may write
$$L(X^1 + X^2) = L(X^1) + L(X^2) = \Theta + \Theta = \Theta.$$

Thus, $X^1 + X^2$ is in K. Similarly, $L(cX^1) = cL(X^1) = c\Theta = \Theta$. Thus, cX^1 is in K. It follows (Theorem 2.1) that K is a subspace of the domain of L.

6 If $k = n$ (kernel dimension = domain dimension), then the kernel of L is all of V. This means that $L(X) = \Theta$ for all X, from which it follows that the range of L consists of the zero-vector alone. Thus, $m = 0$, and we still have $n = m + k$.

SECTION 3.3

1 $f(t) = t^2 + \sin t + c$.

2 $f(t) = ce^t + (-3t + 2)$.

3 a) $f(t) = 100/\pi$ b) $f(t) = b \cos t$ c) $f(t) = b \cos t + 100/\pi$

4 c) $X = a \begin{bmatrix} 1 & 0 \\ -1 & 0 \end{bmatrix} + b \begin{bmatrix} 0 & 1 \\ 0 & -1 \end{bmatrix} + \begin{bmatrix} 25 & 31 \\ 0 & 0 \end{bmatrix}$

5 c) The row rank is 1; the column rank is 1 because each of the last 3 columns is a multiple of the first one.

6 b) Join the 4 columns into a single matrix A. The row rank of A is obviously 2. It follows that its column rank is 2. Thus, the dimension of V is 2.

SECTION 3.4

1 a) $Y_e = \begin{bmatrix} 1/3 & 1/2 & 1 \\ 1/3 & -1/2 & 1 \end{bmatrix} X_\mathfrak{B}$ b) $\frac{19}{6}t + \frac{37}{6}$

2 a) $Y_\mathfrak{B} = \begin{bmatrix} 2 & 3 \\ -3 & 2 \end{bmatrix} X_\mathfrak{B}$ b) $13 \sin t$

3 a) $Y_e = \begin{bmatrix} 5 & 4 & 1 \\ 0 & -1 & 2 \end{bmatrix} X_\mathfrak{B}$ b) $\begin{bmatrix} 10 \\ 1 \end{bmatrix}_e = [10, 11]$

4 a) $\begin{bmatrix} 1 & 2 \\ 4 & 7 \end{bmatrix}$ b) $\begin{bmatrix} 2 & 3 \\ 1 & 2 \end{bmatrix}$

c) $\begin{bmatrix} 2 & 5 \\ 7 & 18 \end{bmatrix}$ d) $\begin{bmatrix} 3 & 8 \\ 2 & 5 \end{bmatrix}$

5 a) $6t^2 + 2t + 2$ b) $36t^2 + 36t - 52$

6 a) $Y_e = \begin{bmatrix} 1 & 0 \\ 0 & 1 \end{bmatrix} X_\mathfrak{B}$ b) $Y_\mathfrak{R} = \begin{bmatrix} -4 & 2 \\ 3 & 0 \end{bmatrix} X_\mathfrak{R}$ c) $\begin{bmatrix} 2 \\ 6 \end{bmatrix}$

SECTION 3.5

1 $\begin{bmatrix} 6 & 9 \\ 15 & 18 \end{bmatrix} + \begin{bmatrix} 3 & 4 \\ 7 & 8 \end{bmatrix} = \begin{bmatrix} 9 & 13 \\ 22 & 26 \end{bmatrix}$

2 a) $(a + c)t^2 + (4a + b)t + \left(\dfrac{13a}{3} + \dfrac{5b}{2} + 2c\right)$

ANSWERS TO SELECTED EXERCISES   317

b) $L(at^2 + bt + c) = \begin{bmatrix} 1 & 0 & 1 \\ 4 & 1 & 0 \\ \frac{13}{3} & \frac{5}{2} & 2 \end{bmatrix} \begin{bmatrix} a \\ b \\ c \end{bmatrix}_{\mathcal{B}}$

$= \begin{bmatrix} (a+c) \\ (4a+b) \\ \left(\dfrac{13a}{3} + \dfrac{5b}{2} + 2c\right) \end{bmatrix}_{\mathcal{B}} = (a+c)t^2 + \cdots.$

3 Let $\mathcal{C} = [t^2, t, 1]$ and let $\mathcal{D} = [1]$, and let Mat assign to each L in T its $\mathcal{DC}$-representation. Then $\text{Mat}(L_1) = [0, 0, 1]$, $\text{Mat}(L_2) = [0, 1, 0]$, and $\text{Mat}(L_3) = [1, 0, 0]$. Since these last 3 matrices form a basis for R^{13}, it follows—note that Mat^{-1} is an isomorphism from R^{13} onto T—that $\mathcal{B}$ is a basis for T.

4 a) Since T is isomorphic to R^{22}, a basis for T is easily obtained by finding the linear transformations whose $\mathcal{BB}$-representations belong to

$$\left\{ \begin{bmatrix} 1 & 0 \\ 0 & 0 \end{bmatrix}, \begin{bmatrix} 0 & 1 \\ 0 & 0 \end{bmatrix}, \begin{bmatrix} 0 & 0 \\ 1 & 0 \end{bmatrix}, \begin{bmatrix} 0 & 0 \\ 0 & 1 \end{bmatrix} \right\}.$$

Letting $f(t) = a \cos t + b \sin t$, the resulting four transformations are given by

$$L_1(f) = a \cos t, \qquad L_2(f) = b \cos t,$$
$$L_3(f) = a \sin t, \qquad L_4(f) = b \sin t.$$

b) $D = L_2 \oplus [(-1) \odot L_3]$.

SECTION 3.6

1 a) $\begin{bmatrix} 3 & 0 & 6 \\ 9 & 4 & -2 \end{bmatrix}$ b) $\begin{bmatrix} 3 & 0 & 6 \\ 33 & 12 & 6 \end{bmatrix}$

2 a) $\begin{bmatrix} 9 & 0 & 0 \\ -12 & 6 & 0 \\ 3 & -3 & 3 \end{bmatrix}$ b) $\begin{bmatrix} 12 & 0 & 0 \\ -4 & 6 & 0 \\ 1 & -1 & 2 \end{bmatrix}$ c) $\begin{bmatrix} 12 & 0 & 0 \\ -6 & 6 & 0 \\ 3 & -2 & 2 \end{bmatrix}$

3 a) $L(at^2 + bt + c) = (a + 2c)t + (3a + b + c)$

b) $\begin{bmatrix} a \\ b \\ c \end{bmatrix}_{\mathcal{B}} = \begin{bmatrix} (a+c) \\ (b-2c) \\ (a+2c) \end{bmatrix}_{\mathcal{B}^*} \xrightarrow{L} \begin{bmatrix} (7a + 2b + 4c) \\ (10a + 3b + 5c) \end{bmatrix}_{e^*}$

$= \begin{bmatrix} (a+2c) \\ (3a+b+c) \end{bmatrix}_{e}$

$= (a + 2c)t + (3a + b + c)$

4 Following the hint, we are led to the equation

$$\text{Mat}(L^{-1} \circ L) = \text{Mat}(L^{-1}) \times \text{Mat}(L).$$

But $L^{-1} \circ L$ is the identity transformation; its $\mathcal{BB}$-representation is I. The desired result follows.

318 ANSWERS TO SELECTED EXERCISES

5 a) $\begin{bmatrix} 1 & 0 & 0 \\ -6 & 1 & 0 \\ 9 & -3 & 1 \end{bmatrix}$ b) $\begin{bmatrix} 1 & 0 & 0 \\ 6 & 1 & 0 \\ 9 & 3 & 1 \end{bmatrix}$ c) Same as (b).

d) $\begin{bmatrix} 1 & 0 & 0 \\ 6 & 1 & 0 \\ 9 & 3 & 1 \end{bmatrix} \begin{bmatrix} a \\ b \\ c \end{bmatrix} = \begin{bmatrix} (a) \\ (6a+b) \\ (9a+3b+c) \end{bmatrix} = at^2 + (6a+b)t + (9a+3b+c)$

$= a(t+3)^2 + b(t+3) + c$

6 Write $Z = \begin{bmatrix} 2 & 3 \\ -5 & 7 \end{bmatrix} Y$ and $Y = \begin{bmatrix} 3 & -2 & 1 \\ 2 & 7 & 0 \end{bmatrix} X$. It follows that

$$Z = \begin{bmatrix} 12 & 17 & 2 \\ -1 & 59 & -5 \end{bmatrix} X.$$

Or,

$z_1 = 12x_1 + 17x_2 + 2x_3$ and $z_2 = -x_1 + 59x_2 - 5x_3$.

The same result can be obtained by direct replacement of y_1 and y_2.

7 a) $Y_{\mathfrak{N}} = \begin{bmatrix} 1 & 0 & 2 \\ 3 & 1 & 0 \\ 2 & 1 & 2 \end{bmatrix} X_{\mathfrak{N}}$ b) $Y_{\mathfrak{B}} = \begin{bmatrix} 3 & 1 & 0 \\ 10 & 12 & -5 \\ 26 & 28 & -11 \end{bmatrix} X_{\mathfrak{B}}$

SECTION 4.1

1 One possibility:

$$a_n = n^2 + \frac{\sqrt{17} - n^2}{n!}(n-1)(n-2)(n-3)(n-4)(n-5).$$

2 The numbers $a_1, a_2, \ldots, a_{40}$ are all primes, but $a_{41} = 41^2$. Thus, the assertion is false.

3 For $n = 1$, the formula reduces to $1 = 1^2$. A true statement. Now suppose that the formula is valid for $n = k$. Then we can write the following:

Sum of $(k+1)$ terms = sum of k terms + $(k+1)$st term
$= [1 + 3 + 5 + \cdots + (2k-1)] + (2k+1)$
$= k^2 + (2k+1) = (k+1)^2$.

This calculation shows that the formula is also true for $n = k+1$.

4 If the formula is assumed valid for $n = k$, then we can write

$$\begin{bmatrix} 1 & 1 \\ 0 & 1 \end{bmatrix}^{k+1} = \begin{bmatrix} 1 & 1 \\ 0 & 1 \end{bmatrix} \begin{bmatrix} 1 & 1 \\ 0 & 1 \end{bmatrix}^k = \begin{bmatrix} 1 & 1 \\ 0 & 1 \end{bmatrix} \begin{bmatrix} 1 & k \\ 0 & 1 \end{bmatrix} = \begin{bmatrix} 1 & (k+1) \\ 0 & 1 \end{bmatrix}.$$

5 We show only that truth for k implies truth for $k+1$:

a) $A^{k+1}B = A^k(AB) = (A^kB)A = (BA^k)A = BA^{k+1}$.
b) $(AB)^{k+1} = (AB)^k(AB) = A^k(B^kA)B = A^k(AB^k)B = A^{k+1}B^{k+1}$.

c) Take $A = \begin{bmatrix} 1 & 1 \\ 0 & 1 \end{bmatrix}$ and $B = \begin{bmatrix} 1 & 0 \\ 0 & 0 \end{bmatrix}$.

6 a) 23   b) 10   c) $15 - [6 - (9-3)] = 15$

ANSWERS TO SELECTED EXERCISES

SECTION 4.2

1 a) 755   b) 6   c) 60   d) 4
2 $\Delta(A) = 11$
3 $\Delta(B) = 11 = \Delta(A)$. For all A, $\Delta(A') = \Delta(A)$.
4 $\Delta(B) = 1/11 = 1/\Delta(A)$. If A is nonsingular, then $\Delta(A^{-1}) = 1/\Delta(A)$.
5 $\Delta(B) = 77 = 7\Delta(A)$. If $E = [i \to (c)i]$, then $\Delta(EA) = c\Delta(A)$.
6 $\Delta(B) = -11 = -\Delta(A)$. If $E = [i \leftrightarrow j]$, then $\Delta(EA) = -\Delta(A)$.
7 $\Delta(B) = 11 = \Delta(A)$. If $E = [i \to i + (c)j]$, then $\Delta(EA) = \Delta(A)$.
8 $\Delta(B) = -10$, $\Delta(AB) = -110$, and $\Delta(A) \cdot \Delta(B) = (11)(-10) = -110$. For all nth-order matrices A and B we have $\Delta(AB) = \Delta(A) \cdot \Delta(B)$.

SECTION 4.3

1 a) 70   b) 0
2 a) -99 b) 0
3 All 3 determinants have the same value: 1. Let $A = [a_{ij}]_{nn}$, where

$$a_{ij} = \binom{i+j-2}{i-1}.$$

Each of the 3 matrices fits the formula for A. Such matrices are so special that the value 1 for each determinant hardly seems to be a coincidence. Thus, we conjecture: $|A| = 1$ for all n.

4, 5, 6, 7 Expand each side using the formula following Example 2 of Section 4.2. Then manipulate each side so as to reveal the equality.

SECTION 4.4

1 $-6 + 38 = 32$
3 Suppose that $|A| = 0$ for any nth-order matrix having a row of zeros, and let A^* be an $(n+1)$st-order matrix whose pth row is a row of zeros. Then, by Definition 4.2,

$$|A^*| = (-1)^{p+1} a_{p1} |A^*_{p1}| + \sum_{k \neq p} (-1)^{k+1} a_{k1} |A^*_{k1}| = 0 + 0.$$

(The first term is 0 because a_{p1} is from the pth row; the second term is 0 because each A^*_{k1} is an nth-order matrix having a row of zeros.)

4 Suppose that each entry in the ith row of A is c times ($c \neq 0$) the corresponding entry in the jth row. Then the matrix $A^* = [i \to (1/c)i] \cdot A$ has identical ith and jth rows. Thus, by the corollary to Theorem 4.3, $|A^*| = 0$. But also, by Theorem 4.2, $|A^*| = (1/c)|A|$. It follows that $|A| = 0$.

5 a) $\begin{vmatrix} x & 3 & 4 \\ y & 2 & 7 \\ 1 & 1 & 1 \end{vmatrix} = 0$ b) $\begin{vmatrix} (x^2+y^2) & 5 & 58 & 26 \\ x & 2 & 3 & -1 \\ y & 1 & 7 & 5 \\ 1 & 1 & 1 & 1 \end{vmatrix} = 0$

6 a) -1 b) 1
c) Successive applications of Theorem 4.3 allow us to write: $|E_1 I| = -1$,
$$|E_2(E_1 I)| = (-1)^2, \ldots, |E_n(E_{n-1} \cdots E_1 I)| = (-1)^n.$$

SECTION 4.5

1 $|A| = 5$, $|B| = 7$, $|AB| = 35$

2 a) Apply Theorem 4.6: $AA^{-1} = I$, $|AA^{-1}| = |I|$, $|A||A^{-1}| = 1$, $|A^{-1}| = 1/|A|$.
b) Apply Theorem 4.6: $|BA| = |B||A| = |A||B| = |AB|$.
c) $\frac{1}{5}$, 35

3 S is dependent if $\lambda^3 - 6\lambda - 4 = 0$, that is, for $\lambda = -2, 1 \pm \sqrt{3}$.

4 The algorithm fails to work only if we run into a row of zeros. In such an event the determinant of the given matrix is 0. It follows (by Theorem 4.8) that the given matrix has no inverse.

5 Let $E = [i \leftrightarrow j]$. Then $B = AE$ and
$$|B| = |AE| = |(AE)'| = |E'A'| = -|A'| = -|A|.$$

6 a) If $AB = I$, then (by Theorem 4.6) $|A||B| = 1$. Thus, $|B| \neq 0$ and (by Theorem 4.8) B^{-1} exists.
b) If $AB = I$, then $(AB)B^{-1} = B^{-1}$. Thus, $A = B^{-1}$.

7 $AX = \Theta$ has nontrivial solutions if and only if A^{-1} does not exist. To guarantee this we have only to set $|A| = 0$, obtaining the values $\lambda = 1, 2, 3$.

SECTION 4.6

1 a) $-8 + 40 + 33 = 65$, $-24 - 20 + 44 = 0$
b) $0 + 33 + 32 = 65$, $-35 + 11 + 24 = 0$

2 a) $-\dfrac{1}{2} \begin{bmatrix} 4 & -2 \\ -3 & 1 \end{bmatrix}$
b) $\dfrac{1}{65} \begin{bmatrix} 11 & -8 & 6 \\ 5 & 20 & -15 \\ -7 & 11 & 8 \end{bmatrix}$

3 $x_2 = \frac{4}{13}$

4 $p(t) = \begin{vmatrix} (t^2 + 2t + 2) & 3 \\ (2t^2 + 7) & 4 \end{vmatrix} \Big/ \begin{vmatrix} 2 & 3 \\ 3 & 4 \end{vmatrix} = 2t^2 - 8t + 13$
$q(t) = -t^2 + 6t - 8$

5 b) $x_1 = \begin{vmatrix} a_2 & a_3 & a_4 \\ b_2 & b_3 & b_4 \\ c_2 & c_3 & c_4 \end{vmatrix}$, $x_2 = -\begin{vmatrix} a_1 & a_3 & a_4 \\ b_1 & b_3 & b_4 \\ c_1 & c_3 & c_4 \end{vmatrix}$, ...

SECTION 4.7

2 a) $\lambda^3 - 8\lambda^2 + 17\lambda - 10 = 0$
b) $\lambda = 1, 2, 5$

c) The characteristic vectors associated with $\lambda = 1$ are given by
$$X = s[-1, -1, 1]',$$
with $\lambda = 2$ by $X = s[-3, -2, 2]'$, with $\lambda = 5$ by $X = s[-2, 0, 1]'$.

d) $B = \begin{bmatrix} 1 & 3 & 2 \\ 1 & 2 & 0 \\ -1 & -2 & -1 \end{bmatrix}$

3 a) $\begin{bmatrix} 1 & 1 \\ \sqrt{5} & -\sqrt{5} \end{bmatrix}$ b) $\begin{bmatrix} 1 & 1 & 0 \\ -1 & 0 & 1 \\ 0 & -1 & 2 \end{bmatrix}$ c) Such a B does not exist.

4 b) $\{[-1, 1, 0, 0]', [-1, 0, 1, 0]', [-1, 0, 0, 1]'\}$

5 a) Let $P = \begin{bmatrix} 1 & -2 \\ 1 & 1 \end{bmatrix}$. Then $P^{-1} = \frac{1}{3}\begin{bmatrix} 1 & 2 \\ -1 & 1 \end{bmatrix}$ and $P^{-1}AB = B$. Thus, $B \stackrel{s}{=} A$.

b) Since $I^{-1}AI = A$, $A \stackrel{s}{=} A$. Suppose next that $B \stackrel{s}{=} A$. This means that there exists a matrix P such that $P^{-1}AP = B$. It follows that $A = PBP^{-1}$. Thus, $A \stackrel{s}{=} B$. Finally, suppose that $C \stackrel{s}{=} B$ and $B \stackrel{s}{=} A$. Then there exist matrices P and Q such that
$$B = P^{-1}AP \quad \text{and} \quad C = Q^{-1}BQ.$$
It follows that $C = Q^{-1}(P^{-1}AP)Q = (PQ)^{-1}A(PQ)$. Thus, $C \stackrel{s}{=} A$.

c) Let $P = \begin{bmatrix} 1 & -1 \\ 1 & 5 \end{bmatrix}$. Then, $P^{-1}CP = B$. Thus, $B \stackrel{s}{=} C$. By part (a), $B \stackrel{s}{=} A$. It follows, since similarity is an equivalence relation, that $C \stackrel{s}{=} A$.

d) $|B - \lambda I| = 0$, $|P^{-1}CP - \lambda I| = 0$, $|P^{-1}(C - \lambda I)P| = 0$,
$$|P^{-1}| \, |C - \lambda I| \, |P| = 0, \, |C - \lambda I| = 0.$$
Since this argument is reversible, we may conclude that the first and last equations are equivalent. That is, they have the same roots.

e) Let X be any vector in V, and let $Y = L(X)$. Then $Y_\mathcal{C} = L_{\mathcal{CC}}X_\mathcal{C}$. Now, let P be the matrix that changes $\mathcal{B}$-columns into $\mathcal{C}$-columns. Then we can write $PY_\mathcal{B} = L_{\mathcal{CC}}PX_\mathcal{B}$, or $Y_\mathcal{B} = (P^{-1}L_{\mathcal{CC}}P)X_\mathcal{B}$. It follows that $P^{-1}L_{\mathcal{CC}}P = L_{\mathcal{BB}}$.

6 a) $\begin{bmatrix} -1 & 0 & 0 \\ 0 & 1 & 0 \\ 0 & 0 & 2 \end{bmatrix}$

b) Let λ be the characteristic value associated with B^j. Then
$$L(B^j) = AB^j = \lambda B^j = \lambda[0, \ldots, 0, 1, 0, \ldots, 0]'_\mathcal{B}.$$

7 a) $L_{\mathcal{BB}} = \begin{bmatrix} 1 & 0 & 0 \\ -2 & 2 & 0 \\ 1 & -1 & 1 \end{bmatrix}$; $\lambda_1 = 1$, $\mathcal{B}_1 = \{1\}$; $\lambda_2 = 2$, $\mathcal{B}_2 = \{t - 1\}$.

8 b) Your sketch should suggest that L transforms arrows in the same way that they would be transformed if the plane of arrows were stretched in a direction parallel to the x_1-axis. Thus, $L(X)$ and X will have the same direction only if X lies in the x_1-axis.

c) $\lambda = 1$, $\mathcal{B} = \{[1, 0]'\}$. Note that the space spanned by $\mathcal{B}$ corresponds to the space of arrows making up the x_1-axis.

SECTION 4.8

1 $x_1 = -c_1 e^{-t} + c_2 e^{5t}$, $x_2 = 2c_1 e^{-t} + c_2 e^{5t}$

2 $x_1 = c_2 e^t + c_3 e^{2t}$, $x_2 = c_1 e^{-t} + 2c_2 e^t + c_3 e^{2t}$,
$x_3 = c_1 e^{-t} + 2c_2 e^t + 2c_3 e^{2t}$

3 a) $x_1 = (c_1 - c_2)e^{2t} + c_3 e^{3t}$, $x_2 = c_1 e^{2t} + 2c_3 e^{3t}$,
$x_3 = 2c_2 e^{2t} + 3c_3 e^{3t}$

b) $x_1 = -10e^{2t} + 12e^{3t}$, $x_2 = -26e^{2t} + 24e^{3t}$,
$x_3 = -32e^{2t} + 36e^{3t}$

c) $x_1 = \frac{4}{3}e^{3t}$, $x_2 = -\frac{3}{2}e^{2t} + \frac{8}{3}e^{3t}$, $x_3 = -3e^{2t} + 4e^{3t}$

d) $x_1 = 2e^{2t}$, $x_2 = 3e^{2t}$, $x_3 = 2e^{2t}$

4 a) Kernel given by

$$X^K = c_1 \begin{bmatrix} e^t \\ 0 \\ 0 \end{bmatrix} + c_2 \begin{bmatrix} e^{2t} \\ e^{2t} \\ 0 \end{bmatrix} + c_3 \begin{bmatrix} e^{3t} \\ 2e^{3t} \\ 2e^{3t} \end{bmatrix};$$

its dimension is 3.

b) $L(A) = B$, where $A = [-2t + 1, t + 1, 5t - 2]'$.

c) $X = X^K + A$

5 a) $X = \begin{bmatrix} -c_1 e^{-t} + c_2 e^{5t} + 2 \\ 2c_1 e^{-t} + c_2 e^{5t} - 5 \end{bmatrix}$

b) $x = ce^{3t} - \frac{3}{10} \sin t - \frac{1}{10} \cos t$

c) Let $L(X) = \dot{X} - AX$. The equations $L(X) = \Theta$ and $\dot{X} = AX$ are equivalent; and so are the equations $L(X) = B$ and $\dot{X} = AX + B$. Thus, the kernel of L is given by X^*, and a particular solution of $L(X) = B$, by B^*. It follows, by Theorem 3.7, that the complete solution of $L(X) = B$ is given by $X = X^* + B^*$.

SECTION 4.9

1 $\begin{bmatrix} 2 & 1 & 0 \\ -1 & 0 & 1 \\ 1 & 0 & 0 \end{bmatrix}$
2 $\begin{bmatrix} 1 & 0 & 0 & 0 \\ 0 & 1 & -1 & 1 \\ 0 & 0 & 1 & -2 \\ 0 & 0 & 0 & 1 \end{bmatrix}$
3 $\begin{bmatrix} 0 & 1 & 0 & 0 \\ 0 & 0 & 1 & -1 \\ -1 & 0 & 0 & 1 \\ 1 & 0 & 0 & 0 \end{bmatrix}$

4 $\begin{bmatrix} 0 & 0 & 0 & 1 \\ 0 & 0 & 1 & 0 \\ 0 & 1 & 0 & 0 \\ 1 & 0 & 0 & 0 \end{bmatrix}$
5 $\begin{bmatrix} 0 & 0 & 0 & 0 & 1 \\ 0 & 0 & 0 & 1 & -3 \\ 0 & 0 & 1 & -2 & 3 \\ 0 & 1 & -1 & 1 & -1 \\ 1 & 0 & 0 & 0 & 0 \end{bmatrix}$

6 $x_1 = 2c_1 e^{3t} + c_2 e^{5t} + c_3 t e^{5t}$, $x_2 = -c_1 e^{3t} + c_3 e^{5t}$, $x_3 = c_1 e^{3t}$

SECTION 4.10

1 a) $|A_4| = (a_1 - a_0)(a_2 - a_0)(a_2 - a_1)(a_3 - a_0)(a_3 - a_1)(a_3 - a_2)$
$\times (a_4 - a_0)(a_4 - a_1)(a_4 - a_2)(a_4 - a_3)$

b) Let

$$p(t) = \begin{vmatrix} 1 & 1 & 1 & 1 \\ t & a_1 & a_2 & a_3 \\ t^2 & a_1^2 & a_2^2 & a_3^2 \\ t^3 & a_1^3 & a_2^3 & a_3^3 \end{vmatrix} \quad \text{and} \quad c = -\begin{vmatrix} 1 & 1 & 1 \\ a_1 & a_2 & a_3 \\ a_1^2 & a_2^2 & a_3^2 \end{vmatrix}.$$

Then $p(t) = c(t - a_1)(t - a_2)(t - a_3)$
$= -(a_2 - a_1)(a_3 - a_1)(a_3 - a_2)(t - a_1)(t - a_2)(t - a_3).$

It follows that $|A_4| = p(a_0)$ has the desired factorization.

2 a) $|B_5| = (2 - 1)[(3 - 1)(3 - 2)][(4 - 1)(4 - 2)(4 - 3)]$
$\times [(5 - 1)(5 - 2)(5 - 3)(5 - 4)]$
$= (1!)(2!)(3!)(4!)$

b) The matrix B_n is a Vandermonde matrix. Adopting the notation of Definition 4.10, we take $a_j = b_{2, j+1} = j + 1$. Thus, applying Theorem 4.17, we have

$$|B_n| = \prod_{j=1}^{n-1}\left[\prod_{i=0}^{j-1}\{(j+1) - (i+1)\}\right].$$

Now,

$$\prod_{i=0}^{j-1}(j - i) = (j)(j - 1)(j - 2)\cdots(1) = j!.$$

Thus,

$$|B_n| = \prod_{j=1}^{n-1}(j!).$$

3 *Case 1:* Suppose that none of the variables is 0. Then we can write

$$|A| = \frac{1}{abcd}\begin{vmatrix} abcd & abcd & abcd & abcd \\ a & b & c & d \\ a^2 & b^2 & c^2 & d^2 \\ a^3 & b^3 & c^3 & d^3 \end{vmatrix} = \begin{vmatrix} 1 & 1 & 1 & 1 \\ a & b & c & d \\ a^2 & b^2 & c^2 & d^2 \\ a^3 & b^3 & c^3 & d^3 \end{vmatrix}.$$

Because of the special factorization of a Vandermonde determinant, we see that $|A| \neq 0$ (hence A is nonsingular) if and only if a, b, c, and d are all distinct.

Case 2: Suppose that one of the variables is 0, say $a = 0$. Then

$$|A| = \begin{vmatrix} bcd & 0 & 0 & 0 \\ 1 & 1 & 1 & 1 \\ 0 & b & c & d \\ 0 & b^2 & c^2 & d^2 \end{vmatrix} = bcd\begin{vmatrix} 1 & 1 & 1 \\ b & c & d \\ b^2 & c^2 & d^2 \end{vmatrix}.$$

It follows that $|A| \neq 0$ if and only if b, c, and d are all distinct and all different from 0.

4 a) The ith row of the determinant defining $W(t)$ is $[0, 0, \ldots, 0, D^i(t^i)$, $D^i(t^{i+1}), \ldots, D^i(t^n)]$, where the first nonzero entry is in the ith column. It follows that the determinant is upper triangular and that the diagonal entries are given by $a_{ii} = D^i(t^i) = i!$. Thus, $W(t) = (1!)(2!) \cdots (n!)$.

b) By part (a), $W(t) \neq 0$. By Theorem 4.19, S is independent.

c) Set
$$a_0 e^t + a_1 e^t + \cdots + a_n t^n e^t = \Theta.$$
Dividing by $e^t \neq 0$, we obtain
$$a_0 + a_1 t + \cdots + a_n t^n = \Theta.$$
Now, since S is independent, it follows that
$$a_0 = a_1 = \cdots = a_n = 0.$$
But this shows that S^* is independent.

5 b) $W(t) = e^{3t} \begin{vmatrix} 1 & 1 & 1 & 1 \\ 0 & 1 & e^t & 2 \\ 0 & 1 & 2e^t & 4 \\ 0 & 1 & 3e^t & 8 \end{vmatrix} = e^{4t} \begin{vmatrix} 1 & 1 & 2 \\ 1 & 2 & 4 \\ 1 & 3 & 8 \end{vmatrix} \neq 0$

for any t. By Theorem 4.19, S is independent.

c) $f(t) = a + be^t + cte^t + de^{2t}$.

SECTION 5.1

1 a) 90° **b)** 30° **c)** 60°

2 a) The cosines of both angles are the same: $1/\sqrt{7}$.

3 a) No. Since $B \cdot C$ and $A \cdot B$ are numbers rather than vectors, expressions like $A \cdot (B \cdot C)$ and $(A \cdot B) \cdot C$ do not even make sense.

4 a) dot $(X + Y, Z)$ = dot $(Z, X + Y)$ = dot (Z, X) + dot (Z, Y)
= dot (X, Z) + dot (Y, Z)

b) dot $(X, Y) = x_1 y_1 + \cdots + x_n y_n = y_1 x_1 + \cdots + y_n x_n$ = dot (Y, X)

5 $X \cdot X \geq 0$, with equality holding only for the case $X = \Theta$; $X \cdot Y = Y \cdot X$; $(cX) \cdot Y = c(X \cdot Y)$; $X \cdot (Y + Z) = (X \cdot Y) + (X \cdot Z)$

6 a) $\cos \alpha_j = a_j/5$. Thus, Dir $(A) = (\frac{2}{5}, \frac{1}{5}, -\frac{2}{5}, \frac{4}{5})$.

b) Dir $(cX) = \left(\dfrac{cx_1}{|cX|}, \dfrac{cx_2}{|cX|}, \ldots, \dfrac{cx_n}{|cX|} \right)$
$= \left(\dfrac{cx_1}{c|X|}, \dfrac{cx_2}{c|X|}, \ldots, \dfrac{cx_n}{c|X|} \right) =$ Dir (X)

c) The direction cosines of the vector X are identical with the natural coordinates of the vector $Y = X/|X|$.

d) $|B| = 10$, $B/|B| = [\frac{2}{5}, \frac{1}{5}, -\frac{2}{5}, \frac{4}{5}]'$, Dir $(B) = (\frac{2}{5}, \frac{1}{5}, -\frac{2}{5}, \frac{4}{5})$

7 a) $|X^2 - X^1| = \sqrt{6}$, $|X^3 - X^2| = 2\sqrt{6}$, $|X^3 - X^1| = 3\sqrt{6}$

b) Dir $(X^3 - X^2) = \left(\dfrac{1}{\sqrt{6}}, \dfrac{2}{\sqrt{6}}, 0, -\dfrac{1}{\sqrt{6}} \right) =$ Dir $(X^2 - X^1)$

8 a) $\sqrt{34}$ **b)** $\frac{1}{2}\sqrt{687}$ **c)** $\frac{1}{6}\sqrt{894}$

SECTION 5.2

1 a) $|p|^2 = \int_0^1 (6t^2 - 12t + 6)\, dt = 2 = \int_0^1 (25t^4 - 60t^3 + 36t^2)\, dt = |q|^2$

b) Since $|p| = |q|$, it is enough to show that $p \cdot r = q \cdot r$. (Why?)

$$p \cdot r = \int_0^1 [5\sqrt{6}\, t^3 + (6 - 11\sqrt{6})t^2 + (-12 + 6\sqrt{6})t + 6]\, dt = \tfrac{7}{12}\sqrt{6} + 2$$

$$q \cdot r = \int_0^1 [25t^4 + (-60 + 5\sqrt{6})t^3 + (36 - 11\sqrt{6})t^2 + 6\sqrt{6}\, t]\, dt$$

$$= \tfrac{7}{12}\sqrt{6} + 2$$

c) $r \cdot s = \int_0^1 (-25t^4 + 60t^3 - 30t^2 - 12t + 6)\, dt = 0$

2 An appropriate sketch of the natural coordinate system for E can be based upon an arrow N^1 of length 1 pointing to the right and an arrow N^2 of length 2 that is 120° counterclockwise of N^1.

3 The matrices under (a) and (d) define dot products. The matrix under (b) does not. For let $X = [0, 2, 3]'$ and $Y = [0, 1, 1]'$. Then $f(X, Y) = 13$, but $f(Y, X) = 11$. The matrix under (c) does not. For let $X = [0, -1, 1]'$. Then $f(X, X) = -1$.

4 (Return to part (2) of the proof of Theorem 5.2.) We have equality if and only if $0 = (bX - aY) \cdot (bX - aY)$. By property 1 of Definition 5.4, this condition is equivalent to having $bX - aY = \Theta$; or, since $a = X \cdot X \neq \Theta$, to having $Y = \dfrac{b}{a} X$.

5 a) Let $X = A$. Since $A \cdot A = 0$, then $A = \Theta$.

b) $A \cdot X = B \cdot X$ is equivalent to $(A - B) \cdot X = 0$. By part (a) we may conclude that $A - B = \Theta$. That is, that $A = B$.

6 Properties 2 and 4 of Definition 5.4 allow us to write

$$(X + Y) \cdot (X + Y) = X \cdot X + 2(X \cdot Y) + (Y \cdot Y).$$

It follows that $|X + Y|^2 = |X|^2 + 2(X \cdot Y) + |Y|^2$. Thus, $|X + Y|^2 = |X|^2 + |Y|^2$ implies that $X \cdot Y = 0$; and conversely.

7 a) As in Exercise 6, use the fact that $|X|^2 = X \cdot X$. Then compute.

b) Sketch on the same figure arrows representing X, Y, $X + Y$, and $X - Y$. Then consider the following theorem from geometry: The sum of the squares of the diagonals of a parallelogram is equal to the sum of the squares of its sides.

SECTION 5.3

1 a) Let $B = \begin{bmatrix} 1 & 0 & 0 & 0 & 0 \\ 0 & 1 & 0 & 0 & 0 \\ -1 & 0 & 1 & 0 & 0 \\ 0 & 0 & 0 & 1 & 0 \\ 0 & 0 & -1 & 0 & 1 \end{bmatrix}$. Then $BAB' = \begin{bmatrix} 1 \\ & 2 \\ & & 2 \\ & & & 4 \\ & & & & 2 \end{bmatrix}$.

It follows that A is positive definite. A is also symmetric. Thus, by Theorem 5.3, f is a dot product.

b) $|N^1| = 1, |N^2| = \sqrt{2}, |N^3| = \sqrt{3}, |N^4| = 2, |N^5| = \sqrt{5}$

3 Identifying the symbols of Example 1 with those occurring in the proof of Theorem 5.5, we have

$$B = \begin{bmatrix} 1 & 0 \\ 0 & 1 \end{bmatrix} \quad \text{and} \quad C = \begin{bmatrix} 1 & \frac{1}{2} \\ \frac{1}{2} & 1 \end{bmatrix}.$$

Then take

$$P = I_2 \quad \text{and} \quad Q = \begin{bmatrix} 1 & -\frac{1}{2} \\ 0 & 1 \end{bmatrix},$$

so that

$$P'BP = \begin{bmatrix} 1 & 0 \\ 0 & 1 \end{bmatrix} \quad \text{and} \quad Q'CQ = \begin{bmatrix} 1 & 0 \\ 0 & \frac{3}{4} \end{bmatrix}.$$

Finally, take $R = \begin{bmatrix} \sqrt{1} & \\ & \sqrt{4/3} \end{bmatrix}$ and let L be the transformation from E onto F such that

$$L_{\mathfrak{F}\mathcal{E}} = QRP^{-1} = \begin{bmatrix} 1 & -1/\sqrt{3} \\ 0 & 2/\sqrt{3} \end{bmatrix}.$$

4 a) $|N^1| = |-1| \, |X^1| = 1.$
$|N^2|^2 = (-X^1 - X^2) \cdot (-X^1 - X^2) = |X^1|^2 + |X^2|^2 = 4.$
Thus, $|N^2| = 2.$

b) $N^1 \cdot N^2 = (-X^1) \cdot (-X^1 - X^2) = 1.$ Thus, $\cos \phi = \frac{1}{2}$ and, consequently, $\phi = 60°.$

d) We use Theorem 5.4, obtaining $A = \begin{bmatrix} 1 & 1 \\ 1 & 4 \end{bmatrix}.$

e) $X \cdot Y = X'_{\circledR} \begin{bmatrix} 1 & 0 \\ 0 & 3 \end{bmatrix} Y_{\circledR}$

SECTION 5.4

2 b) $X_{\circledR} = [-1, \sqrt{3}/3]'$ \qquad c) $|X| = \frac{2}{3}\sqrt{3}$

d) The method leads to $\circledR$. (There are, of course, other orthonormal bases for E.)

3 $\dfrac{1}{\sqrt{3}}[1, 1, 1]'$, $\dfrac{1}{\sqrt{6}}[2, -1, -1]'$, $\dfrac{1}{\sqrt{2}}[0, 1, -1]'$

4 $\dfrac{1}{\sqrt{2}}[1, 1, 0, 0, 0]'$, $\dfrac{1}{\sqrt{6}}[1, -1, -2, 0, 0]'$,

$\dfrac{1}{\sqrt{5}}[0, 0, 0, 1, -2]$, $\dfrac{1}{\sqrt{5}}[0, 0, 0, 2, 1]$

5 $b_1 = B^1 \cdot X = 10$, $b_2 = B^2 \cdot X = -5$, $b_3 = B^3 \cdot X = 8.$

6 Set $a_1 X^1 + a_2 X^2 + \cdots + a_i X^i + \cdots + a_n X^n = \Theta.$ Multiply both sides by X^i obtaining $a_i = 0.$

ANSWERS TO SELECTED EXERCISES   327

SECTION 5.5

1 a) $\dfrac{1}{\sqrt{2}}\begin{bmatrix} 1 & -1 \\ 1 & 1 \end{bmatrix}$ 	 b) $\dfrac{1}{\sqrt{2}}\begin{bmatrix} 1 & -1 \\ 1 & 1 \end{bmatrix}$

c) $\begin{bmatrix} -\dfrac{1}{\sqrt{3}} & \dfrac{1}{\sqrt{6}} & -\dfrac{1}{\sqrt{2}} \\ -\dfrac{1}{\sqrt{3}} & \dfrac{1}{\sqrt{6}} & \dfrac{1}{\sqrt{2}} \\ \dfrac{1}{\sqrt{3}} & \dfrac{2}{\sqrt{6}} & 0 \end{bmatrix}$ d) $\begin{bmatrix} -\dfrac{1}{\sqrt{2}} & -\dfrac{1}{\sqrt{6}} & \dfrac{1}{\sqrt{3}} \\ \dfrac{1}{\sqrt{2}} & -\dfrac{1}{\sqrt{6}} & \dfrac{1}{\sqrt{3}} \\ 0 & \dfrac{2}{\sqrt{6}} & \dfrac{1}{\sqrt{3}} \end{bmatrix}$

e) $\begin{bmatrix} 1 & 0 & 0 & 0 \\ 0 & 1 & 0 & 0 \\ 0 & 0 & -\dfrac{1}{\sqrt{2}} & \dfrac{1}{\sqrt{2}} \\ 0 & 0 & \dfrac{1}{\sqrt{2}} & \dfrac{1}{\sqrt{2}} \end{bmatrix}$

2 a) We use Theorem 5.11 in both directions: If A and B are orthogonal, then
$$(AB)^{-1} = B^{-1}A^{-1} = B'A' = (AB)'.$$
It follows, since $(AB)^{-1} = (AB)'$, that AB is orthogonal.
b) $|A|^2 = |A||A'| = |AA'| = |AA^{-1}| = |I| = 1$.
c) By Theorem 5.12, there is a matrix B such that
$$B^{-1}AB = B'AB = D,$$
where D is a diagonal matrix whose entries are the characteristic values of A. By Theorem 1.14, A is positive definite if and only if the entries of D are all positive.

3 By Theorem 5.6, $X \cdot Y = X'_\circledB Y_\circledB$. Substituting for the $\circledB$-columns, we obtain $X \cdot Y = (A^{-1}X_e)'(A^{-1}Y_e) = X'_e Y_e$, where we have used the fact that $A^{-1} = A'$. Again applying Theorem 5.6, the converse part, we may conclude that $\mathcal{C}$ is orthonormal.

4 a) $B^1 = [1/\sqrt{2}, 1/\sqrt{2}]'$ and $B^2 = [-1/\sqrt{2}, 1/\sqrt{2}]'$. It follows that $B^i \cdot B^j = \delta_{ij}$.
b) It is the same as the given change-of-basis matrix.

SECTION 5.6

1 a) Center the circles about the origin. Take the tangent line as the vertical line through $(1, 0)$; then its equation is $x_1 = 1$. Solving with $x_1^2 + x_2^2 = 4$ gives $x_2 = \pm\sqrt{3}$. It follows that the length of the chord is $2\sqrt{3}$.
b) Proceed as in part (a). The equations of the tangent line and outer circle are no longer obvious. Since $A = -N^1 + N^2$ is vertical, our tangent

line may be described by $X = t[-1, 1]' + N^1$ or $x_1 = -x_2 + 1$. The outer circle is described by $|X| = 2$ or $x_1^2 + 2x_1x_2 + 2x_2^2 = 4$. Solving simultaneously we obtain

$$X^1 = [1 - \sqrt{3}, \sqrt{3}]' \quad \text{and} \quad X^2 = [1 + \sqrt{3}, -\sqrt{3}]'.$$

Then the length of the chord is $|X^2 - X^1| = 2\sqrt{3}$.

2 Let $A = \begin{bmatrix} 7 & -\sqrt{3} \\ -\sqrt{3} & 5 \end{bmatrix}$. Its characteristic values are 4 and 8. Thus, there exists an orthogonal matrix B such that $B'AB = \begin{bmatrix} 4 & 0 \\ 0 & 8 \end{bmatrix}$. It follows that there is a u_1u_2-coordinate system with perpendicular axes in which the given ellipse is described by the equation

$$4u_1^2 + 8u_2^2 = 8, \quad \text{or} \quad \tfrac{1}{2}u_1^2 + u_2^2 = 1.$$

From this last equation it is clear that the lengths of the major and minor axes of the ellipse are $2\sqrt{2}$ and 2, respectively.

4 The characteristic values of the matrix associated with the given quadratic form are 2, 3, and 6. Thus, there exists an orthogonal coordinate system in which the ellipsoid has the equation $2u_1^2 + 3u_2^2 + 6u_3^2 = 6$. It follows that the axes of this ellipsoid have lengths $2\sqrt{3}$, $2\sqrt{2}$, and 2.

GLOSSARY OF SYMBOLS

$[a_{ij}]_{mn}$	The m by n matrix with entries a_{ij}, 1
$[x]$	Greatest integer in x, 2
δ_{ij}	Kronecker delta
I	Identity matrix, 5
Θ	Zero matrix, 5
A^{-1}	Inverse of the matrix A, 13
$[i \leftrightarrow k]_n$	Row interchanger, 14
$[i \to (c)i]_n$	Row multiplier, 14
$[i \to i + (c)k]_n$	Row adder, 14
$A \stackrel{(\text{row})}{=} B$	Row equivalence, 21
A'	Transpose of the matrix A, 31
$X \oplus Y$	Vector sum of X and Y, 52
$c \odot X$	Scalar multiple of X, 52
$(V, R, \oplus, \odot)$	Vector space, 52
R^{mn}	Space of m by n matrices, 60
P^n	Space of polynomials of degree not exceeding n, 60
$\text{Id}(X)$	Vector space isomorphism, 62
$\{a, b, c, \ldots\}$	Set of objects $a, b, c, \ldots$, 71
$\mathfrak{B}, \mathfrak{C}, \ldots$	Ordered bases, 92
$X_{\mathfrak{B}}$	$\mathfrak{B}$-column for X, 99
$\begin{bmatrix} a \\ b \\ \vdots \end{bmatrix}_{\mathfrak{B}}$	$\mathfrak{B}$-label for a given vector, 100
$N^1, N^2, \ldots$	Natural basis vectors, 101
L	Linear transformation, 123
$L_{\mathfrak{C}\mathfrak{B}}$	$\mathfrak{C}\mathfrak{B}$-representation for L, 144
$L_1 \circ L_2$	Composition of L_1 and L_2, 149
$|A|$	Determinant of the matrix A, 176

GLOSSARY OF SYMBOLS

$\text{cof}(a_{ij})$ Cofactor of a_{ij}, 191
A_{ij} The matrix obtained by deleting the ith row and the jth column of A, 191
$B \stackrel{s}{=} A$ Matrix similarity, 206
$|X|$ Length of the vector X, 235
$X \cdot Y$ Dot product of X and Y, 236

SUGGESTED REFERENCES

For a more advanced treatment of linear algebra:

Cullen, C. G., *Matrices and Linear Transformations*, Addison-Wesley, Reading, Mass., 1966.

Ficken, F. A., *Linear Transformations and Matrices*, Prentice-Hall, Englewood Cliffs, N. J., 1967.

Finkbeiner, D. T., *Introduction to Matrices and Linear Transformations*, W. H. Freeman, San Francisco, 1960.

Marcus, M. and H. Minc, *Introduction to Linear Algebra*, Macmillan, New York, 1965.

For applications of linear algebra:

Crowell, R. and R. Williamson, *Calculus of Vector Functions*, Prentice-Hall, Englewood Cliffs, N. J., 1968.

Kaplan, W., *Operational Methods for Linear Systems*, Addison-Wesley, Reading, Mass., 1962.

Karlin, S., *Mathematical Methods and Theory in Games, Programming, and Economics*, Addison-Wesley, Reading, Mass., 1959.

Noble, B., *Application of Undergraduate Mathematics in Engineering*, Macmillan, New York, 1967.

Pipes, L. A., *Matrix Methods for Engineering*, Prentice-Hall, Englewood Cliffs, N. J., 1963.

For numerical methods that are useful in solving matrix problems:

Fadeev, D. K., and V. N. Fadeeva, *Numerical Methods in Linear Algebra*, W. H. Freeman, San Francisco, 1963.

Householder, A. S. *The Theory of Matrices in Numerical Analysis*, Blaisdell, New York, 1964.

Varga, R. S., *Matrix Iterative Analysis*, Prentice-Hall, Englewood Cliffs, N. J., 1962.

INDEX

INDEX

Abstract mathematical structure, 51
Analytic geometry applications, 182, 184, 236, 241, 270
Angle between two vectors, 235
Arrow space, 50
Augmented matrix, 25

Basis, 75
  infinite, 85
  natural, 101
  ordered, 92
  orthonormal, 258
Beginner variable, 26

Cauchy-Schwarz inequality, 246, 249
Cayley-Hamilton theorem, 208
Change-of-basis matrix, 92
Change-of-coordinates formula, 108
Characteristic equation, 205
Characteristic value (vector), 205
  for symmetric matrices, 264
Coefficient matrix, 25
Cofactor, 191
Column equivalence, 22
Column space, 88, 90
Congruence, 48, 112
Coordinate equation, 120, 242
Coordinate-free methods, 238
Coordinate system, 97, 99
  natural, 101
Cramer's rule, 194
Cross product, 133

Determinant function, 169
  evaluation of, 169, 170, 174
  formula for A^{-1}, 193
  a geometric interpretation, 288
  properties of, 173, 181, 185, 186
  the traditional approach, 285
Diagonalization problem, first, 32
  application to quadratic forms, 39
Diagonalization problem, second, 197, 263
  application to differential equations, 212
  application to geometry, 275, 276
Differential equation applications, 85, 134, 210, 215, 216, 228
Dimension, 65, 84
Dimension theorem, 129
Direction cosines, 240
Distance between vectors, 240
Dot product, 236
  matrix for, 253

Elementary matrices, 14
Elementary matrix applications,
  determinant theory, 186, 187
  determining a basis, 88
  diagonalizing A, 32
  finding A^{-1}, 17
  solving systems, 23
Equivalence of matrices, 159
Equivalence relation, 45, 279

Equivalent systems, 25
Euclidean space, 244
　of infinite dimension, 244

"Golden rule," 15, 16
Gram-Schmidt process, 259
　an alternative for, 263

Identification function, 62
Inductive definition, 165
Inductive set, 163
Isomorphism of, Euclidean spaces, 252
　linear algebras, 156
　matrices to linear transformations, 149
　vector spaces, 62

Kernel, 127
k-plane, 114
　parallelism of, 118, 120

Length of a vector, 235
Linear algebra, 156
Linear combination, 68
Linear dependence (independence), 72, 74
Linear equation, 127
　homogeneous (nonhomogeneous), 127, 133
Linear system, 25
Linear transformation, 123
　an application to calculus, 291
　matrix representation for, 144
　operations with, 149

Mathematical induction, 164
Matrices, diagonal, 10
　elementary, 14
　Jordan, 217
　lower (upper) triangular, 10
　orthogonal, 266
　positive definite, 40
　row echelon, 26
　singular, 13
　skew symmetric, 36
　symmetric, 10
Matrices with other than real numbers, 91, 94, 95
Matrix operations, 2, 3, 5
　properties of, 6
Matrix representation, 144

Numeral system, 96

Orthogonal vectors, 236
Orthonormal basis, 258

Parallelogram law, 249
Projection, 160, 224, 241

Quadratic form, 37
　positive definite, 40
　negative definite, 41
　indefinite, 41

Rank, 138
Rotation of axes, 276;
　see also Exercise 6, p. 270
Row echelon form, 26
Row equivalence, 21, 48
Row (column) rank, 88
Row (column) space, 87

Similarity of matrices, 206
Spanning set, 70
Subspace, 57

Transpose of a matrix, 31
Triangle inequality, 249

Vandermonde determinant, 225
Vector projection, 241
Vector space, 52
　subspace, 57
　real versus complex, 80

Wronskian, 229